MW00710145

Security Operations Management

Security Operations Management

Robert D. McCrie

Boston • Oxford • Auckland • Johannesburg • Melbourne • New Delhi

Recognizing the importance of preserving what has been written, Butterworth–Heinemann prints its books on acid-free paper whenever possible.

Butterworth–Heinemann supports the efforts of American Forests and the Global ReLeaf program in its campaign for the betterment of trees, forests, and our environment.

Library of Congress Cataloging-in-Publication Data

McCrie, Robert D.
Security operations management / Robert D. McCrie.
p. cm.
Includes bibliographical references and index.
ISBN 0-7506-7087-8 (alk. Paper)
1. Private security services--Management. 2. Security systems--Management. I. Title.
HV8290 .M39 2001
363.28'9'068--dc21 00-045498

British Library Cataloguing-in-Publication Data
A catalogue record for this book is available from the British Library.

The publisher offers special discounts on bulk orders of this book.
For information, please contact:
Manager of Special Sales
Butterworth-Heinemann
225 Wildwood Avenue
Woburn, MA 01801-2041
Tel: 781-904-2500
Fax: 781-904-2620

For information on all Butterworth–Heinemann publications available, contact our World Wide Web home page at: http://www.bh.com

10 9 8 7 6 5 4 3 2 1

Printed in the United States of America

TABLE OF CONTENTS

PREFACE

What does an organization expect from its managers, directors, and executives concerned with protection of assets from loss? It expects leadership, analytical ability, relevant knowledge to solve problems, flexibility to confront new situations, and sufficient experience to enhance sound judgment. At the core of that expectation is the need for effective action.

Security Operations Management is written for practitioners, students, and general managers who are involved with or interested in managing security operations more effectively. The purpose of this book is not immodest: it seeks to bring order to the sometimes chaotic task of protecting people, physical assets, intellectual property, and opportunity. It endeavors to provide a structure to operate programs for the benefit of the organization, and it wishes to relate such principles clearly and directly to readers.

Security programs in the workplace are growing robustly, a development that can be traced back several years. Meanwhile, security trade and professional organizations have grown in numbers, specializations, and aggregate membership in recent decades. Protection-related issues appear regularly on board agendas, and security-related matters are covered in the media on a daily basis. However, the actual number of security directors and managers decreased in the late 1980s and 1990s. The *need* for security programs and services did not decrease, but several managerial positions, including security, declined in total numbers. Middle management positions were squeezed by the impetus to create tighter, more cost-effective organizations.

This book is written with the implications of this trend in mind. It explores both the problems and opportunities for protection management in contemporary organizations, and the ways in which security operations managers constantly must demonstrate their programs' value.

A curious feature of organizational security leadership is how little effect it has had on undergraduate or graduate education within criminal justice, economics, and business programs. This book seeks to integrate the nascent but growing academic discipline of security management with the more mature, directed studies encountered both in undergraduate and graduate schools of business administration, as well as in academic programs in criminal justice. Some of the book's material is based on the academic framework of business school management courses—indeed, syllabi from general management courses at leading schools of business administration were evaluated in the preparation of the early chapters. Further, the book is written within the context of the security management academic program at John Jay College of Criminal Justice, a liberal arts institution focusing on public service. John Jay was located briefly in its earliest days within Baruch College, now a highly regarded business and public affairs–oriented liberal arts institution whose library, like John Jays, was particularly helpful for this volume. Hence, elements of criminal justice, business and public administration have influenced some aspects of the content.

This book also integrates many relevant papers from *Security Journal* and criminal justice and business-oriented publications that have served as primary sources for research on security operations management. Additionally, findings and recommendations from academic/practitioner symposia, sponsored by the American Society for Industrial Security, held at Webster University in 1997, John Jay College the next year, and the University of Nevada at Reno in 2000, have been incorporated into the text.

ACKNOWLEDGMENTS

A book of this sort is long in the making and incurs many debts along the way. In a general way, the 450 or so authors of the papers of *Security Journal*, which I edited from 1989 to 1998, provided inspiration as to the content of this book. Additionally, the readers of *Security Letter* have informed me of operational issues of concern to them, which find expression throughout this work.

In a more specific way, many talented security practitioners and academics have provided me with inspiration—knowingly or unknowingly. Those persons include, but are certainly not limited to, J. Kirk Barefoot; Ronald V. Clarke; Charles "Sandy" Davidson; Minot B. Dodson; John G. Doyle, Jr.; David H. Gilmore; Robert A. Hair; Raymond F. Humphrey; Robert V. Jacobson; Ira A. Lipman; Robert F. Littlejohn; Bonnie S. Michelman; Lawrence J. O'Brien, Jr.; Hans Öström; Richard D. Rockwell; Marcel Sapse; Ira S. Somerson; Michael J. Stack; and Steven B. Stein.

Specifically in the writing of this book, Bo N. Sorensen read almost every chapter and invariably made helpful contributions. Walter B. Parker read part of Chapter 8 and offered constructive comments. My associate, Luis A. Javier, saw to the details of the work with considerable faithfulness. And above all, thanks go to Fulvia Madia McCrie, without whom this book would never have been realized. At Butterworth-Heinemann, my appreciation extends to Laurel DeWolf, who originally proposed the theme and suggested the title, and to Jennifer Packard and Jennifer Plumley, who saw this project through to completion.

—*R.D. McC.*

PART 1

GENERAL FUNDAMENTALS AND COMPETENCIES

1

SECURITY OPERATIONS IN THE MANAGEMENT ENVIRONMENT

> Security management is ready and eligible to be considered as a management science.
>
> —Charles H. Davidson in *Security Journal*

To achieve optimal protective goals, security executives, directors, and managers must operate successful programs. The origins of certain management-related words clarify this objective. The word "operate," for example, is derived from the Latin *operatus*, the past participle of the verb "to work"; hence, operations are concerned with exerting power or influence in order to produce an effect. Security operations, therefore, are the processes whereby the protective aims of the organization are to be achieved. Success does not depend upon good intentions alone. Personal effort causes such desired changes to occur. The security practitioner must assume correctly that his or her appropriate involvement is consequential in achieving what needs to be done.

Operating security programs is not easy. Protection is an inherent factor in success and continuity of an operation. Because of this, one might assume—falsely—that efforts to protect assets would receive broad, largely uncritical support from senior management and ownership. That's not necessarily so. A paradox exists within the workplace: freedom results in creativity, spontaneity, and economic development, while at the same time making abuses within the organization easier to occur. Therefore, controls that decrease the possibilities of loss are implemented. However, these same controls may also decrease creativity and efficiency. The art of the security practitioner is thus to encourage expression and achievement, while making the control mechanisms reasonably unburdensome to employees, visitors, vendors, and the public at large so that the organization flourishes without the appearance of constricting security operations.

This book considers the tasks of operating security loss prevention programs in contemporary organizations. The principles involved are applicable in for-profit and not-for-profit corporations and within government units at all levels. Many of thc organizational protective features are common to the concerns of general management. Indeed, security operations are aspects within a broad management context. Therefore, the initial part of this chapter will consider the concepts that have helped form management practices in the 20th century and that are guiding it in the 21st.

ORGANIZATIONS AND MANAGERS

To understand what a manager does, it is essential to consider the ways in which organizations have evolved in modern public and private institutions. Management must be rational in order to achieve long-term success. Therefore, the creation of organizations and their successful achievement of desired objectives must be understandable both to those within and outside the organization. This is true for security departments as well as for every other operating unit.

What Is an Organization?

The word "organization" is derived from the Greek *organon*, meaning organ, tool, or instrument, and is akin to work. Organizations are composed of groups of people bonded by a purpose: a systematic scheme to achieve mutually agreed-upon objectives. Typically, organizations might be divided into a bifurcated scheme: administrators (leaders and planners) and functional members (followers and processors). These roles may be interchangeable according to different circumstances. Organizations are created, therefore, in order to achieve objectives deemed desirable by leaders and planners of the organization, by those who carry out tasks, or in some cases from both.

What Is a Manager or Director?

The word "manage" is derived from the Gaelic *mano*, related to hand, and was related to the handling or training of a horse in graceful or studied action. Thus, the word suggests the concept of controlling, directing, or coping with challenging and constantly diversifying circumstances. A manager is a person who controls or directs an organization in a desired, purposeful direction. The title of director usually outranks that of manager and refers to the person who directs the work of managers and their subordinates.

What Is a Security Manager?

Security is defined as the protection of assets from loss. Each word in this definition carries its own implications. The word "protection" means to cover or defend. The word "assets" encompasses numerous possibilities of tangible and intangible resources of value. Clearly, cash and physical property can be considered assets, but knowledge-related activities and the opportunity to achieve a desired goal due to particular circumstances are also assets. A security manager (or director) is a person who protects identified assets through personnel, procedures, and systems under his or her control. The goal is to achieve objectives—agreed upon with senior management—that also produce minimum reasonable encumbrances to overall operations.

A POINT ABOUT TITLES

Titles within the organization can change according to fashion. For most of the 20th century, the titles "president," "executive," "chief," "director," "manager," and others had specific meanings. They connoted a hierarchy well understood by those within and outside the organization. Such a hierarchy still exists, but title connotations may neither be clear nor consistent and vary from one organization to another. Indeed, sometimes an executive (or manager) creates new titles for structural or motivational purposes (see Chapter 5). Thus, words like "deputy," "associate," "assistant," "managing," "acting," "senior," and "junior" are parts of some titles that may serve to provide the level of significance of the position.

WHAT IS THE PURPOSE OF AN EXECUTIVE?

Executives and those with executive tasks—regardless of their titles—are responsible for the planning and analysis of required programs. They are further responsible for implementation of such programs. Ultimately, the challenge to organizational leaders is to be effective in achieving or surpassing the reasonably set goals of the organization. That is, they *execute* such programs. Peter F. Drucker, a leading management consultant, argues that the primary strategy of work is measured not in the brilliance of its conception, but in how well the desired goals were actually achieved. The nature of work changes constantly, he observes.

According to Drucker, "knowledge workers" are the human capital through which objectives are achieved. Knowledge workers are members of an organization whose effectiveness is realized though the use of information often accessed and partially analyzed through technology. In *The Effective Executive,* Drucker posits that effectiveness is not simply necessary as a managerial attribute, it is vital and can be learned through concerted effort, leading to still greater effectiveness. Drucker writes:

> I have called "executives" those knowledge workers, managers, or individual professionals who are expected by virtue of their positions or their knowledge to make decisions in the normal course of their work that have significant impact on the performance and results of the whole. They are by no means a majority of the knowledge workers. For knowledge work too, as in all areas, there is unskilled work and routine. But they are a much larger proportion of the total knowledge workforce than any organization chart ever reveals.[1]

The effective security executive or manager is a person who can identify the problems and opportunities facing the organization, plan to resolve them, organize resources so that the mission may be successfully achieved, deputize others to follow through on his or her behalf, and then supervise the continuing operation. This is spelled out further in the next section.

WHAT IS THE STRATEGY OF MANAGEMENT?

Management refers to the way in which members of an organization make key decisions on how goods and services are produced. Management can also refer to the process by which such goals may be achieved.

Throughout contemporary organizations, the strategy of management is accomplished via a process of identifying, analyzing, planning, organizing, deputizing, and supervising activities common to the attainment of these goals. This process is systematic in that order and conduct is required to achieve objectives by members of the organization. The manager sees to this process in each link of the chain (see Box 1.1). Specifically, the concatenation of managerial tasks is as follows:

1. **Problem identification.** The first organizational process step identifies the need for desirable and required managerial action. This need may be to commence a new program or initiative, to revise an old one, to solve a problem, to seize an opportunity, to expand or contract operations, or to handle still other options. The management process begins by asking the question "What needs be accomplished and why?" It then grapples with the clarified requirements that emerge from the following stages.

Box 1.1. The Security Management Process

Managers in modern organizations use a simple, logical process to achieve desired goals. The problem or opportunity may differ in significance, and the time required to adequately analyze and plan it also may vary. A major problem or opportunity may require weeks or months to resolve, but the sequence of events remains the same. Here's an example:

Problem identification. Assume that the organization is expanding and must create a new facility to achieve desired production. This new facility will require a security program to protect its assets. How will it be created? Early in the process of planning for such a facility, the security director collects pertinent information so that an optimal security program may be designed. The size, condition, employment, production requirements, environmental issues, potential problems, and other issues will be considered, and the most problematic matters will be isolated. Then the director, often aided by others, completes additional tasks until the program is fully implemented. The process is as follows:

1. **Analysis and planning.** The security director might collect and analyze the following information about the new facility:

   - Function of the new facility (what it does, its size and significance)
   - Site selection (for protective and risk-averse features of the topography)
   - Architectural and engineering involvement
   - Local conditions where the facility is to be located (for example, crime pattern)
   - Local resources available (police, fire, emergency-oriented)
   - Special security features likely to be required at such a facility

   This process involves a fact-finding process in which the manager, or a surrogate, visits the site to determine its potential risks and opportunities so that these may be incorporated into the formal plan. When as much relevant data is collected as possible in the time available, the planning team is ready to prepare the physical security plan for the new facility. Planning for security measures needed by the facility once it begins operating is also undertaken at this time. The manager then discusses the analysis and planning with senior management.

2. **Organizing.** The plan must now be accepted by relevant decision-makers throughout the organization. Resources required for the

Box 1.1. *(Continued)*

security program at the new facility are then mobilized. The steps taken may include:

- Consulting with architectural and engineering personnel about specific security design needs
- Issuing a request-for-proposal (RFP) for the system (Chapter 10)
- Establishing qualified bidders for the security project
- Reviewing submissions and awarding the contract
- Supervising the project's installation
- Assuring adequate training and support materials
- Testing the system under normal and adverse circumstances

At this point, a complex system has been created for the new facility. Meanwhile, a security staff must be hired and procedures for both security and non-security personnel must be prepared and reviewed. The next step assures these goals are met.

3. **Deputize.** As the new security system is configured and the operational commencement for the new facility can be scheduled, a manager must run the enduring, ongoing security program. Consequently, someone is deputized to assume this responsibility on behalf of management at headquarters. He or she will carry out the earlier aspects of the plan through the commencement and subsequent routine operations of the new facility.
4. **Supervise.** The central manager's time commitment for the new facility gradually lessens as the deputy assumes control. That deputy reports regularly on developments. The central manager maintains quality control over the physical and procedural process involved in creating the plan for the new facility.
5. **Constant critical analysis and change.** At this point, the managerial process has been completed. The time it takes to complete the process varies considerably depending on the particular problem to be managed. It may take as little as a few hours by a single individual or as much as months of concerted effort by a managerial team. The process is dynamic; circumstances change constantly, often in ways that could not have been anticipated early in the planning period even by the most conscientious and rigorous planners. Therefore, the manager must be prepared to constantly refine the plan to new circumstances, seizing fresh opportunities for further gains in programmatic objectives whenever possible.

2. **Analyzing and planning.** Analyzing is the process of separating something into its constituent parts or basic principles. This allows the nature of the whole issue to be examined methodically. To analyze a security problem, the practitioner seeks to collect all pertinent information, which then becomes the basis of planning—or formulating—a means to achieve the desirable goals. These are the critical next parts of the managerial process. Wise managers do not proceed generally to the next step in the sequence until the previous one is reasonably completed. *How much planning is enough?* A manager is never likely to have all the knowledge and facts necessary to comprehend every relevant facet to analyze fully and then plan comprehensively without ever looking back. Further, conditions change constantly and create situations with which the manager must contend. Yet at some point the analysis must be summarized when a reasonable quantity of information has been collected and a plan for action evolved. That process of working with finite knowledge and resources is what is fascinating and challenging about the art and method of management.
3. **Organizing.** After the need has been determined, its critical parts have been identified, and a plan has been established to respond to the need, resources must be organized—that is, created or accumulated in order to achieve the objective. Money and personnel must be committed. Technology and software strategies may be required and must be allocated. Impediments must be resolved. Commitments must be assured. Then the plan can be implemented by selecting subordinate managers.
4. **Deputizing.** A manager does not achieve the objectives of the plan solely by his or her actions: a manager works in the company of others. In the management process, the problem has been analyzed and a plan to deal with it has been agreed upon. Resources have been committed firmly. Now the process of assuring that the plan achieves its objectives is shared with persons who will follow through—hopefully to realize the intended goals. Persons deputized to achieve these ends on behalf of the planning managers are themselves managers who are now transferred the responsibility for assuring that the plan will be carried out. The senior planning manager is now free to supervise this person or persons.
5. **Supervising.** The planning manager next supervises the manager who has been given responsibility for achieving the goals set by the plan. Through this process, the manager can assure that goals are reached in the face of constantly changing circumstances. Thus, the principal manager is engaged in controlling the work of others and the allocation of resources in pursuit of the desired objectives. The supervising manager in the hierarchy remains available to critique,

and supports and guides the manager deputized to carry out the plan. The supervising manager now has time to concentrate on other matters, such as identifying another need and planning its resolution or supervising other operating programs.

6. **Constant critical analysis and change.** At this point, the planning process has been completed from inception to realization. The sequence may take as little as a few hours by a single individual or as much as months of concentrated effort by a devoted managerial team. Such a team could include internal managers, contract personnel, and independent consultants retained for the project. However, although the program may be functional, the process is never complete. Circumstances change constantly, often in ways that could not have been anticipated even by the most conscientious and rigorous planning process. Therefore, the manager must refine the plan to fit the new circumstances, seizing new opportunities for further gains in programmatic objectives whenever possible.

THE CHARACTERISTICS OF MODERN ORGANIZATIONS

Contemporary organizations of size and complexity must possess a pertinent structure to achieve operational success. Civilization is about 5,000 years old, but the industrial age arrived in Europe only in the 18th century, arriving decades later in what would become the U.S. The demands of constantly competing, expanding industrialization—coupled with expanding urbanism—created pressures for greater effectiveness on organizations. This process attracted the attention of seminal early observers who first described evolving characteristics of the operational processes. These observations created the basis for methodological observers who sought ways of improving industrial output. Much later still, security practitioners emerged to protect organizations in specific and distinctive ways.

The pivotal figures in this process may be divided into three categories: classical management theorists, scientific management proponents, and some recent distinctive contributors to security management practices.

Classical Management Theorists

Industrialization flourished following principles of expediency and common sense. In time, the processes of production came under analysis and improvement. The first significant and comprehensive codification of management principles was provided by a French mining engineer, Henri Fayol (1841–

1925). He observed workplace processes, which he then categorized into logical and distinct descriptives with broad applications and significance:

- **Division of work.** In an organization of any size, labor is divided into specialized units to increase efficiency. Work within the organization tends to become increasingly specialized as the organization grows in size.
- **Hierarchy.** Organizations disperse authority to managers and employees according to their formal position, experience, and training.
- **Discipline.** Good discipline exists when managers and workers respect the rules governing activities of the organization.
- **Unity of command.** No individual normally should have more than one supervisor. Work objectives concerning tasks should relate rationally among supervisors and subordinates. (Fayol derived this point from his observations of military structure.)
- **Chain of command.** Authority and communication should be channeled from top to bottom in the organization. However, communication should flow from bottom to top as well.
- **Unity of direction.** The tasks of an organization should be directed toward definable and comprehensible goals under the leadership of a competent manager.
- **Subordination of interests.** The goal of the organization should take precedence over individual desires. When personal agendas become paramount, the goals of the organization cannot be achieved effectively.
- **Remuneration.** Pay and the total benefits package should be fair.
- **Equity.** Managers should be just and kind in dealing with subordinates.
- **Stability of tenure.** Management should plan so that positions are stable. Reduction of positions (downsizing; "rightsizing") may be necessary under times of market and production downturn, but often the reduction of previously budgeted positions reflects the failure to plan and execute wisely.
- **Order.** The workplace should be orderly.
- **Initiative.** Employees should be encouraged to show personal initiative when they have the opportunity to solve a problem.
- **Teamwork.** Managers should engender unity and harmony among workers.
- **Centralization.** Power and authority are concentrated at the upper levels of the organization. The advantages of centralization versus decentralization are complex and may be regarded as a cyclical phenomenon in management fashion; that is, despite a penchant for centralization of organizational power, there may be times

when production is best achieved by decentralization of planning and much decision-making.

According to Fayol, all managerial activities can be divided into six functions:

1. Technical (engineering, production, manufacture, adaptation)
2. Commercial (buying, selling, exchanging)
3. Financial (searching for an optimal use of capital)
4. Accounting (stock taking, balance sheets, cost analysis, statistical control)
5. Managerial (goal-setting, analyzing and planning, organizing, deputizing, supervising)
6. Security (protecting physical assets and personnel)

These six functions are always present, regardless of the complexity and size of an organization. Thus, all organizational undertakings involve an interlinking of functions. Note that security is identified as one of these fundamental activities of general management. Fayol observed that the security function "involves exposure identification, risk evaluation, risk control, and risk financing."[2] In a remarkably insightful observation for its time, he also added:

> Quite frankly, the greatest danger to a firm lies in the loss of intellectual property, a loss that the firm may attempt to prevent through patent protection, trade-secret protection, signed agreements (nondisclosures) with key personnel, and access to its innermost secrets on a strictly "need to know" basis.

Fayol's prescient views hold that security of know-how and opportunity take precedence over physical assets, an opinion many contemporary security practitioners readily agree with.

Fayol is regarded as a classical administrative theorist. Other pioneers of his genre include Max Weber (1864–1920) and Chester Barnard (1886–1961). Weber developed the term "bureaucracy," which he described as the most rational form of an organization.[3] According to Weber, large-scale tasks could be pursued by organizing human activity as follows:

1. Activities directed toward meeting organizational goals are constant and officially assigned.
2. Activities are controlled through a hierarchical chain of authority.
3. A system of abstract rules ensures that all operations are treated equally.
4. Bureaucratic officials remain emotionally uninvolved while fulfilling their formal duties.

Barnard, an executive for New Jersey Bell Telephone, emphasized that a "cooperative system" generally is necessary for an organization to reach its goals. In *The Functions of the Executive*, Barnard advanced a concept known as acceptance theory, concluding that subordinates would assent to authority from supervisors and managers when four conditions were met:

1. They could and did understand the communications they received.
2. They believed that the communication was consistent with the purpose of the organization.
3. They believed that it was compatible with their own personal interest.
4. They were mentally and physically capable of complying with the communication.[4]

Thus, Weber underscored the importance of managerial structures to achieve desirable goals and Barnard espoused that the principle that defined communications could result in acceptance of the desires of bureaucracies by workers.

SCIENTIFIC MANAGEMENT PROPONENTS

Exponents of scientific management seek to use data collection and analysis to improve work performance. The costly and time-consuming efforts required to save a few minutes or seconds might seem like a frivolous activity to some; however, improved techniques, when applied to a repetitive process on a large scale, pay generous rewards over time by improving efficiency. Furthermore, the same process of job analysis could offer improvements in safety and work comfort.

Frederick W. Taylor (1856–1915) was a self-taught engineer who had become chief engineer of a steel company by the age of 28. His impressive early climb up the career ladder was related to his ability to study work scientifically and then to apply the results directly. His contributions had enormous influences on the workplace throughout the 20th century.[5] Taylor's principles are summarized as follows:

1. **Determine what's important in a task.** Managers must observe and analyze each aspect of a task to determine the most economical way to put that process into general operation. The use of time studies helps to establish what works best.

   Example: Federal Express couriers delivering or picking up packages knock on a door before ringing the bell. Their studies have revealed that customers respond faster to the knock-first-then-ring sequence. Perhaps regular FedEx customers are conditioned to faster response because they know who is at the door. Similarly, security officers responding to an incident can be more productive and thorough by following a form that sequentially prompts them to collect the essential facts about the event.

2. **Select personnel scientifically.** Taylor believed that all individuals were not created equal. Training could help modify differences in behavior and performance, but still some persons would be more effective than others in performing the same tasks. It stands to reason, therefore, that operations will be improved when managers concentrate on selecting only those who show the best capacity to perform the job required.

   Example: An organization may determine that its security personnel must write clear, cogent reports of incidents. Therefore, a pre-employment (vetting) test may be designed to ascertain how saliently candidates for employment express themselves on paper.

3. **Offer financial incentives.** Selecting the right worker for the right task does not by itself assure optimal effectiveness. Workers need motivation, and hourly pay and benefits alone may not be sufficient to achieve that goal. Taylor ascertained that providing a differential piece-rate form of incentive can produce higher worker output than what would ordinarily be expected.

   Example: The manager of an investigative department provides incentive payments for those staff investigators who are able to complete more investigations than the baseline expectation. Quality control assures that such investigations meet or exceed expected standards of quality for the assignments undertaken so that investigators seeking to achieve additional payments may not sacrifice standards to achieve higher benefits.

4. **Employ functional foremanship.** Taylor argued that responsibility should be divided between managers and workers. Managers primarily would plan, direct, and evaluate the work; the individual worker was responsible for completing the designated tasks. This permitted a worker to take orders from a functional foreman regardless of the stage of work because all manager-foremen would understand the same work processes.

   Example: Assume that a new security supervisor replaces another normally responsible for a work unit. The goals of the workers being supervised are identical. Since procedures to achieve these objectives are understood by all workers, a new supervisor reasonably should be able to achieve the same objectives with the workers as the regular supervisor would have.

Frank Gilbreth (1868–1924) and Lillian Gilbreth (1878–1972) were a husband and wife team who translated Taylor's scientific management approach and applied it to specific tasks. Taylor often tried to identify means to help workers get their jobs done faster. The Gilbreths further sought to increase the speed of attaining production objectives by eliminating useless motions. They noted that efficient procedures also led to less fatigue and chances of error by workers.[6] Their research underscores the importance of designing systems and tasks that support them carefully. As a result, errors are less apt to occur or may be less frequent and serious after such analysis than in systems that are not established with empirical methods.

Example: On March 28, 1979, at Three Mile Island, near Harrisburg, Pennsylvania, a near-meltdown of a nuclear power facility almost occurred. It resulted in a limited evacuation of the area. As a result of the fear generated by this emergency, the nuclear industry in the U.S. was stigmatized, and additional construction of nuclear power facilities ceased. In subsequent investigations, many factors explained why the nuclear accident at Three Mile Island occurred. One significant issue was that critical gauges and controls were not within the line of sight of engineers at the control consoles. An investigation of the Three Mile Island facility by the Nuclear Regulatory Commission determined that an inadequate quality assurance program to govern construction and monitor quality "resulted in the construction of a facility of indeterminate quality."[7] Failure to design a facility properly may explain why losses occur; conversely, good design system may be more important than marginal differences in human competency in explaining the achievement of desired effects.

SECURITY MANAGEMENT PRECEDENT SETTERS

The craft of operating security programs effectively is a recent one, when judged by contemporary standards. The principal professional association in the field, the American Society for Industrial Security-International, was founded in 1955. The Security Industry Association began in 1967; the National Council of Investigation and Security Services was organized in 1975; and the International Security Management Association began in 1976. Surely, informal private security operations existed prior to the founding of these groups, and thousands were employed in security positions in the 19th century and the first half of the 20th century. But only in the last half of the 20th century did security emerge as a defined, usual, respectable, and visible part of management. In the process, security operations have been enhanced by the writings and practices of those who have directed successful programs. In particular, four practitioners and researchers are among those who have contributed notably to the conceptual and operational framework of the discipline. They are Dennis R. Dalton, Richard D. Paterson, J. Kirk Barefoot, and Ronald V. Clarke.

In the last decade of the 20th century, the process of contracting-out was an option increasingly elected by organizations employing security forces. That is, in-house (proprietary) guard or alarm monitoring services were replaced by professional security contract workers. Dennis R. Dalton evaluated and streamlined the process by which organizations could make such a transition. In *Security Management: Business Strategies for Success*, Dalton presents a strategy for outsourcing by creating a "strategic alliance" with the service vendor.[8] Outsourcing by for-profit businesses and privatization of services by government have helped shape contemporary security services. Dalton's work helped make the process rational and efficient.

Investigations are an important option for organizations, both for external and internal loss control and management purposes. Failure to institute a fact-finding inquiry may result in unchecked losses or other vulnerabilities. Richard D. Paterson and J. Kirk Barefoot removed the mystery of undercover operations by establishing a school that trained students to be effective fact-finders for internal and external deviance. The process encouraged managers, in appropriate circumstances, to consider the regular use of undercover operations as an ethical, reasonable, and efficient means of detecting and deterring crime victimization and the flouting of recognized performance standards. Barefoot further detailed the process in *Undercover Investigations*.[9]

Although his research career largely has been rooted in studies aimed at aspects of community crime mitigation and funded by various governmental agencies, Ronald V. Clarke has contributed exceptionally to the philosophical and research basis of private sector security practices. Clarke, a professor at the Rutgers University School of Criminal Justice, was an early social science researcher who helped develop the field of situational crime prevention. Other pioneers in this field include Paul and Patricia Brantingham, L.E. Cohen, D.B. Cornish, and Marcus Felson. These researchers postulate that three separate factors are involved in determining whether a criminal act will be successful or thwarted. These factors are a motivated offender, a suitable reward or goal for the offender's actions, and the absence of appropriate controls that could check such action by the offender (see Box 1.2). A fourth component, often mentioned, is the potential creation of shame or image problems for a perpetrator.

By intervening with any one of these three primary factors—which is often possible at low or no substantial cost—measurable crime should decrease. Situational crime prevention does not envision situations in which an environment will entirely be free of crime. Rather, it seeks to engineer practical measures that will permit a normal pattern of human and commercial activity while reducing violent acts and property offenses to a tolerable level.

HOW ORGANIZATIONS ARE STRUCTURED

Fayol stated that a formal structure naturally evolves over time to achieve efficiency. This view was new when it was first propounded. Yet organiza-

Box 1.2. Situational Crime Prevention: Key Elements, Possible Controls, or Mitigating Factors

Key Elements	Possible Controls or Mitigating Factors
A motivated offender	• Deny access to sensitive areas • Warn of punishment for illegal behavior • Prosecute apprehended offenders
A suitable reward or goal	• Decrease available assets that might be stolen from a potential victimization site • Render vulnerable assets less attractive to thieves or alter behavior of potential victims so that they might be less likely to be victimized. • Make vulnerable assets impossible for thieves or other offenders to benefit from.
Absence of appropriate control	• Assign security officers ("place minders") to protect a location or increase their numbers. • Install or upgrade a security system. • Educate non-security employees and others to participate willingly in loss prevention strategies.

Note: Situational crime prevention posits that all three elements may be assessed to determine the crime vulnerability of a location or situation. A fourth element sometimes mentioned relates to image-risk to the perpetrator by shame or embarrassment. By changing any one element, the possibilities of increasing or decreasing violent or property crime change.

tions have always used ranks, grades, classes, or other categorizations to reflect significance and authority. While ranks and titles may change and considerable variation may exist within characteristics of the organization, the structures of modern corporations and institutions fit general patterns. A review of the two major types of non-governmental organizations will illustrate where security management may be found.

For-Profit Corporations

Most corporations are established at the behest of private investors who seek a return on their invested capital; that is, they are for-profit corporations. Such corporations are considered perpetual entities performing the activities described in their charters. Corporations issue common stock to investors, who hope to generate a profit (through dividends and growth of value) from the capital they put at risk.

Individuals or institutions who purchase common stock are termed the corporation's shareholders or stakeholders. These shareholders own the corporation, and their degree of ownership (equity) depends on the number of shares they own relative to the total number of shares authorized to be outstanding in the organization. Large corporations with thousands of shareholders are not democratic organizations. They are in no position to hear from all shareholders individually on corporate matters, and modern shareholders expect to have no voice in routine operations or planning. However, shareholders are not without representation. The board of directors legally represents total ownership—that is, the shareholders of common stock. Figure 1.1 shows a corporate organizational chart showing related security functions. In publicly held corporations, in which shares are traded on public stock exchanges, investors exercise their factual ownership by casting votes for directors annually and approving any major changes in the financing, structure, and governance of the entity. A chairman or chairwoman of the board heads the board of directors. This person may also hold other executive duties within the corporation or has held such responsibilities in the past.

The board may be composed of two classes of directors. One category is inside directors, who are currently employed by the corporation. This will include the chief executive officer (CEO) (who may also hold other titles). The CEO's role is self-evident: he or she is the person most concerned with executive responsibilities, being in charge of all planning, growth, and operations. Usually immediately subordinate to the CEO is the chief operating officer (COO), who is the main officer concerned with managing day-to-day operations and who reports to the CEO. Formerly, the title of president was equivalent to CEO, but that is no longer the case in most large, complex organizations. The board may also include one or more vice presidents (sometimes titles of executive or senior vice president are used). These vice presidents may be responsible for a variety of corporate tasks, including financing, manufacturing and production, marketing, legal affairs, and research and development. Other functions can include information (data operations), human resources, and international operations. Most vice presidents will not be members of the board. They often are referred to in large corporations as senior staff officers. They constitute the executive cadre in large for-profit businesses and variously have responsibility for finance, human resources, research and development, legal affairs, and information systems.

Other staff officers may be included as board members, depending on the nature of the corporation. Outside directors also may be included as

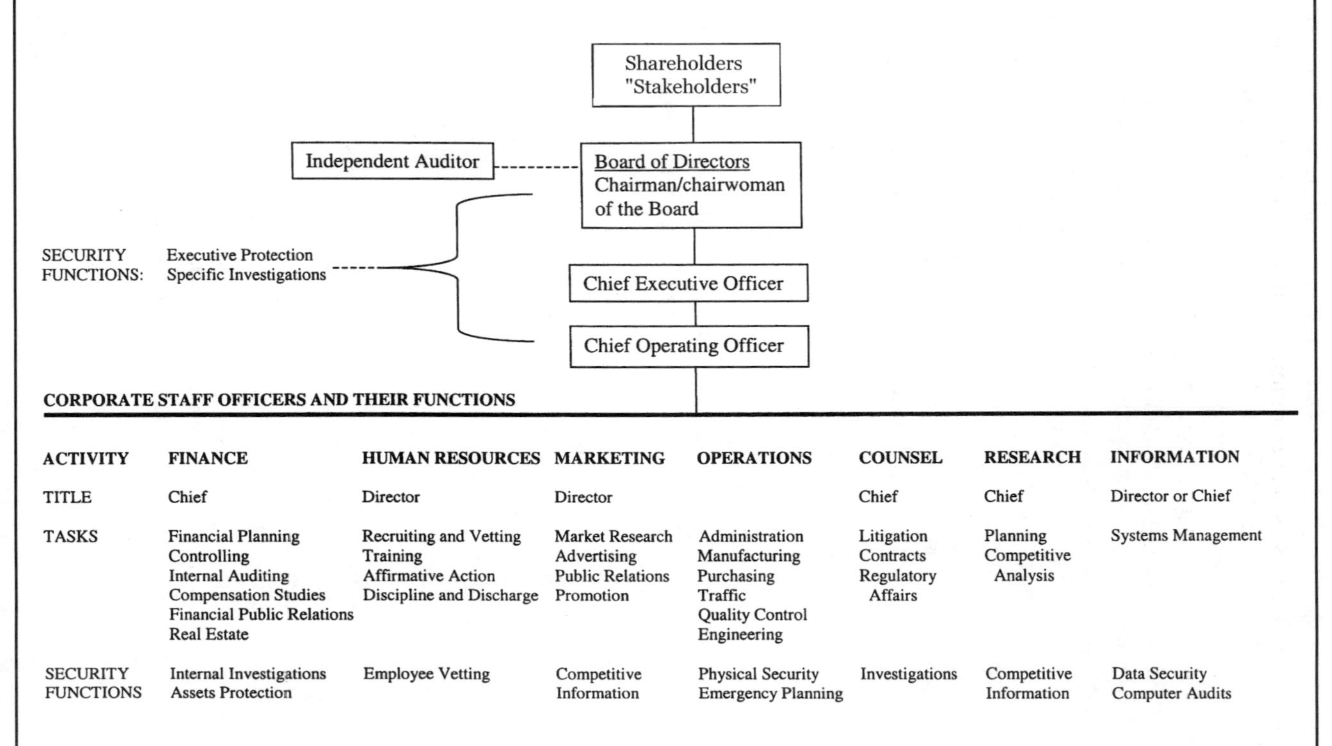

Figure 1.1. Corporate Organizational Chart Showing Related Security Functions

board members. Although they are not employees of the corporation, they do possess skills and experience believed to be valuable in directing the strategic affairs of the corporation. Sometimes, an outside director represents, or is personally, a major shareholder, or such a director may own or represent significant debt obligations of the corporation. Other outside directors may be executives of other non-competing corporations. They may thus be enlisted for board membership because of the experience they may offer to business decision-making. Still other outside directors may be academics, public figures, or diverse leaders with insight and professional connections that can aid board decision-making.

Boards meet with the frequency set in the bylaws of the corporation. In addition to full board meetings, members often serve on committees, which conduct deliberations on specific issues and make recommendations to the whole board. Typically, the board committees include executive (daily operations), public affairs, executive compensation, and audit committees. In large, publicly held corporations, executive compensation and audit committees usually are composed exclusively of outside directors. This particular composition of the board committees enables fiscal or ethical irregularities at the senior level to reach independent fact-finders for evaluations.

The audit committee receives prepared financial reports from the independent auditor, an external firm of accountants which audits financial records of the institution and reports on their soundness to the board. While serving the interests of the shareholders and corporate operations, the audit report also meets reporting requirements of the Securities and Exchange Commission. From the security standpoint, should dishonesty or ethical deviance be occurring by a senior staff officer or officers, a whistle-blower—defined as an employee who reports illegal activities of his or her employer or fellow employees to outside authorities—can contact the independent auditors, who would have a legal duty to evaluate the charge. Often, whistle-blowers have already condemned the illegal activities inside their organization, but to no avail. Thus they turn to outside authorities as a last resort. In other cases, the whistle-blower may be motivated to reveal information for personal or financial reasons.

The highest ranking executive concerned with security may interact with the board and senior corporate officers in several ways, one of which is shown in Figure 1.2. In some organizations, security directors present periodic reports to the board on significant protection issues and their implications for the organization. Additionally, the security director is likely to supervise executive protection measures, if relevant, and efforts to safeguard proprietary information at the board level, as well as elsewhere in the organization. Finally, security may be involved in specific investigations at the request of the board or in cooperation with the auditors or other senior corporate officials.

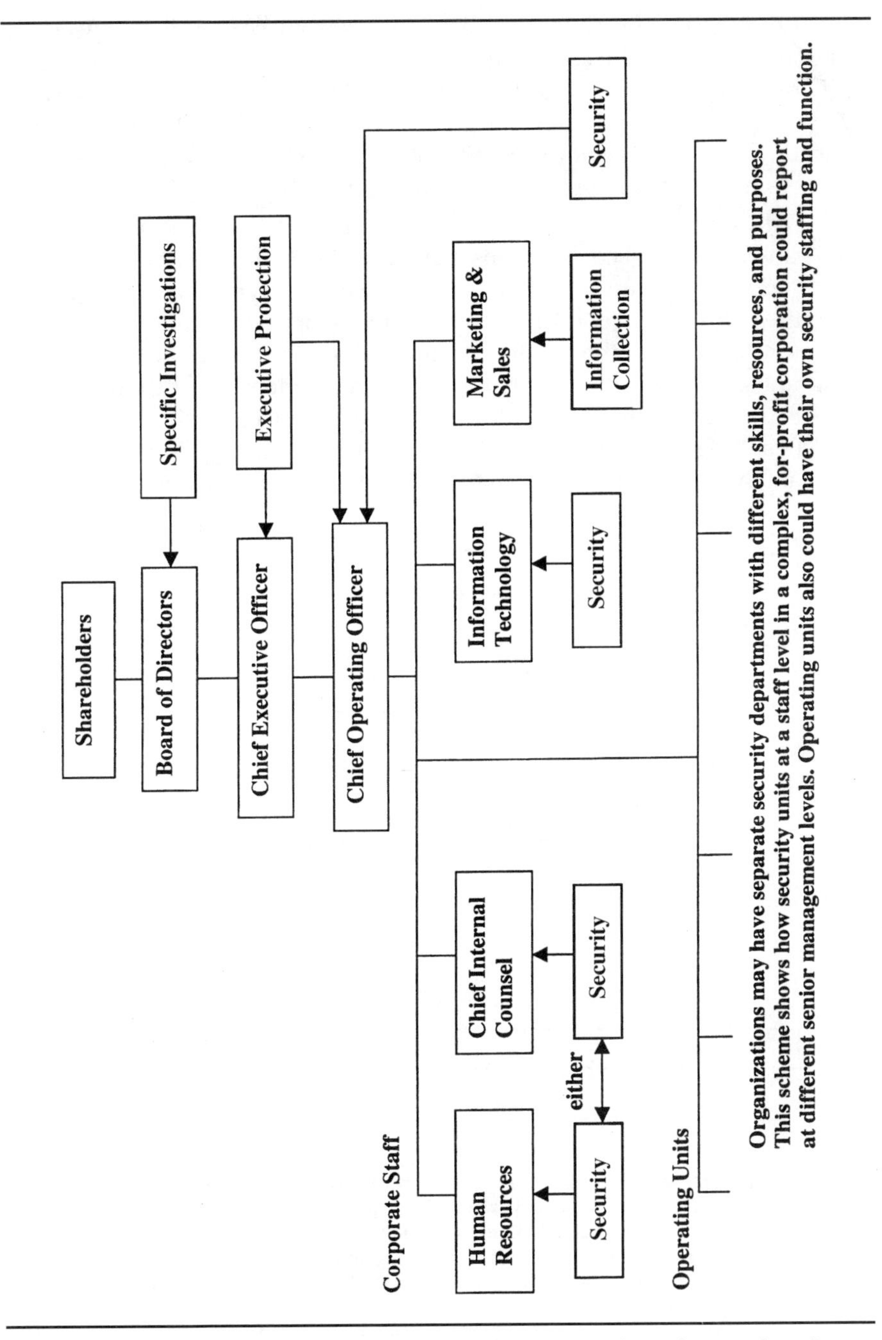

Figure 1.2. Possible Reporting Structure for a Security Director in a Large For-Profit Corporation

The organizational chart of a large for-profit corporation reflects the relationship among the corporate staff at headquarters. It may be described as hierarchical and somewhat like a pyramid, as suggested by Fayol's earlier observations. The trend in recent decades has been to shrink the headcount at headquarters. The senior executive cadre in such organizations sets policy and objectives and often provides internal consulting.

Daily operations management is less frequently found at headquarters. Large and diversified corporations may replicate the headquarters hierarchy with various operating units possessing a similar pyramidal management structure to the parent corporation. An example of such a hierarchy is shown in Figure 1.3. These subordinate corporations or companies are called operating units. The operating units function independently of headquarters to achieve their goals, though headquarters may retain a planning, monitoring, and consulting role. Thus, a diversified corporation may have a board, a CEO, a COO, and other senior staff officers at headquarters, but also numer-

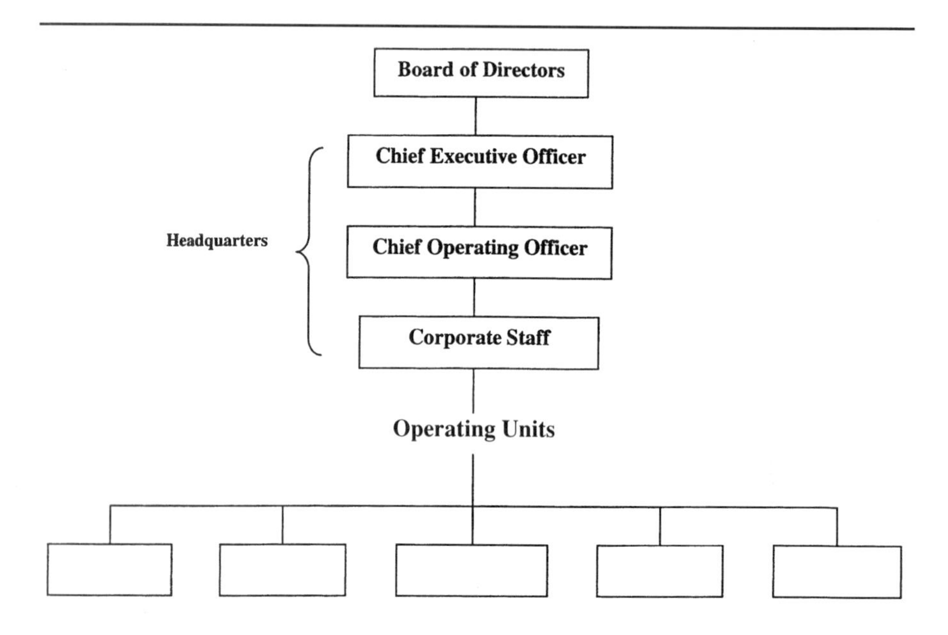

The corporate staff is small in many large contemporary operations. A corporate security director may or may not exist at the corporate level. Operating units may be divided into subordinate divisions or subsidiaries based on the nature of their work or geography. These divisions or operating units will have their own staffing requirements met, possibly with separate security developments for each one.

Source: T. Moore (December 21, 1987). "Managing: Goodbye, Corporate Staff," *Fortune*, p. 65.

Figure 1.3. How Corporate Staff Relates to Operating Units

ous operating companies within the structure, all of which may replicate the hierarchical structure at the staff level. This structure of a small headquarters senior staff followed by member operating units with varying degrees of independence from staff operations constitutes the usual situation currently encountered in large for-profit entities.

Not-for-Profit Corporations

For-profit corporations generally are what the public thinks of when reflecting on the nature of corporate structure. Many organizations do not have as their goal the necessity of returning dividends and increased value to their shareholders. These are not-for-profit (NFP) organizations. They include educational, healthcare, and research institutions, as well as charities and professional associations. NFPs possess much of the same hierarchical and reporting structure as for-profit organizations. However, titles may differ; instead of a president or CEO, the leader might be called a director or administrator. The board of directors may be equivalent to a board of trustees, governors, or supervisors. No shareholders exist because the board represents the public at large, which the non-profit corporation is chartered to serve through its endeavors.

Many NFP groups are large, diversified, well known to the public, and operate with the same reporting structures and operating practices as for-profit businesses. While profit is not the motive for NFPs, the accumulation of losses is not an objective either. In reality, NFPs face most of the same kinds of management issues common to for-profit organizations. Therefore, a director of security possesses analogous responsibilities and creates similar types of programs in NFPs as in for-profit entities.

GOVERNMENT SECURITY OPERATIONS

Government has an obligation to the public to operate effectively. This includes reducing losses, waste, error, and risks to the lowest practicable level. Depending on the size and complexity of such units, government may achieve its goals with a variety of resources. They may include law enforcement personnel delegated to internal protective functions or independent police or security units. Also, many large government units possess inspectors-general to investigate internal allegations of improper behavior.

LAYERS OF MANAGEMENT

The management structure of large organizations appears on paper like a pyramid. This reflects the hierarchical structure of the organization. For operations to operate efficiently, management often is divided into several categories: senior management (includes the staff officers most concerned with strategy, planning, and consolidation of results from subordinate units); middle management (includes numerous support roles with more restricted planning and strategizing, while operational tasks are greater); and first-line management (includes those most concerned with the daily work product of the organization and who have diminished planning activities).

SECURITY IN THE ORGANIZATIONAL HIERARCHY

In a large, diversified organization, the highest officer concerned with protection of assets from loss may have the title vice president. He or she is usually categorized as working within middle management. The security vice president reports directly to a senior officer, who may differ by title according to the type of industry involved. For example, in research-oriented businesses, the security chief often reports to the chief internal counsel; in manufacturing firms, reporting tends to be with the function concerned with operations or production; in service businesses, reporting generally occurs to the director of human resources. These reporting relationships are variable and other reporting structures are common.

While the top corporate security director usually is classified in middle management, this categorization should not be regarded as inconsequential or unimportant. Security directors frequently provide reports to the board of directors and may routinely interact with all senior officers of the corporation in providing pertinent services.

STRUCTURE OF A COMPLEX SECURITY DEPARTMENT

Security or loss prevention departments can possess considerable variation. Further, the structure of such departments is likely to change over time. For example, if security officer services are contracted-out, supervision of the contract is still required, though the total number of proprietary employees required will be reduced considerably by the out-contracting process. A typical security department is apt to oversee propriety personnel, contract staff, and internal consulting services, as shown in Figure 1.4.

A security department may incorporate considerable breadth and diversification in its resources and duties. It reflects the guarding, alarm monitoring, and asset moving and protection found in most organizations, as shown in Table 1.1. Additionally, it reflects the internal consulting, risk management, data protection, investigation, and human resources tasks often performed by security departments. Security operations also audit programs to determine how loss prevention efforts can be improved.

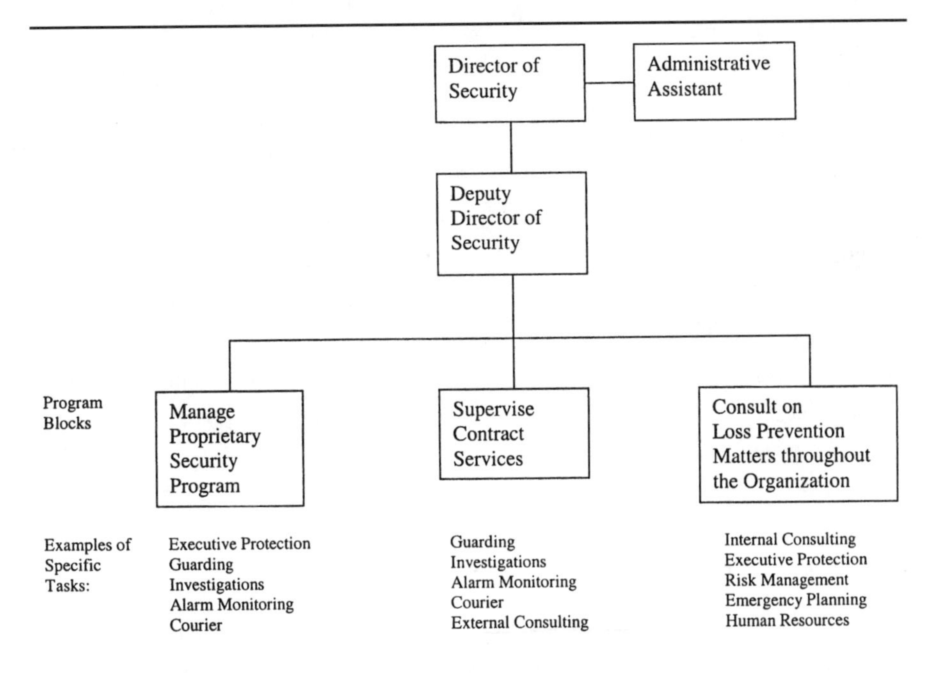

Considerable variation exists in the way security units are organized. In this example, two managers aided by an administrative assistant are responsible for proprietary (in-house) staff and supervising contract security services. They also act as internal consultants for other management issues where their skills could be valuable.

Figure 1.4. Work Relationships of a Security Program

Table 1.1. Types of Services Potentially Offered by Large, Complex Security Programs

Services	Frequency
Alarm Monitoring	High
Computer Security	High
Competitive Intelligence	Low
Emergency Planning	Moderate/High
Executive Protection	Moderate
Facilities Management	Moderate
Guarding—Propriety or Contract	High

(continues)

Table 1.1. *(Continued)*

Services	Frequency
High Security Courier	Low/Moderate
Investigations	High
Loss Prevention Consulting	Moderate
Polygraph	Low
Regulatory Compliance	Low/Moderate
Pre-Employment Screening	High
Risk Management	Low/Moderate
Safety Audits	Low/Moderate
Security Training, Awareness	High

Related to this function is risk management, which is concerned primarily with property, casualty, and liability insurance of an organization. In this case, as risks are reduced, the organization may benefit from lower insurance premiums, fewer restrictions, and benefits in coverage achieved through improved security operations.

ETHICS AND SECURITY OPERATIONS

Ethics relates to moral actions, conduct, motive, and character. It is professionally the right or befitting action within its context. While a criminal act generally is also a breach of moral conduct, ethics includes numerous behaviors that fall short of breaching criminal or civil laws. The widely heard cliché is that "ethics start at the top" in any organization. As Ira Somerson, an industry consultant, noted: "When busy CEOs take time to discuss ethical issues in their work, the message soon filters down."[10]

Seminal research on workplace deviancy was conducted by academics John P. Clark and Richard C. Hollinger.[11] Over 9,500 employees at all levels were queried in three geographical areas, representing numerous types of public and private workplaces. Results from the Clark-Hollinger study show that the level of self-reporting workplace deviance differs widely and generally is not related to income. Surely not all protection employees are above reproach ethically. Yet security personnel were assessed in all employment segments and ranked among the highest in ethical standards. This finding may be due to the fact that security personnel tend to be selected for having higher ethical behavior. Another explanation could be that security practitioners have less opportunity for workplace deviance due to the nature of their job design.

In many organizations, operational security personnel are regarded as ethical arbiters, or are normally part of the facility's ethical resources. At such organizations, managers are likely to be involved in setting, promoting, and managing ethical programs. They may:

- Draft a corporate ethics policy and disseminate it broadly.
- Emphasize the importance of ethical standards at new employee orientations.
- Provide new employees with a workplace ethics statement they may sign.
- Establish mechanisms whereby someone with an ethical concern may be heard confidentially and non-judgmentally.
- Investigate promptly and thoroughly all allegations of unethical behavior and refer the results of such efforts to appropriate authorities.

The motivation for the growing emphasis on ethics has many bases. Some executives claim that ethical behavior is morally proper and that is why they believe in it. Others would agree and discreetly add that voluntary ethical standards decrease public censure and chances of unwelcome litigation and legislation. But more than this is at stake. Perhaps the biggest factor behind the wave of ethical enlightenment is that such behavior is good business. Put differently, if only one part of an organization is perceived as being unethical, the entire organization can be tainted and potentially devastated in the process.

The American Society for Industrial Security (ASIS) promulgates a Code of Ethics (see Appendix A). Violators who come to the attention of the ASIS Ethical Standards Committee are given the opportunity to explain their perceived misconduct. Expulsion from ASIS is one of the consequences for those persons who deviate from the code and whose cases are considered by the Ethical Standards Committee and found in violation of established practices.

Other professional and trade organizations concerned with loss prevention also possess codes of ethics and good conduct. Some of these are the Academy of Security Educators and Trainers, the Business Espionage Controls and Countermeasures Association, the International Association for Healthcare Security and Safety, the National Burglar and Fire Alarm Association, and the National Council of Investigation and Security Services. This list is not meant to be comprehensive. The point is that security practitioners generally take ethics as a serious, profound reflection of their responsibilities to their colleagues, employees, and clients—and to society as a whole. Such ethical structures usually permit censure, suspension, and expulsion as possible sanctions for errant members. Normally, the person accused of unethical behavior has an opportunity to respond to the charges at a specially convened board to hear charges and responses. The appointed group then collects the facts in the situation, arrives at a conclusion, and may report its findings to the full group for a final consideration.

SUMMARY

Organizational concerns of corporations only became the object of research in the late 20th century. Security operation as a discipline is thus less than two generations old. Successful security operations are critical to the growth and stability of organizations of any size and complexity. While usually a part of middle management, security operations are concerned with performance throughout the entire organization. The functions of the executive charged with security operations are diverse and subject to change according to the primary operation of the organization. The ethical nature of the chief executive often influences the behavior of subordinate employees and others concerned with the operation. Security practitioners generally are viewed as exponents of an organization's ethical policy and program and frequently are involved in establishing and managing such a policy.

DISCUSSION AND REVIEW

1. What is the essence of "the art" of contemporary security practice?
2. When did the era of modern management emerge? When did protection management appear as a distinct managerial function?
3. Briefly describe the purpose of an executive within contemporary organizations.
4. The managerial process involves a sequence of interrelated activities. What are they and why does each have significance?
5. What are the similarities on Henri Fayol's categorizations of the workplace and a typical operation today? How are Fayol's descriptives similar to contemporary organizational structure and activity? What differences exist between his observations and the present place of work?
6. What were the contributions of scientific management to the contemporary workplace?
7. How does "outsourcing" affect current security practices?
8. Describe the connections between situational crime prevention and research applications for loss problems or concerns.
9. Explain how the structure of the organization permits recourse to investigate and respond to allegations of improper behavior, even at the highest level.

10. Describe the role of security managers in establishing policies and maintaining standards in ethical issues within the workplace.

ENDNOTES

[1] P.F. Drucker (1985). *The Effective Executive.* New York, NY: HarperBusiness, p. 8. Also: Nancy Stone, (Ed.) (1998). *Peter Drucker on the Profession of Management.* Cambridge, MA: Harvard Business School.

[2] H. Fayol (1984). *General and Industrial Management,* revised by Erwin Gray, New York, NY: Institute of Electrical and Electronics Engineers, p. 11.

[3] M. Weber (1947). *The Theory of Social and Economic Organization,* trans. A.M. Anderson and T. Parsons, ed. T. Parsons. New York, NY: Free Press.

[4] C.I. Barnard (1938). *The Functions of the Executive.* Cambridge, MA: Harvard University Press. Also: L.A. Hill (1992). *Becoming a Manager.* Boston, MA: Harvard Business School Press.

[5] F.W. Taylor (1911). *Principles of Scientific Management.* New York, NY: Harper & Brothers.

[6] F.B. Gilbreth (1972). *Motion Study.* Easton, PA: Hive Publishing Company.

[7] J.P. Tomain (1987). *Nuclear Power Transformation.* Bloomington and Indianapolis, IN: Indiana University Press, p. 36

[8] D.R. Dalton (1995). *Security Management: Business Strategies for Success,* Boston, MA: Butterworth-Heinemann.

[9] J.K. Barefoot (1995). *Undercover Investigations.* 3rd ed. Boston, MA: Butterworth-Heinemann.

[10] W.C. Cunningham, J.J. Strauchs, and C.W. Van Meter (1990). *Private Security Trends 1970 to 2000: The Hallcrest Report II.* Boston, MA: Butterworth-Heinemann, p. 49.

[11] J.P. Clark and R.C. Hollinger (1983). *Theft by Employees.* Lexington, MA: Lexington Books.

ADDITIONAL REFERENCES

E.J. Criscuoli, Jr. (1988). "The Time Has Come to Acknowledge Security as a Profession." *Annals AAPSS* 498:99.

C.H. Davidson (1989). "Toward a New Discipline of Security Management: The Need for Security Management to Stand Alone as a Management Science." *Security J.,* 1(1):3–13.

C.D. Shearing and P.C. Stenning (1983). "Private Security: Implications for Social Control." *Social Problems* 30(5):503–504.

2

CORE COMPETENCIES TO INITIATE EFFECTIVE PROTECTION PROGRAMS

> Private security is more than twice the size of federal, state, and local law enforcement combined.
>
> —*The Hallcrest Report II*

Security activities for an organization are often centered within a department concerned with delivering value to the organization through services. As the previous chapter indicated, much flux occurs in the nature of organizations themselves and within various departments providing such services. Still, some generalizations can be made that will be appropriate for various types of managerial situations. This chapter will examine the means whereby organizations with dedicated security departments are organized to serve the entire operation. The chapter further looks at the relationship between organizations that contract-out for routine security services. We begin by examining core competencies of security operations.

CORE COMPETENCIES OF SECURITY OPERATIONS

Core competencies refer to the fundamental abilities a protective program needs in order for it to deliver services. These needs will vary according to the type of organization, its size and geography, recent history, criticality of resources, vulnerability to losses, and other factors. No single executive is expected to be competent in all demands required of the position, but the list below serves as a means of generating thought as to what a protective operation's value to the organization is or should be. This list is dynamic and reflects the changing nature of the requirements of security programs and of the expectations of people heading them.

Initiating and managing security programs. As discussed in the previous chapter, problems and opportunities require a response by appropriate programs. The identification of these situations, their analysis after fact-finding, the organizing of an appropriate program, the appointing of a deputy to operate the new program, and its supervision and constant improvement are reasonable expectations of a high-performing operations professional. Three skills are mentioned first in terms of core competencies:

1. **Initiating new programs.** This was discussed in the previous chapter.
2. **Operating existing programs.** The ability to initiate a successful program is a strategic skill, whereas the operation of existing programs is less challenging. Nonetheless, this is the basis of most daily work and includes opportunities for creativity and constant program improvement, much like what occurs in the initiating of such activities. Another core skill is the ability to collect information that is critical to the operation and assess the success of ongoing programs (see Box 2.1). The manager or director for such operations is expected to manage the budget for these activities (see Chapter 8).

Box 2.1. Collecting and Measuring What's Important

Once goals are set, data are needed to evaluate how successfully aims are being reached. Relevant data collection can also point to other issues that require more attention than was initially apparent. The use of computer programs helps make the capture of incidents and services easier than the previous manual form of collecting inputs. Analysis also is improved. This is partially why security operations have improved over the years. At a minimum, incidents must be collected so that they can be measured for any relevant trend. At the maximum, all service events performed by the security operation may be collected and measured for the same reason. The following is the information collected to assess significant programmatic developments.

Number of crime incidents, including:

- Robbery
- Aggravated assault
- Other assault

Box 2.1. *(Continued)*

- Burglary
- Felony-theft
- Motor vehicle theft
- Forgery and counterfeiting
- Fraud and embezzlement
- Vandalism
- Trespassing
- Other

Number of non-criminal emergencies, including:

- Accidents (within the facility)
- Accidents (automotive)
- Accidents (other)
- Computer service response
- Dangerous behavior
- Fire and smoke conditions
- False alarms
- Loss of utilities
- Malfunctions of critical equipment
- Water and flood damage
- Wind damage
- Other

Number of service activities, including:

- Complaints and miscellaneous
- Employee records checks
- Escort service
- Executive protection
- Investigations (internal)
- Investigation (external)
- Key runs
- Lock or key service
- Lost and found

---

3. **Handling personnel administration.** The recruiting, screening, hiring, training, supervising, promoting, disciplining, terminating, and conducting of other personnel-related activities are expectations of high-performance security operations (see Chapters 3 through 7).

Initiating new programs, operating existing ones, and dealing with personnel issues are expectations of all managers, not just those concerned with assets protection. However, some tasks are specific to loss prevention staff:

- **Contract services management.** Since an increasing portion of security services are provided by contract personnel, operations must be able to select, motivate, supervise, and discipline contract vendors and their personnel so that goals are met (see Chapter 9). Such investigations may be contracted to investigators or consultants, but the manager in charge of the program is likely to monitor the assignment to assure that objectives are pursued diligently.
- **Assess security technology.** Security practitioners are not expected to be engineers. However, they are required to be familiar with current technologies to serve the protective objectives of the organization. They should further be able to procure such technology and services under favorable terms for management (see Chapter 10).
- **Other expectations.** As indicated above, security programs have considerable variations in their operational goals. Therefore, some organizations will have such core competency objectives as executive protection, international affairs, risk management, competitive intelligence, data security, emergency planning and response, and other topics.

Some general personal characteristics are also critical for all high-performance executives:

- **Communications.** Security leaders and their programs obtain and retain support by successfully serving various "customer bases" (management, employees, visitors). Those responsible for security programs constantly must enunciate through in-house educational efforts and other means what security does and why it is relevant, without being repetitious and boring.
- **Leadership.** Security programs often require various groups to take—or not take—actions against their will. Personal leadership by persons responsible for the program helps retain the credibility and support such programs require (see Chapter 11).

HOW CONTEMPORARY SECURITY SERVICES HAVE EVOLVED

Many state governments have used licensed security guard companies and investigators for decades. But by the late 1960s, no independent research of the growing security industry had taken place. That was soon to change. In 1970, contemporary security practices in the U.S. were described and evaluated by the scathing and influential *Rand Report*. This document represented the first time the burgeoning security industry received a systematic analysis from a disinterested research group. With a grant from the Law Enforcement Assistance Administration (LEAA), the Rand Corporation began, in 1970, a 16-month investigation of "private police" in the U.S. The authors, James S. Kakalik and Sorrel Wildhorn, were lawyers trained as policy analysts and employed by Rand in Santa Monica, California, to assess private security personnel as they affected public policy.

For starters, the *Rand Report* impugned correctly the level of employment standards then common for private security personnel. The authors observed:

> The typical security guard is an aging white male, poorly educated, usually untrained, and very poorly paid. Depending on where in the country he works, what type of employer he works for (contract guard agency, in-house firm, or government), and similar factors, he averages between 40 and 55 years of age, has had little education beyond the ninth grade, and has had a few years of experience in private security. . . . He often receives few fringe benefits; at best, fringe benefits may amount to 10 percent of wages. But since the turnover rate is high in contract agencies, many employees never work the 6 months or 1 year required to become eligible for certain of these benefits.[1]

The *Rand Report* continued with its litany of harsh observations of private security practices: weak pre-employment screening, high turnover in the industry, and a lack of meaningful licensing standards. That depiction has not changed much in the opinions of many contemporary critics, but such views tend to be unanalytical and episodic. Much evidence indicates that security practices have improved, though the persistence of numerous substandard protection service providers and programs remains a reality. In the decades following publication of the five-volume *Rand Report*, considerable advancement in the industry occurred, though at a measured, slow rate.

The next significant official scrutiny of private security services also emanated from LEAA funding. In 1972, LEAA created the National Advisory Committee on Criminal Justice Standards and Goals. This group undertook a number of detailed, analytical reviews of various issues connected with criminal justice. One was the Private Security Task Force (PSTF). Following a series of discussions and inquiries stretching over 18 months, the PSTF issued its comprehensive report in 1976[2] (see Appendix B). The PSTF was composed of a variety of individuals, including law enforcement officials, corporate security directors, and an executive of a major security services company. The report identified almost 80 goals and standards for private security. The list encompassed such areas as licensing, regulations, consumer services, personnel training, crime prevention systems, and conduct and ethics. The *Report of the Private Security Task Force* was not intended as an impetus to achieve federal legislation to regulate aspects of the security industry; rather, its intention was to identify significant issues that would stimulate local and state laws and codes to be passed or strengthened. Also, it would serve as an industry guide to improvement in procedures.

These two documents—from Rand and the PSTF—served to inform legislators, regulators, general management, the security industry, the media, and the public at large about issues relating to private security. In some ways, changes have occurred in almost all aspects of security services delivery; in other aspects, change has been barely discernable. Yet the industry and the performance of security services—both proprietary and contract—have experienced steady growth from a period beginning at least since the end of World War II to the present. Why? Several reasons exist for the growth of private security. Senior executives do not accept their subordinates' recommendations for increased security expenditures—or any other kinds of financial allocation—without due justification. Such asset allocations are generally predicated upon defined needs that the organization has identified and that, therefore, make the existence of security expenditures an informed imperative, rather than a capricious decision.

WHAT DRIVES SECURITY OPERATIONS?

Security operations normally do not exist within an organization for a single reason. Typically, numerous factors interweave to justify commitments to fund security operations. These will vary in significance according to a wide variety of factors relating to the degree of risk appetite, demand for internal services, and the value of assets to be protected in the workplace. Here are the leading factors that underpin the reason for being and growth of contemporary security programs and that drive their growth and vitality:

- **Cost savings.** An operating security program may reduce losses to an organization that will in turn offset the apparent cost of the security services. For example, employees may be unwilling to work certain shifts because they feel unsafe. Their replacement could be costly. The presence of a security patrol could make the perilous shift a possibility.
- **Risk mitigation.** Security is a fundamental necessity for corporate endurance and success. Lack of adequate protection could lead to devastating results. Security programs identify weaknesses and seek to reduce risk (see Box 2.2).*

* The goal of security management programs generally is not to reduce risks as low as possible. That would be excessively burdensome and costly. Rather, it is to reduce risks to an acceptable or practicable level.

Box 2.2. A Case of Inadequate Security: The Demise of Pan Am World Airways

---

The advertisement proclaimed: "Pan Am Makes the Going Great!" Pan Am World Airways was the first transatlantic carrier to provide regularly scheduled flights. For most of the 20th century, Pan Am possessed its own distinctive cachet. Pilots, flight attendants, ground crew, and passengers were attracted to the carrier for its élan and quality services, and the airline prospered. In fact, one of the major mid-town skyscrapers constructed in Manhattan in 1960s was named for the airline (now the MetLife building).

In the late 1960s, airlines became aware of their vulnerabilities to breaches of security. Numerous planes were skyjacked, and pre-board screening became a routine requirement. The impetus for international air carriers to improve security had become a priority. Almost all international air carriers saw their business decline as a result of travelers' fears of skyjacking.

Most airlines developed security programs to provide for their own needs. For some, attractiveness to contract-out protective services to other airlines became enticing. One of these was Pan Am, which created, in 1986, a wholly owned subsidiary, Alert Management Systems, Inc., to provide services to Pan Am and other customers. It was financed, in part, by a surcharge of $5 per ticket on each transatlantic flight.

Pan Am's Alert Management Systems was presumably a high-visibility service provider and revenue generator for the parent company. Yet the security "was more for show than genuine security," according to Steven Emerson and Brian Duffy, authors of *The Fall of Pan Am 103*. When Alert Management Systems began operations at New York's J.F. Kennedy Airport, for example, Alert personnel paraded dogs throughout Pan Am's check-in counters. However, according to Alert's first president, Fred Ford, they were not dogs trained to sniff for bombs; they were merely "well-behaved German shepherds."

Pan Am retained the services of a security consultancy, Ktalav Promotion and Investment Ltd. (KPI), to critique its performance and to review operations at Frankfurt and 24 other airports. Isaac Yeffet, a former security chief for El Al Airlines, then with KPI, wrote to the airline that: "Pan Am is highly vulnerable to most forms of terrorist attack," despite the existence of Alert Management, and that "a bomb would have a good chance of getting through security" at the Frankfurt Airport. Yeffet concluded: "It appears, therefore, that Pan Am is almost totally vulnerable to a mid-air explosion through explosive charges concealed in the cargo." But Yeffet's report and Ford's request for more resources for Alert Management were ignored by Pan Am's senior management. The price of inadequate security would be

Box 2.2. *(Continued)*

---

high. On December 21, 1989, Pan Am flight 103, a Boeing 747 jet, was blown apart over Lockerbie, Scotland, killing all 259 people aboard and an additional 11 on the ground.

Pan Am's decline as a viable business did not begin with Lockerbie, but instead started in 1973 when the Arab oil embargo pushed up fuel prices at the same time as a sharp recession began. From then through the 1980s, Pan Am lost over $2 billion and only survived by selling its Pacific routes to United Airlines in 1986. But Lockerbie substantially sealed the fate of Pan Am. By 1994, the airline was bankrupt. A jury held that Pan Am and its Alert Management Systems, because of the numerous security deficiencies, were guilty of "willful misconduct" in permitting a security breach that allowed a bomb to be placed aboard the craft.

Sources: S. Emerson and B. Duffy, *The Fall of Pan Am 103*, New York: G.P. Putnam's Sons, 1990; R. Stuart, "Pan Am Ads Touting Security Plan Stir a Debate," *New York Times*, June 10, 1986; John Greenwald, "Fallen Emperors of the Air," *Time*, January 7, 1991, p. 71; and R. Sullivan, "Court Upholds Pan Am 103 Awards," *New York Times*, February 1, 1994, p. D2.

- **Income generation.** A security program is often thought by managers not involved with protection to be a "cost" to the operation, not a source of "profit." However, in some circumstances, security departments perform services that can generate income for the organization. For example, some organizations provide their own security services to other businesses or institutions and charge for them accordingly. Hence, they can become a profit center for the parent organization (see Box 2.3).

Box 2.3. Making Security a Profit Center

Profit centers are workplace activities that bring rewards from income derived by the enterprise from an unrelated entity.

Profit centers provide security services to non-competitive businesses for fees. Such activities include guarding, investigations, alarm monitoring, computer back-up services, and consulting. The process can be profitable for the service provider. The customer or client derives benefits from resources with demonstrable performance characteristics and ongoing management attention. Not all security operations can or should possess profit centers, but for some, opportunities exist and may be pursued.

- **Crime.** Violent and property crime that could occur within or near a facility or property can be deterred by the presence of security personnel, the installation and functioning of an alarm system, and good security design. This is supported by research derived from situational crime prevention studies, which confirms that pertinent measures may reduce losses from crime and other risks.
- **Fear.** The presence of trained security personnel and state-of-the-art systems make employees, vendors, and visitors feel safer at the workplace. For example, the availability of a parking lot security patrol may reduce trepidation while it lowers actual risk. In this sense, security provides a desirable service to those who use the parking lot.
- **Litigation.** The failure to have an adequate security program may leave the owner and operator of a facility vulnerable to a successful tort action for negligent security in the event that a crime or related loss occurs. The defense burden is greater if the facility has a weaker protective program than do comparable operations within the region. The existence of a security program by itself, however, does not protect the facility from successful litigation in the event an actionable offense for negligence takes place.
- **Insurance.** Organizations often are required to provide security services and systems for themselves because their property and casualty liability insurance coverage— or other specific insurance policies—mandate certain minimum protective measures.
- **Legal mandates.** In some cases, specific litigation directly requires the presence of security operations. For example, financial institutions face general obligations to maintain a security program subsequent to the Bank Security Act of 1968 and as subsequently changed.
- **Bureaucratic requirements.** Numerous governmental agencies create regulations that mandate the existence of security programs. Usually, these are the outgrowth of federal laws that contain broad language and leave the specifics to be developed by a designated federal agency. For example, the Federal Aviation Administration requires airport managers and airlines to institute a variety of protective measures, including pre-board screening of airline passengers and personnel. These regulations were the result of the passage of a measure to protect the air-traveling public.
- **Accreditation requirements.** Institutions that meet the general standards of their appropriate accreditation body sometimes also face the specific demands of security measures from such an accrediting association. For example, the Joint Commission on the Accreditation of Healthcare Organizations (JCAHO) promotes high-quality patient care through a voluntary process of accredita-

> tion, encompassing thousands of healthcare organizations. JCAHO has no specific security standards at present; however, in practice, the desirability of an appropriate security program is expressed through the "Plant, Technology, and Safety Management" section of the JCAHO 2000 *Comprehensive Accreditation Manual for Hospitals*, which requires a safe environment for institutions desiring to meet the criteria of JCAHO.

Clearly, the need for security programs and services in commerce and institutions is not derived from a single requirement, but rather from a combination of factors. The individual reasons for having a particular level of security are affected by geography, time, financing, available personnel, legal precedents pending legislation and litigation, and other considerations. Ultimately, security programs exist due to the conviction that any vulnerability eventually will lead to unfavorable consequences. This explains why security services continue to grow.

A BRIEF HISTORY OF A GROWING FIELD

Security has always been important for the protection of people and property. However, the security industry emerged as a modern business activity within the U.S. in the second half of the 19th century.[3] During this period, investigations, guarding, executive protection, consulting services, alarm monitoring and response, and armored courier services all had their origins. By the mid-20th century, corporations themselves had established proprietary security programs that initially were concerned with loss prevention and order maintenance. Managers for these programs in industrial applications often reported to engineering, maintenance, and general administrative or operational units.

The era of security as a modern management function did not arrive until the second half of the 20th century. At the end of the 1950s, a resurgent economy and the implications of Cold War protectionism vastly increased the importance of security as an organized business practice. Industry was serving the needs of military preparedness and expanding commercial inventiveness: both required adequate security measures, though of differing sorts. Such diverse interests in proprietary security were met, in 1955, with the founding of the American Society for Industrial Security (ASIS). Full members were employed usually as loss prevention directors serving for-profit and institutional organizations. Typically, they would be concerned about physical security, emergency response planning and coordination, and internal investigations. Members who worked for industrial corporations that provided products, systems, services, and research for government, especially the military, monitored extensive compliance requirements to protect information and production know-how from possible compromise. ASIS members in those years included many retired military officers. Membership also consisted of retired police officers, special agents of the Federal Bureau of Investigation and other law enforcement organizations, and persons who became responsible for security without having had any previous formal preparation in the military or law enforcement.

Security directors in these organizations often were responsible for identifying and assigning security classification to information and materials requiring protection in the national interest.[4] With the fall of the Berlin Wall in 1989, the military threat between the superpowers in the East and West diminished rapidly. The need for protection of intellectual property and of physical developments related to military requirements declined, but did not disappear. Meanwhile, other security priorities and duties emerged.

While the modern origins of professional security are related to military and industrial concerns, protection was needed in other organizations where theft, vandalism, and employee safety were issues. Retailing, distribution, general manufacturing, and many types of service businesses—especially banking—added security services at the place of work. Most security programs in the 1950s and 1960s concentrated on anti-theft and information protection measures. But by the late 1970s, numerous security programs began to absorb other management and administrative duties, including safety. In the 1990s, data security, emergency planning, and organizational ethical concerns became significant management issues.

Today, the security industry is not one entity but a series of commercial activities that overlap each other. These activities include services such as guarding, investigations, alarm monitoring, escorting, and consulting; electronics (including companies that manufacture, distribute, and add value to systems); and hardware (encompassing non-electronic, high-quality products and materials that serve above-standard protective needs). These services and products compose thousands of independent entities.

Laws That Influenced Growth

The Occupational Safety and Health Act (OSHA) of 1970 (29 USC 651 *et seq.*) was passed to develop and promulgate occupational safety and health standards. It required and established a bureaucracy to develop and issue regulations, conduct investigations and inspections to determine the status of compliance with safety and health standards and regulations, and issue citations and propose penalties for noncompliance with safety and health standards and regulations.[5] Many security directors also serve as OSHA compliance officers at their workplaces.

Other specific laws calling for increased commercial security measures were passed. Notable among these was the Bank Security Act of 1968.[6] This law was passed because bank crimes had grown steadily during the decade. The increased incidence of bank robberies, burglaries, and extortions prompted Congress to require all federally chartered banking institutions to undertake specific security measures to reduce the risk of successful criminal acts. In retrospect, the measure in itself did not reduce the growing pattern of violent and property crimes against financial institutions covered by the Act: in fact, they kept increasing for years after passage of the law.[7] This Act was significantly modified years later to make requirements more reflective of changing circumstances. For example, the original law of 1968 produced regulations for Minimum Security Devices and Procedures [12CFR21]. These included specific language related to anti-theft technology, which became less relevant with the introduction of new types of cameras and recording media. By the 1990s, bankers possessed greater latitude with regards to designing security systems most appropriate for their needs. While the Bank Security Act is not directly associated with decreasing bank crime patterns, common sense argues that such institutions must implement reasonable security measures to protect employees and the public. Perhaps without the vigorous measures that were instituted by this law, bank crime would have increased. This act is one of the few examples of government passing a measure that created a distinctive security response and in the process helped the role of security management grow in a particular sphere, in this case financial industries.

Another—and demonstrably more successful—example of how a legislative initiative causes the creation of security strategy relates to skyjacking, or the criminal high-jacking of commercial airplanes.[8] This type of crime became an issue in November 24, 1971, when a man who called himself Dan B. Cooper commandeered a Northwest Orient flight from Portland, Oregon, en route to Seattle. The passengers were released in Seattle and a ransom and parachutes were taken aboard. Cooper ordered the flight to take off for Reno and to fly at a minimum speed at low altitude. When the plane landed in Reno, Cooper, some parachutes, and $200,000 in small, used bills were gone. Within six months, six other attempted skyjackings with parachute demands occurred. Most were unsuccessful, but the crime pattern threatened the air transportation industry.

Historically, Cooper's was not the first skyjacking: that occurred in Peru in the 1930s. Yet from 1930 to 1967, only nine incidents of air piracy occurred, of which only four were successful. In 1968, 17 skyjacking incidents took place, of which 13 were successful. The next year, the total jumped to 40 incidents, of which 33 were successful. For the next three years, as the private and public sectors fought the trend, the number of attempted skyjackings on U.S.-scheduled aircraft held steady, while the number of successful incidents dropped from 17 in 1970, to 11 the next year, and then eight in 1972. With implementation of control measures, skyjacking attempts dropped for the period 1973–1979 to 31, with only three being successful. The measures succeeded in reducing the risks of air piracy and restored confidence in travelers. This is a convincing example of how a coordinated security policy can reduce an international problem.

While loss clearly helped increase the formation of security programs in banking and aviation, many other industries were developing security programs without the force of directive legislation. For example, no legislation requires the retail industry to engage in security measures; however, substantial measurable losses in retailing have produced a cadre of managers who evaluate losses and seek to mitigate them. The same holds true for other aspects of commerce and industry—public, private, and not-for-profit sectors alike. Non-aviation transporters, distributors, mining and processing facilities, and a myriad of service organizations have all enhanced their security policies and systems.

These factors stimulated the growth of proprietary security programs. In the process, this trend stimulated the emergence of the contemporary security industry, which expanded to fill the growing services required by this commercial impetus. Meanwhile, many proprietary organizations partially or fully shifted security activities to outside services and contractors. We examine this issue next.

THE GROWTH OF THE MODERN PROTECTIVE INDUSTRY

The security industry that grew from the 19th century to the present developed not as a single category of commercial activity but as a group of independent interrelated services and security product manufacturers. Although little convergence in ownership occurred over the various businesses, the industry did grow. The U.S. security industry surged at a rate considerably exceeding that of the nominal gross domestic product (GDP) for the same period of time. (The GDP is the aggregate measure of economic activity excluding income from foreign investments.)

For example, for the years 1987 to 1989, the U.S. security industry grew at a rate of more than 11 percent annually, from $15.9 billion to $19.8 billion. This compares with an average annual growth rate of 7.5 percent for the GDP during the same period.[9] During this period, gross revenues from security services—security guarding and related services, central station monitoring, and armored car services—grew from $11.1 billion to $13.8 billion, an increase of 24.3 percent. Meanwhile, the retail value of electronic security equipment sales grew from $4.8 billion to almost $6 billion, an increase of 25 percent. Within these categories of products and services, numerous product lines grew or emerged to serve the varied interests of proprietary security operations.

Between 1992 and 1996, growth of the industry continued, but at a slower rate. Total revenues increased about 8 percent, from $29.1 billion to $39.3 billion. Meanwhile, the nation's GDP grew at an average annual rate of

4.2 percent.[10] Therefore, this increase actually represented an improvement relative to the period 1987 to 1989. Indeed, security revenues as a percentage of GDP grew from 0.46 percent in 1992 to 0.52 percent in 1996 and are projected to reach 0.69 percent in 2004, as shown in Table 2.1.

Historically, most organizations directed security operations through a proprietary department. This implied that the organization "owned" the unit that provided its security services. Proprietary security directors and associates remained central to the planning, organizing, and management of such services. However, as the previous chapter observed, security departments have been diminished in terms of proprietary personnel as more resources have been provided by outside service suppliers. The proprietary organization became the client, customer, or contractee of the supplier or contractor. Meanwhile, organizations generally turned to outside suppliers for security products and systems—hardware, software, and electronics. Therefore, this relationship has not reflected much change.

Leading Edge Group, a market research firm, divides the security industry into two major components: services and electronics. Robust growth in the services and electronic security equipment segments occurred in the 1990s and is expected to continue to 2004 and beyond, as shown in Table 2.2. The reasons for and extent to which these industry segments are changing reflects changes taking place within the security industry's clients, namely, proprietary organizations, which contract or purchase such services, products or systems.

Security Services

Three out of four dollars expended for purchased protection needs are allocated for personnel-based services. These activities require the support of equipment and technology, but the bulk of the expenditures are for direct

Table 2.1. Security Service Revenues and Equipment Sales by Product Function

Factor	1992	1996	2000	2004
GDP (in $ Billions)	$6,244	$7,576	$8,869	$10,147
Security Revenues as a Percentage of GDP	0.46%	0.52%	0.60%	0.69%
Total Security Revenues (in $ Millions)	$29,109	$39,397	$53,424	$70,271
Security Service Revenues	24,013	31,651	41,410	53,600
Electronics Security Equipment Sales	5,096	7,746	12,014	16,671

Source: U.S. Department of Commerce; Leading Edge Reports, Study LE907

Table 2.2. Security Service Revenues and Equipment Sales by Type (in $ Millions)

Services/Equipment Purchased	1996	2000	2004
Security Services Revenues	**$31,651**	**$41,410**	**$53,600**
Central Station Services	13,190	17,610	22,020
Security Guard Services	13,300	17,000	22,700
Consultant Services	661	1,000	1,480
Private Investigation Services	3,500	4,600	6,000
Armored Car Services	1,000	1,200	1,400
Electronic Security Equipment Sales	**$7,746**	**$12,014**	**$16,671**
Vehicle Security Systems	1,596	2,187	2,699
Intrusion Detection Equipment	1,840	2,560	3,030
Fire Detection Equipment	1,100	1,415	1,690
CCTV Equipment	818	1,528	2,534
Electronic Article Surveillance	703	1,100	1,690
Metal Detection Equipment	55	84	123
X-ray Inspection Equipment	76	100	111
Bomb Detection Equipment	53	80	121
Secure Telephone Equipment	112	157	205
Access Control Equipment	548	900	1,480
Computer Security Equipment	845	1,903	2,858
Total	**$39,397**	**$53,424**	**$70,271**

Source: Leading Edge Reports, Study LE907

compensation and benefits of personnel and their support. Personnel-based services may be divided into five categories:

1. **Security guard services.** This category absorbs a number of protection workers with varying responsibilities, including extensive public service contact ("officers"); assets protection specialists ("guards"); receptionists; patrol officers; executive protection personnel; watchmen; timekeepers; and others. These personnel provide both the important "visible security" presence required by many protective objectives and the less apparent, behind-the-scenes securing of physical assets by deterring, detecting, and reporting activity related to threats to people or property.

2. **Central station services.** The computing revolution has permitted organizations to monitor people, places, and events with efficiency and accuracy. Originally, central stations provided burglar alarm, fire notification, and messenger-requesting signals. Today, burglar and fire alarms remain the core of such services, but numerous other monitoring functions are also available. Expenditures for such services have grown steadily in the last 25 years, and are close to the amount paid for security guard services.
3. **Private investigation services.** In the past, investigators were linked to the resolution of specific losses. That's still true, but the assignments faced by investigators are broader. Investigators today are much more likely to conduct evaluations to make sure that corporate policies are maintained, such as by assuring that licensing fees and payments are properly documented. Investigators and consultants are frequently integral in due diligence (that is, the vigilant care needed in a given situation) prior to acquiring an asset or related to litigation involving the organization. This category of contracted service continues to grow in importance for proprietary security programs.
4. **Armored car services.** Ever since Washington Perry Brink started his transport business with a horse and a wagon in the 19th century, customers have turned to outside organizations to physically move assets. In addition to transporting cash, these services provide activities such as the servicing of automated teller machines (ATMs) and the transporting of high-value non-cash assets like jewelry and computer tapes and documents. They also handle aspects of cash management for financial organizations.
5. **Consultant and other services.** When particular problems emerge, security operations managers often turn to consultants with special expertise in particular activities. Most major industries and most types of specific security concerns—for example, data protection and financial investigations—can retain the services of such persons or their organizations for a defined period of time to achieve the desired goals. Due to the shrinking of central staff management in the last 25 years, managerial resources were reduced and the use of outside consultants replaced in-house management capabilities. Meanwhile, other services are equally significant to organizations. For example, in the event of a loss of computer processing capability, an organization may turn to facilities that have compatible hardware that may be commandeered for immediate use.
6. **Electronic security equipment.** To provide more control at less cost, managers turn to electronic security equipment, which can augment, supplement, and verify the actions of security personnel. Though smaller than personnel-intensive services, capital outlay for such

equipment is growing steadily. Here are the ways these funds are allocated:

- **Intrusion detection equipment.** These devices and systems indicate the unauthorized passage of individuals into a protected area.
- **Vehicle security systems.** These products deter and detect the theft or removal of cars, trucks, vans, and other mobile conveyances.
- **Computer security equipment.** Because of the importance of data protection, systems to protect information assets are growing at a substantial rate and play an important role in loss deterrence for data systems.
- **Closed Circuit Television (CCTV) equipment.** Security operations increasingly use CCTV to monitor and record activities.
- **Fire detection equipment.** Facilities are required to use fire detection systems by codes, standards, insurance requirements, or common sense.
- **Electronic Article Surveillance (EAS) systems.** These systems are primarily used to control the losses of retail merchandise. However, EAS systems also may be applied to broader applications of assets control.
- **Access control equipment.** Systems to efficiently allow or deny entrance make an important contribution to operating a secure facility. This category is closely related technically to intrusion detection equipment.
- **Secure telephone equipment.** This equipment protects telephone and data transmissions from unauthorized interception.
- **X-ray inspection equipment.** X-ray impressions may identify the presence of weapons or contraband material hidden in packages or on persons.
- **Metal detection equipment.** These systems identify metal content hidden within packages or on persons. Wide applications are found at airport pre-board screening and the checking of people and hand parcels entering at-risk locations.
- **Bomb detection equipment.** Still another way of identifying contraband is the use of technology to identify the chemical signatures of such materials. These may result in an alarm being sounded or in a security officer setting aside a suspicious object for further evaluation.

7. **Security hardware.** In addition to personnel and electronic devices, systems, and software, security operations often require the purchase and use of particular products and materials. Hence, any object or material that is not electronic is categorized as hardware. This category also is substantial, placed at about $3.775 billion in

1996, and projected to grow at a rate of 4.8 percent to $4.775 billion by 2001.[11] A door that is required to be intrusion-resistant would be included in this category; a normal door would not be. Locking devices including door locks, padlocks, deadbolts, latches, as well as security storage equipment like safes and vaults and fire extinguishers, and related products are also included in this category.

HOW SECURITY MANAGERS RANK PRIORITIES

Security practitioners deploy proprietary and contract services and electronic security systems in order to achieve a wide range of objectives. Principally, tasks relating to personnel matters, budgeting, training, and program planning and administration comprise most of the time available in a manager's week. However, in addition to routine program management, numerous security-related threats require consideration and action. These change with the times, geography, and the nature of particular industries.

The security threats and management issues listed by respondents to a *Fortune* 1000 study are various, as shown in Table 2.3. (These issues are discussed in greater detail in the final chapter of this book.) The following is a list of these issues:

- **Workplace violence.** For most people, the workplace is a safe place. But occasionally, violence intrudes into an otherwise nonviolent environment. Despite the rarity of the events, the issue cannot be dismissed by managers; indeed, it is at or near the top of any list of workplace concerns.[12] Diligent managers must take measures to make employees feel safe on the job, and at the same time deter disgruntled employees, terminated workers, enraged customers or clients, and others from untoward action.
- **Crisis management** (also called **contingency management**). Emergencies in organizations can be due to nature-based, people-based, design-based, and technology-based factors, among others.[13] The crisis or contingency manager seeks to avoid a crisis from occurring by establishing contingency plans.
- **Executive protection.** The protection of senior officials from risk has become a highly evolved skill that draws upon management planning and analysis.[14] Protection choreography, advance security preparations, domestic and international travel assessment, and physical training are all involved in the process.
- **Fraud.** The investigation and prevention of fraud (that is, false representation or intentional perversion of truth to induce another to part with something valuable) is an important task in any organization.[15]

Table 2.3. Most Important Security Threats and Management Issues

Rank	Security Threat
1.	Workplace violence
2.	Crisis management/executive protection
3.	Fraud/white-collar crime
4.	Employee screening concerns
5.	Computer crime (hardware/software theft)
6.	General employee theft
7.	Computer crime (Internet/intranet security)
8.	Drugs in the workplace
9.	Unethical business conduct
10.	Property crime
11.	Sexual harassment/EEOC
12.	Business espionage/theft of trade secrets
13.	Litigation liability ("inadequate security")
14.	Terrorism
15.	Litigation (negligent hiring)
16.	Insurance/workers' compensation fraud
17.	Cargo/supply-chain theft
18.	Kidnapping/extortion
19.	Political unrest/regional instability
20.	Product diversion/transshipment
21.	Product tampering/contamination
22.	Organized crime

Source: Sixth annual survey, "Top Security Threats Facing Corporate America" (1999), Westlake Village, CA: Pinkerton. (See also Chapter 11.)

- **White-collar crime.** The term "white-collar crime" was coined by the sociologist Edwin Sutherland, and signifies unlawful, nonviolent conduct committed by corporations and individuals. It includes theft, fraud, and other violations of trust, including embezzlement. Embezzlement is the fraudulent appropriation of property by one lawfully entrusted with its possession. Frequently, this crime is committed by someone with a fiduciary responsibility within the organization, which he or she then abuses.

- **Employee screening concerns.** Security directors frequently are involved in directing, assessing, and improving ways by which new employees may be screened (vetted) before an offer of employment or a significant promotion is made (see Chapter 3).
- **Computer crime.** The protection of assets relating to misuse of information technology (IT) or electronic data processing (EDP) systems is a large task. Security is involved with the protected data and its output. To achieve this end, hardware, software, communications, and the physical environment of the IT/EDP facility must be considered.[16] Similarly, security procedures need to be evaluated for their adequacy.
- **General employee theft.** An omnipresent concern for security practitioners is larceny by employees. Security operations must seek to decrease the opportunity for such acts and to manage investigations.
- **Drugs in the workplace.** Employee drug testing is widely used in industry. Policies, procedures, consent forms, checklists, and training materials are needed for such programs.[17]
- **Unethical business conduct.** In many circumstances, one of the tasks for security operations is to play varying roles in drafting, monitoring, and enforcing an organizational code of ethical conduct.
- **Sexual harassment/Equal Employment Opportunity Commission (EEOC).** Complaints of harassments by co-workers or others at the workplace have been recurrent in recent years. Similarly, alleged violations of the EEOC guarantees have affected security operations. Both types of allegations require investigation and response.
- **Business espionage/theft of trade secrets.** For many organizations, the loss of crucial information is of greater importance than the fraudulent disappearance of products or supplies. These transgressions include espionage or theft of trade secrets, including developmental procedures and know-how.[18]
- **Litigation: inadequate security and negligent hiring.** The reality or fear of legal action for negligence is one of the driving features in security management today. Security program operators are involved in a variety of activities related to these risks, including instituting procedures and controls to reduce such risks and preparing actions or defenses for legal cases.[19]
- **Terrorism.** The bombings of the World Trade Center, the Oklahoma City federal building, and public and private assets in various locales suggest the importance of measures to deter such attacks.[20] While the chances of being a victim of a domestic terror-

ist attack are remote, a security director must evaluate such risks nevertheless. In certain international markets where an organization may operate, the risks can be higher.

- **Insurance/workers' compensation fraud.** Organizations that provide workers' compensation insurance sometimes encounter abuse, which requires investigation and response.
- **Cargo/supply-chain theft.** The stealing of goods that are being transported or stored is an ongoing concern for product manufacturers and distributors.[21] Outright theft requires prompt investigation and response.
- **Kidnapping/extortion.** Kidnapping (that is, the forcible abduction of a person from his or her residence or business) for profit via ransom payment is rare in the U.S. Extortion (that is, obtaining property by threatening to injure or commit any other criminal offense) is more common and sometimes involves features of kidnapping. Security operations must seek to minimize the chances of these offenses from occurring.
- **Political unrest/regional instability.** Many organizations operate in nations where risks to employees and assets require constant monitoring. Optimal security operations must seek to assess the risks in various nations. They must further stay abreast of any changing conditions that could lead to threats to employees or expropriations of assets. In the event of an emergency, security planners must attempt to remove employers safely from harm's way.
- **Product diversion.** Sometimes, products are sold abroad at a lower price than what is charged to distributors and retailers in other markets. The manufacturer loses revenues when these products intended to be sold into the lower-priced market are surreptitiously sold back into the domestic distribution channels by a third party. Security operations must seek to prevent such transshipment and to investigate any suspected incidents that may be encountered.
- **Product tampering or sabotage.** The integrity of a product is vital to its maker. Sometimes, however, products are intentionally adulterated. In extreme situations, this willful and malicious destruction of property can lead to injury and death. Additionally, the organization can have its valuable market position threatened by the results.
- **Organized crime.** Criminal activity that is perpetuated by individuals who are systematized and concerted to common goals is defined as organized crime. It is not solely a law enforcement problem, but is also directly of importance to any organizations affected by it.[22]

SPECIFIC CONCERNS FOR DIFFERENT INDUSTRIES

The security threats and management concerns listed in the preceding sections reflect responses from managers in numerous types of industries. Naturally, when responses from a particular industry are disaggregated from the total, the rank and pattern of priorities change. The following sections present four large industry groups from the sample as well as their respective top five concerns.

Manufacturing (30 Percent of Total in Response in the Survey)

Security directors of *Fortune* 1000 manufacturing firms cite the top two concerns mentioned above as their primary management concern. The distinctive variance from the overall profile was seen in the high rank accorded to the security of intellectual property. This reflects the importance of safeguarding proprietary technology and processes that create a competitive difference.

Top Security Threats: Manufacturing

1. Workplace violence
2. Crisis management/executive protection
3. Employee screening concerns
4. Business espionage/theft of trade secrets
5. Computer crime (hardware and software theft)

Business Services (21 Percent of Total)

Business services in the survey included computer and data service firms, financial institutions, and insurance companies. This segment placed two computer-related security concerns among the top five. The theft of PCs and laptops, for example, can represent a loss of current value and future opportunity.

Top Security Threats: Business Services

1. Computer crime (hardware/software theft)
2. Fraud/white-collar crime
3. Employee screening concerns
4. Crisis management/executive protection
5. Computer crime (Internet/intranet security)

Retail Trade (13 Percent of Total)

Unlike other major industry groups, retailers and related companies in the *Fortune* 1000 group placed employee theft as their paramount concern. Employee screening concerns also ranked higher with this group than with other industry groups, which reflects the concerns that security executives in the retail trade industry have about obtaining and managing ethical employees.

Top Security Threats: Retail Trade

1. General employee theft
2. Employee screening concerns
3. Fraud/white-collar crime
4. Workplace violence
5. Drugs in the workplace

Utilities (9 Percent of Total)

Consistent with the broad survey group, the utilities industry placed workplace violence as its top concern. Also, computer-related security issues rank high for this group.

Top Security Threats: Utilities

1. Workplace violence
2. Computer crime (hardware/software threats)
3. Crisis management/executive protection
4. Computer crime (Internet/intranet security)
5. Fraud/white-collar crime

SUMMARY

Security operations must possess competence in a "core" set of skills in order to run successfully. These skills include program initiation, ongoing monitoring, and constant endeavors to improve performance. Corporate security endeavors are still new, having been the focus of independent research only in recent decades. Yet the reasons for the growth of such services involve diverse psychological, legal, social, and financial requisites. Politics and federal and state laws have helped shape security services. The changing priorities of security practitioners in large corporations focus on workplace violence, crisis management, and executive protection, although the nature of those concerns differs according to the particular industry of the corporation.

DISCUSSION AND REVIEW

1. Why are some operational contingencies considered "core"? What would be an example of a non-core competency?
2. Why was the *Rand Report* so influential on security practices? Is it pejorative to refer to private security services as "private police"?
3. What factors are most important in driving the growth of security services and programs?
4. How has the ending of the Cold War changed security practices?
5. Describe federal laws that have helped form security practices.
6. What are the differential rates of growth of security services, electronics, and hardware? Why would these grow at different rates?
7. Corporate security managers place workplace violence at or near the top of their concerns. Is this likely to change? What factors could influence different types of industries to report different security priorities?

ENDNOTES

[1] J.S. Kakalik and S. Wildhorn (1972). *Private Police in the United States: Findings and Recommendations*, Washington, DC: Government Printing Office, Vol. 1, p. 30.

[2] National Advisory Committee on Criminal Justice Standards and Goals (1976). *Private Security: Report of the Task Force on Private Security*. Washington, DC: Government Printing Office.

[3] R.D. McCrie (July 1988). "The Development of the U.S. Security Industry," *Annals, AAPSS* 498:23–33.

[4] The specialized nature of information identification and assignment of security classifications produced a group of managers who founded the National Classification Management Society in 1964. Classification management may be part of the responsibility of a security operative.

[5] An Assistant Secretary for Occupational Safety and Health reports to the Secretary of Labor. OSHA regulations do not pertain to the federal or state governments or to mining. A separate act, the Federal Coal Mine Health and Safety Act of 1969 (30 USC 801 *et seq.*) is concerned with safety and health issues in that industry.

[6] The Bank Protection Act of 1968 (PL 90-389) embraced the jurisdiction of four federal banking supervisory agencies: the Comptroller of the Currency, the Board of Governors of the Federal Reserve Systems, the Federal Deposit Insurance Corporation, and the Federal Home Loan Bank Board. Nothing requires a bank to install a surveillance system. However, if it is installed, it must meet Title 12 of the U.S. Code.

[7] Bank crime has fluctuated over the years. In 1932, there were 609 holdups across the country. By 1943, it had declined to 24, following passage of the 1934 Bank Robbery Statute. (R.F. Cross [1981]. *Bank Security Desk Reference*, Boston: Warren, Gorham & Lamont, p. 1–4.) In 1968, the year the Bank Protection Act passed, 1,769 bank robberies occurred. The following year, this number was 1,793. In 1997, bank robberies reached 7,840.

[8] The Federal Aviation Act of 1958, as amended. Section 315 (a) required the Federal Aviation Administration to report on the effectiveness of the Civil Aviation Security Program to Congress on a semi-annual basis. In the 1980s domestic airline security measures were supplemented with assessments of foreign airports, conducted pursuant to the International Security and Development Cooperation Act of 1985 (public law 99-83).

[9] Security Products and Services, Cleveland Heights, OH: Leading Edge Reports, Study LE7302, 1990.

[10] Security Industry Markets: Products and Services, Commack, NY: Leading Edge Group, Study LE907, 1997.

[11] Commercial and Industrial Security, Cleveland, OH: Freedonia Group, Study 923, 1997.

[12] M.D. Kelleher (1996). *New Arenas for Violence.* Westport, CT: Praeger. Also: J.W. Mattman and S. Kaufer (1997). *Complete Workplace Violence Prevention Manual.* Costa Mesa, CA: James Publishing; and M.D. Southerland, P.A. Collins, and K.E. Scarborough (1997). *Workplace Violence.* Cincinnati, OH: Anderson Publishing.

[13] A.M. Levitt (1997). *Disaster Planning and Recovery.* New York, NY: John Wiley & Sons.

[14] R.L. Oatman (1997). *Art of Executive Protection.* Baltimore, MD: Noble House. Also: M.J. Braunig (1993). *Executive Protection Bible.* Aspen, CO: ESI Education Development Corporation.

[15] J.T. Wells (1997). *Occupational Fraud and Abuse.* Austin, TX: Obsidian Publishing Company.

[16] J.M. Carroll (1996). *Computer Security,* 3rd edition. Boston, MA: Butterworth-Heinemann.

[17] J. Fay (1991). *Drug Testing*. Boston, MA: Butterworth-Heinemann.

[18] J.J. Fialka (1997). *War by Other Means.* New York, NY: W.W. Norton & Company. Also: L. Kahaner (1996). *Competitive Intelligence,* New York, NY: Simon & Schuster.

[19] F.E. Imbau, B.J. Farber, and D.W. Arnold (1996). *Protective Security Law,* 2nd edition. Boston, MA: Butterworth-Heinemann.

[20] P.H. Heymann (1998). *Terrorism and America.* Cambridge, MA: MIT Press.

[21] L.A. Tyska (1989). "Transportation-Distribution Theft and Loss Prevention," in L.J. Fennelly, *Handbook on Loss Prevention and Crime Prevention,* 2nd edition. Boston, MA: Butterworth-Heinemann.

[22] D.F. Pace and J.C. Styles (1983). *Organized Crime.* 2nd edition. Englewood Cliffs, NJ: Prentice Hall.

Additional References

M. Chaiken and J. Chaiken (June 1987). Public Policing—Privately Provided. Washington, DC: National Institute of Justice.

L. Johnston (1992). *The Rebirth of Private Policing*. London & New York: Routledge.

Report of the National Advisory Commission on Law Enforcement (February 1990). Washington, DC: U.S. General Accounting Office.

C.D. Shearing and P.C. Stenning (Eds.) (1987). *Private Policing*. Sage Criminal Justice System Annuals. Newbury Park, CA: Sage Publications.

N. South (1988). *Policing for Profit: The Private Security Sector.* Sage Contemporary Criminology. Newbury Park, CA: Sage Publications.

3

STAFFING TO MEET PROTECTIVE GOALS

> It is imperative that those exercising supervisory authority over security guards (whether contract or proprietary) initiate affirmative actions to ensure that the guards are physically, mentally, and morally capable and qualified to perform their assigned duties.
>
> —Jennifer F. Vaughan, *Avoiding Liability in Premises Security*

Staffing relates to the recruiting and pre-employment screening (vetting) of individuals required for protective activities. This process is critical for all managers concerned with security operations. Indeed, security directors and their deputies normally spend considerable time and effort on the topics discussed in this and the next three chapters. Failure of any organization to make such screening a priority can lead to regrettable developments (see Box 3.1).

The staffing process must balance the requirements of the organization with the abilities provided by a finite pool of potential employees. The requirements further change depending on the position needing to be filled. This chapter discusses staffing matters relating to entry-level, supervisory, and managerial issues; however, the emphasis will be on entry-level employment. Hiring personnel for security positions is like hiring persons for any other position within the organization; however, security positions are distinctive in some ways and require a level of background screening that extends beyond the scope suitable for most other employees. This vetting is required due to the high level of trust expected in such positions. The means of ascertaining the levels of confidence in new hires are discussed later in this chapter.

Box 3.1. The Perils of Poor Pre-Employment Screening

The failure to vet prospective employees in a comprehensive manner increases the risks that dangerous and dishonest employees may be hired. Employers frequently learn that applicants for employment fail to reveal significant facts of their work or personal history. At BellSouth, for example, some 15 to 20 percent of new applicants conceal a secret.[1] Serious error and omissions in corporate and professional résumés should be identified by a diligent review of "facts" provided on the application form. Lack of appropriate pre-employment screening has led to the following situations:

- An international trade director for the City of San Francisco misled people into believing that he was a lawyer by including on his résumé that he has attended the Hastings College of the Law and had passed the California Bar. Neither was true.[2]
- A former prosecutor, who tried hundreds of cases under two Brooklyn, New York, district attorneys and the city's special narcotics prosecutor, pleaded guilty to practicing law without a license.[3]
- A professor of criminal justice at the University of Nebraska was arrested in Lincoln, Nebraska, on burglary charges. A routine police check of the suspect's fingerprints showed he was a convict mistakenly released from prison 11 years earlier and was also a suspect in the murder of two New York police officers.[4]
- A medical doctor serving as an intern at University Hospital of the School of Medicine at the State University at Stony Brook, New York, failed to disclose on his application that he had spent time in prison for sprinkling arsenic-laced ant poison on co-workers' donuts and coffee while working as a paramedic in Illinois. While at Stony Brook, the doctor treated 147 patients, one of whom lapsed into a coma shortly after the doctor's treatment.[5]

These examples involve non-security professionals. The same dangerous deception occurs among security personnel. In one example, a security guard working at the civic auditorium in Albuquerque, New Mexico, was employed to maintain order during a wrestling match. A bystander claimed he was observing a disturbance, but the contract security officer concluded otherwise. The guard handcuffed the observer, removed him to a separate area, and beat him. At trial, the guard's earlier conviction for violent crime was revealed. An appellate level decision found that the security contractor was negligent. The court ruled, in part, that the security firm did not adequately investigate the background and character of individual guards prior to hiring them.[6]

Box 3.1. *(Continued)*

References

[1] E.A. Robinson, "Beware—Job Secrets Have No Secrets." *Fortune*, December 29, 1997, p. 285.

[2] J.E. Rigdon, "Workplace: Deceptive Resumes Can Be Door-Openers but Can Become an Employee's Undoing," *Wall Street Journal*, June 17, 1992, p. B1.

[3] P. Hurtado, "Prosecutor Was Unlicensed," *Newsday*, November 27, 1990, p. 18.

[4] T.J. Knudson, "Professor and His Past Stun Nebraska Campus," *New York Times*, November 28, 1986, p. A14.

[5] J.T. McQuiston, "Lawyer Ties Man's Coma to Intern Stoney Brook Dismissed," *New York Times*, October 28, 1993, p. B7.

[6] J.F. Vaughan (Ed.) (1997). *Avoiding Liability in Premises Security*. Atlanta, GA: Strafford Publications, p. 157.

At all employment levels, with security being no exception, complex concerns relating to the size of the employment pool, geographical considerations, time of the year, legal constraints, reputation and industry of the potential employer, and other factors affect the choices an employer is likely to have.

PERSONNEL PLANNING

In order to achieve optimal performance, organizations must plan to fulfill personnel needs for three basic requirements. First, ongoing programs require personnel to replace those who retire, resign, or leave for cause. Next, personnel may be needed on a temporary basis to staff a short-term requirement. And finally, personnel may be needed for new programs. The processes involved in this planning activity are as follows:

1. **Identify personnel resources.** The number of employees required for current operations as well as any new organizational goals must be assessed. Further levels of skilled and experienced personnel needed for any internal growth or contraction or new venture must be estimated.
2. **Monitor current internal personnel resources.** At least some of the personnel required for a new venture may be found among existing employees. Therefore, management must have a database of existing personnel to draw upon when openings arise.
3. **Estimate labor resources available for a particular market.** The characteristics of the workers available (quantity, skills and educational level, wage rates, and unemployment trends) need to be identified for each particular market.
4. **Analyze future personnel requirements.** The program planner should be reasonably assured of the capacity of the market to meet

needs in the foreseeable future for the variety of staffing demands to be encountered.

5. **Create personnel strategies.** Issues needing to be addressed include when new employees will be hired and what proportion will be proprietary or contract.

JOB DESCRIPTIONS

After the quantity of personnel needed for the project or program has been determined, job descriptions for these positions are prepared. These are summaries of the basic tasks required for the position. They serve to establish minimum desired criteria for persons who will be recruited and vetted for the available positions. Job descriptions may include title, tasks to be performed, work conditions, persons reported to, and hours and wages. Wages and benefits are usually determined by the compensation office, which is often located within the finance department in large organizations. Wages are set relative to local pay ranges, corporate pay policy, and the urgency with which the positions must be filled.

Job descriptions clarify organizational structure and briefly summarize the key performance standards required for the position. The job description also is a basis of comparing the eventual performance of an employee with the original written expectation. Job descriptions vary markedly according to the position. Job descriptions for an entry-level position emphasize the minimum skills required (see Box 3.2), while those for supervisory positions attempt to identify persons who have sufficient experience and personnel skills to supervise the work of others even if they previously were employed in a non-protective position. Job descriptions for managerial posts normally emphasize years of experience and the nature of responsibility held.

Box 3.2. Example of a Job Description for a Security Officer

Title: Security Officer
Grade Level: (employer may set different grades according to skill and pay)
Hours: (day, evening, night shifts indicated)
Salary Range: (usually quoted on an hourly basis for this position)

Major Duties and Responsibilities

Working with people:

- Provides necessary assistance to the public in case of injuries or emergencies
- Deters unauthorized persons from entering restricted areas by ensuring that all entering such areas display proper identification

Box 3.2. *(Continued)*

- Provides information and supportive assistance courteously in all contacts with employees and the public
- Is trained or trainable in cardiopulmonary resuscitation and emergency first aid

Working with systems:

- Operates a computer terminal; capable of entering data and writing own reports
- Can use Deister, Detex, or other data-inputting systems
- Monitors access control, alarm, and CCTV system
- Can operate a fire alarm system in an emergency
- Issues temporary identification documents to visitors

Routine procedures:

- Patrols assigned areas in accordance with procedures and instructions located in the post orders
- Conducts periodic inspections of facilities to identify hazards and prepare reports
- Investigates notable incidents observed and reported by him or her and prepares concise factual written reports

Qualifications Required

- Communicates orally in English clearly and courteously
- Writes reports factually and grammatically correctly
- Possesses minimum five years' experience in law enforcement, fire safety, or with private security where systems experience was included; or minimum of two years higher education from an accredited academic program, preferably in security management, criminal justice, communications, or related fields
- Meets physical standards necessary to perform assigned duties such as:
    - Responding to emergencies, including assisting a person or persons to safety
    - Carrying and operating a fire extinguisher
    - Restraining psychologically or physically removing a person or persons from the premises
    - Driving a vehicle (must have a driver's license)
    - Standing, walking, or sitting for an extended period of time
- Possesses no record of felony convictions or convictions for relevant misdemeanors
- Passes a medical examination including, but not limited to, blood and urine analysis to detect the presence of any illegal substances.

NEGLIGENT HIRING LITIGATION

The primary reason to implement excellent staffing programs is to attract and retain a cadre of competent, motivated, and productive employees who are capable of performing the necessary work. However, another important factor influencing such programs is the fear of a lawsuit for an untoward personal action. Less than diligent personnel procedures could allow the hiring of a rogue employee whose behavior, for example, could hurt another person or result in loss of assets. This could lead to litigation for negligent hiring. Prudent operational management can reduce the possibility of such a plaintiff's action succeeding. Such an action is usually lodged in civil courts against the employer-defendant and can be based on constitutional principles, case law, or federal or state statutes.

The pressure to take shortcuts in an effort to save money and time—sometimes because personnel slots need to be filled quickly—can result in the wrong person being hired and retained until an actionable offense occurs. But most employment errors take place because the organization has not created a rigorous human resources vetting program in the first place. Regardless of the cause, organizations must take a reasonable degree of care in selecting employees and determining if they are fit for the positions to which they are assigned.[1] If any actionable occurrence results, the specific charges against the defendant-employer may be negligent hiring and negligent retention, which are separate causes of action based on employer negligence.[2] Negligent hiring occurs when the employer knew or should have known of the employee's unfitness prior to offering that individual a position within the organization. Negligent retention takes place when the employer becomes aware of the employee's unfitness for duty and does not terminate the employee promptly.

In addition, an employer is responsible for certain acts of the employee under the doctrine of *respondeat superior*. This is a concept of vicarious liability in which the employer assumes responsibility in certain cases for wrongful acts of the employee acting on behalf of the employer. However, if the employee acts outside of the course and scope of his or her employment—even when on the job and being compensated at the time—such an incident may not necessarily result in a judicial verdict against the employer. Certainly, if the employer could have determined something about the employee from a routine background investigation that would have prevented hiring the person, but did not conduct such an investigation, the employer may be held liable for negligence in hiring.

Other reasons for vetting employees are to assure that certain frequently occurring hiring-related problems do not occur.[3] Some examples of these are:

- Hiring people who have too much in common with their interviewers

- Hiring friends or repaying favors
- Settling for mediocre candidates because of the pressure to fill positions
- Hiring someone for political reasons or for the feeling that one cannot afford the best candidate
- Not probing for limitations, lies, or relevant details in the prospect's work or education history
- Talking instead of listening while conducting pre-employment interviews
- Overselling the position

For whatever reasons, the consequences of failing to vet employees properly prior to making an offer of employment can lead to substantial losses by the employer in a trial (see Box 3.3). The legal principles pertaining to negligent hiring are not fixed and are thus subject to change, from court decisions to appeals, to changing conditions and new laws. Generally, the main requirements of a plaintiff to establish civil liability in which an allegation of injury or damage occurred are:

1. The existence of a legal duty by the plaintiff to protect the defendant,
2. A breach (failure) of that duty, and
3. Damage, harm, or injury as the proximate cause of that breach.

Box 3.3. Negligent Hiring: Verifying Indications of Integrity

---

The burden of appropriate vetting is incumbent upon both proprietary employers and employers of contract workers. In such cases, an employer is obliged to conduct a reasonable investigation into the employee's work experience, background, character, and qualifications. The standard of care does not vary though the greater the risk of harm, the greater the degree of care necessary to constitute ordinary care.

This is demonstrated in the case *Welsh Manufacturing, Division of Textron, Inc.* v. *Pinkerton's, Inc.* [474 A.2d 436 (1984)]. Welsh Manufacturing sued Pinkerton's for losses sustained due to three major thefts connected to a Pinkerton's security guard. Welsh claimed that Pinkerton's was negligent in the hiring, training, supervising, and assigning of a guard who was later found to have been a co-conspirator in such losses, and that the negligence was proximate cause to the losses. A Rhode Island civil court returned the verdict in favor of Welsh. Pinkerton's appealed. The appeals court reviewed Pinkerton's contention that insufficient evidence existed relative to the theories of liability raised in the lower court. The security firm requested that the lower court's decision should be set aside.

Box 3.3. *(Continued)*

The Background

For 30 years, Pinkerton's had a contract to provide security services for Welsh, which manufactured gold sunglasses frames for the U.S. government. The Welsh manufacturing complex contained sizeable quantities of gold required for these manufacturing purposes. Pinkerton personnel were aware of the gold stored on the premises. Pinkerton's assigned a 21-year-old part-time employee to patrol the Welsh premises during the night shift. The guard had worked for Pinkerton's less than six months at the time of the assignment. Over the next 45 days, during the night shifts, three thefts occurred at Welsh's facilities, resulting in a loss in excess of $200,000. An investigation identified the new guard as the culpable thief in two of the larcenies. He had resigned by the time the third crime occurred but testified later that he provided vital information to parties who robbed the facility of another $180,000. In that third crime, the former Pinkerton guard put a gun to the head of a replacement guard. The erstwhile guard further testified that he secured the position with Pinkerton's in order to steal from Welch's, to which he expected to be assigned.

The Deficient Pre-Employment Process

Pinkerton's application for employment requested the names of former employers and of three character references. Pinkerton's did not contact the character references. The company did forward reference forms to the applicant's high school principal and to a hospital where the applicant had worked for about one month. The forms did not address questions relating to the applicant's honesty and trustworthiness. The hospital provided an employment-termination form in which a supervisor checked a box indicating the worker was deemed "good" for "honesty." The supervisor also noted that the worker had failed to report for work on two occasions. Pinkerton's also telephoned someone who had been the applicant's superior officer in the navy for about two months, who said that he would recommend the applicant for a job with Pinkerton's.

Despite these measures, the appellate court ruled that "Pinkerton's cursory investigation prior to [the applicant's] employment provided it with little current intelligence on him and could well support an inference of negligence in hiring for such a sensitive assignment as the guarding of gold." The court further noted that the applicant had no prior working experience similar to Pinkerton's business. His prior assignment before being assigned to Welsh did "not evidence ascending levels of sensitivity." The appeals court dismissed Pinkerton's appeal and affirmed the judgment of the lower court.

Box 3.3. *(Continued)*

The Lesson

At the time of this incident, which happened in 1973, vetting standards were lower than the levels reached in later years. Indeed, Pinkerton's procedures did not appear to represent a deviation from the norm of the time. What the case establishes, however, is that an ordinary duty of care requires a reasonable investigation of a security applicant's background, including his or her character. In this case, Pinkerton's did not have a single affirmative statement of the candidate's probity from a supervisor or other independent person. If such statements were obtained, an employee nonetheless might steal on the job. But the employer in such a circumstance would have demonstrated good faith in meeting this appellate court's standard.

Source: D.A. Maxwell (1993). *Private Security Law: Case Studies*. Boston, MA: Butterworth-Heinemann.

In various cases, appellate courts have established criteria for negligent hiring and training. They further identify issues in the failure or omission to take reasonable measures to reduce these risks. Frequently in plaintiffs' actions, the loss prevention director and other individual employees involved in the case may be charged personally as defendants, in addition to the employer. Even if the loss prevention director is defended by attorneys for the employer, such an individual may need to consult independent counsel to protect his or her own interests. Thus, a lawsuit for negligent hiring can require corporate and individual response, involving financial costs and unwanted publicity and misuse of time, which possibly could have been mitigated by the exercise of reasonable vetting measures. The remainder of this chapter will concern such measures.

THE VETTING PROCESS

Hiring the right people can be costly, slow, and frustrating. But optimally performing security operations personnel are not deterred from their goals even during times when unemployment is low, compensation costs are rising, and the pool of potentially desirable candidates appears to be weak. The astute manager will persevere to obtain the best employees as possible under the circumstances.

The entire vetting process is costly. For entry-level and supervisory employees, vetting involves expenditures to recruit candidates, perhaps to pay personnel agencies, costs of staff time to review applications and conduct interviews, and outlays for background investigations. Only then can the additional cost of training be contemplated. For managerial positions, costs may involve management recruiters and the expenses incurred to interview and eventually relocate the successful applicant.

The personnel process can be partially analyzed according to productivity costs; that is, what means of attracting candidates produces the best long-term results. This can vary widely in organizations. One means of attracting candidates for employment can be exhausted or become less efficient with time, much as a vein of gold becomes fully consumed by miners who must then move their search elsewhere. The vetting process in security-oriented employment contains a number of distinct phases, as shown in Figure 3.1. The first step is to obtain as large a pool of prospects as possible at a reasonable cost. This process often involves recruiting.

Recruiting

For entry-level personnel positions, most managers prefer that many candidates be considered before a final selection is made. In cases where specific and demanding criteria are established as a requirement for employment, the number of qualified candidates will be less. Recruiting expenses hopefully should produce a commensurate return with their outlays. A manager is likely to experiment with different means of attracting applicants and determining which methods produced the most desirable results. For example, Table 3.1 shows a simple security recruitment productivity worksheet whereby different recruiting methods can be analyzed for their cost effectiveness. The table identifies cost per completed application, but the final criterion of successful recruiting is the number of successful applicants who are screened and offered a position that is accepted. In addition, it is important to take into account whether the worker makes a positive contribution over an extended period of time. The recruitment method that produces the most acceptable applications may be different from the one that produces the most acceptable employees on a long-term basis. Analysis can help identify the most effective recruitment medium for an employer.

Security program managers must attract applicants so that they can have a large enough pool to select the best candidates for the position. The comparative study divides the dollar cost of acceptable applications by the number of applications received. This yields a cost per application, which can lead to a ranking of the most productive means of recruiting. While this sample report is concerned with measuring completed applications, other analyses could identify actual hires, or those employees who remain on the job for 90 days or more. Such results may differ from those based only on applications completed.

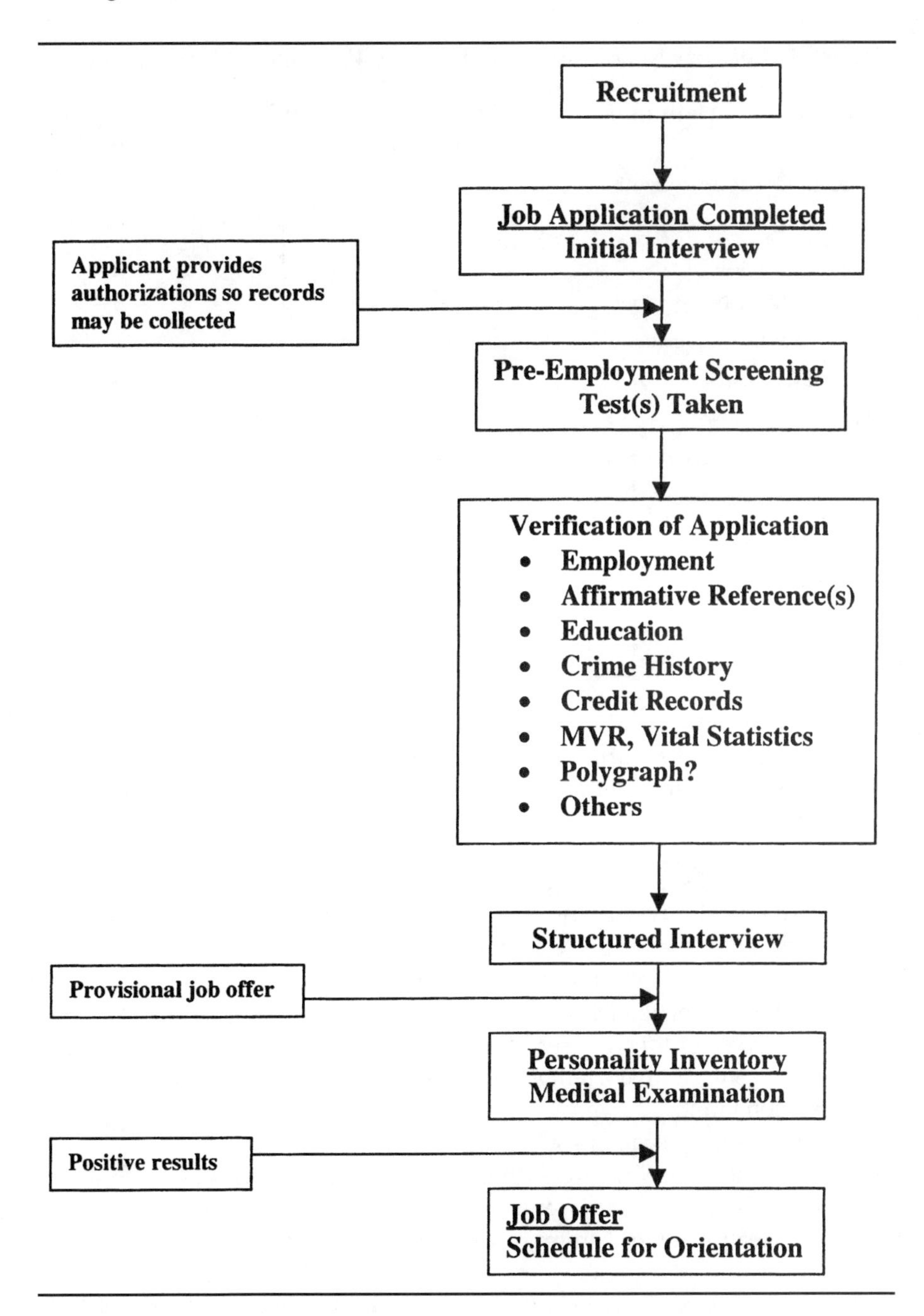

Figure 3.1. Stages of the Security-Oriented Vetting Process

Table 3.1. Security Recruitment Productivity Worksheet Monthly Review

Recruitment Method	Completed Applications	Recruitment Cost	Cost per Application
Classified Advertisements			
Daily Newspaper A	—	$—	$—
Daily Newspaper B	—	$—	$—
Sunday Newspaper	—	$—	$—
Display Advertising	—	$—	$—
E-mail Responses	—	$—	$—
Web Site Inquiries	—	$—	$—
Incentives to Current Employees	—	$—	$—
Job Fair Attendance	—	$—	$—
Personnel Agencies	—	$—	$—
Radio Spots	—	$—	$—
TV Spots	—	$—	$—
Other	—	$—	$—

Employers of security personnel use a number of strategies to attract applicants. The following are means used for entry-level and supervisory personnel:

- Classified advertisements
- Display advertisements
- Radio or television advertisements
- Internet employment sites
- Personnel agencies
- State employment services
- Incentives to current employees
- Job fairs
- College placement services
- Military or Veterans Administration employment services
- Association job placement services
- Health club or community centers member recruitment

If some types of recruitment sources begin to pale in effectiveness, high performance managers turn to other means of obtaining needed applicants. Flexibility on management's part may achieve the desired goal. For example, offering part-time or flexible schedules and unusual incentives may draw applicants who might otherwise not consider applying for full-time work. Similarly, making prospects feel welcome, valuable, and engrossed in interesting and relevant work—like security—may produce qualified applicants who might otherwise pass on the opportunity to work for the organization. Employers may find at times that they must aggressively and imaginatively lure individuals to apply for the positions they have available.[5]

In analyzing the effectiveness of different recruitment sources, a manager may later determine that one type of source is better than another. The results may differ according to the type of security assignments involved and the culture of the employer. For example, a study of newly hired marketing representatives found that employees recruited through college recruitment efforts had better initial levels of performance than did those recruited from newspaper advertisements.[6] The analytical security executive is likely to identify which recruiting sources are most productive for the short- and long-term by monitoring performance of different recruiting sources over time.

The potential employer's Web site, which should contain available job descriptions and an opportunity to apply over the Internet, will appeal to some applicants and is growing as a recruitment tool. Managers are particularly apt to use the Internet when looking for employment opportunities. Applicants for managerial and executive positions may also be recruited by outside personnel recruiters, or headhunters. These share some of the burdens of vetting candidates they propose to the prospective employer. Management and executive recruiters usually charge the employer a percentage of the successful applicant's beginning annual compensation; in other cases, a flat fee is charged, regardless of success.

Pre-screening

Anyone who physically presents himself or herself to a personnel office and requests an application for work should be provided one readily, even if no positions are currently available or will be in the foreseeable future. To do otherwise could risk the perception of discrimination and subject the organization to a civil action for employment discrimination. If no positions are currently available or likely to be in the foreseeable future, a posted notice or a receptionist may inform potential applicants of this situation. At their option, job-seekers may then complete an application that can be filed for possible future consideration when positions become available.

If positions are available for those with particular skills or requirements, these attributes can be indicated on signage apparent to all visitors at the personnel reception office. Upon the receipt of a completed application from such a candidate, a clerk or manager may conduct a brief interview to provisionally ascertain the facts of the application and to inform the applicant of the subsequent steps in the employment process. The brief informal interview also helps establish the communications skills of the applicant.

The Application

The application is management's opportunity to obtain information needed to determine whether the applicant should be considered for further background screening and a possible offer of employment. For most of the past century, applications were completed on paper forms. However, organizations increasingly have the applicant enter the information directly into a computer terminal within a personnel office or send it to the company's e-mail address. The quantity of information required from the applicant on the employment form differs widely based on the employer. The degree of detail and the thoroughness of the evaluation process reflect the employment philosophy of the employer (see Box 3.4).

Box 3.4. The Employment Application: How Much Is Enough?

Some employers of security personnel use simple application forms that require directory-type information and little more. Guardsmark, a nationwide security service firm based in Memphis, Tennessee, presents applicants with a 24-page booklet that requests a detailed educational, work, and social history. The copyrighted application also includes information on the company's philosophy and ethics policy. Applicants are informed that they must successfully pass a polygraph examination for which a release is included in the application. A consent to drug and alcohol screening tests and release is also included.

Vetting methods from the application process to the point of finally offering a position of employment involve aspects of privacy as well as questions of fairness and reliability. A balance between the rights of the applicant and the needs of the employer must be maintained.[7] This is particularly true for employers of protection personnel. For security-related positions, more information generally is requested of applicants than in other types of employment. Regardless of the length of the application, some minimum information is required. Directory-type data—current address, phone, and Social Security number—are required, as well as work history. The specific questions asked of new prospective employees on applications have been shaped in recent years by various federal laws. The main federal laws affecting pre-employment standards are:

- The Fair Labor Standards Act of 1938 (The Wage and Hours Law)
- Title 7, Civil Rights Act of 1964, as amended by the Equal Employment Opportunity Act of 1972
- The Occupational Safety and Health Act (OSHA) of 1970
- The Rehabilitation Act of 1973
- The Vietnam Era Veterans' Readjustment Assistance Act of 1974
- The Pregnancy Discrimination Act of 1978
- Executive Orders 11246, 11375, and revised orders 4 and 14 (Affirmative Action)
- Title 42, United States Code, Section 1983 (Civil). This section guarantees citizens of the U.S. or other persons within its jurisdiction from the deprivation of any rights, privileges, or immunities secured by the Constitution and its laws.
- Title 18, United States Code, Section 242 (Criminal). This section protects citizens and others from discrimination by reason of color or race and allows for criminal penalties against offenders.
- The Fair Credit Reporting Act of 1971
- The Immigration Reform and Control Act of 1986
- The Employee Polygraph Protection Act of 1988
- The Americans with Disabilities Act of 1990. This act prohibits employers with 15 or more workers from discriminating against qualified persons based on their disabilities. It is important to note that knowing what constitutes a disability is not always clear (see Box 3.5).

Box 3.5. Defining What Is a Disability

The Americans with Disabilities Act of 1990 defines a disabled person as someone who has a physical or mental impairment that substantially limits one or more major life activities, has a history of physical or mental impairment, or is regarded as having a physical or mental impairment. The nature of specific disabilities is in a constant flux related to court decisions that modify what employers may or may not exclude in considering the applicant. Sometimes the concerns about excluded positions has led to searching questions for security and Human Resources managers. For example, does wearing glasses constitute a disability? May security employers be forced to hire convicted felons because they are in a protected class?

Box 3.5. *(Continued)*

The ADA modifies the hiring process, but guidelines are compatible, nonetheless, with the desire of achieving a competent and effective workforce. The employer must be aware, however, of the ADA's context. Applicants who are "qualified" and able to perform "essential functions" of a position with or without "reasonable accommodation" may be in a protected status to receive employment consideration. The employer may insist on the following guidelines:

1. The disabled individual must satisfy the prerequisite of the established position. Certain levels of experience and education are typically permitted prerequisites.
2. The individual must be able to perform essential job functions with or without reasonable accommodation. A key factor is whether a particular job function is essential. For example, a position requiring a security officer to communicate with the public verbally could exclude a candidate who is mute. However, if verbal communication was not an essential job function, the employer might make "reasonable accommodation" by permitting a mute employee to communicate by signs or symbols or by using a technological device to respond.
3. In terms of defining "reasonable accommodation," employers only have to provide accommodation if it does not present the organization with "undue hardship." This is further defined as "requiring significant difficulty or expense." The employer is held to ADA standards that are related to size of the corporation, number of employees, and the cost of the accommodation necessary to render the disabled person capable of performing the required work to established standards.

Source: L. Greenhouse, "Justices Wrestling with the Definition of Disability: Is It Glasses? False Teeth?" *New York Times*, April 28, 1999, p. A26.

Existing measures may be further strengthened by subsequent modifications in rules and regulations. Also, state measures may expand upon federal coverage. As a result of these proscriptives, application forms are limited in what they may or may not ask. Such limitations may change at any time based on new rulings in case law and passage of federal and state legislation.

The employer has a legal obligation to verify essential information in the application, particularly for security-related positions. This is a vital service because applicants may provide incorrect or misleading information in their applications and resumes. Verification of references and other relevant information should be completed substantially or fully before a formal interview is scheduled.

References

Most applications request references of applicants. Generally, these may be of two types: employment-related or personal. Of these two, employment-related are more significant in that they relate to workplace attitude and proficiency. Personal references tend to be biased and are usually not job-oriented; therefore, most employers place emphasis on job-related sources for information about the applicant. However, government and private sector positions requiring a high level of integrity in their hires may request character references in addition to those linked to past or present employment. Self-serving personal references from applicants' mothers and best friends are not convincing. Employment verifications and confidential references from applicants' direct supervisors carry much weight.

References may be vetted by clerks telephonically, by written requests, or by a combination of both. Some organizations turn to investigators or contract services to aid them in their reference checking. As noted earlier (Box 3.3), the failure to obtain a single affirmative statement of probity concerning a candidate can lead to a charge of negligence in hiring. This would occur in the event that an employee had a relevant personal history experience, capable of negating consideration for employment, that was not revealed on the application but that a reasonable effort on the employer's part would have discovered.

Employment Verification and Continuity

References and employment evaluations from an applicant's previous direct supervisors are valuable to employers in making a decision on whether to hire that person or not. Some employers are reluctant to permit their supervisors to share candid insight on a departed worker's performance and personal behavior. They fear possible civil suits for slander or liability from workers who may be denied an employment offer based on such negative statements. However, supervisors who speak truthfully and without malice about a former worker, even while revealing negative aspects about the worker's employment history, rarely are targets of litigation that is sustained by the court.

Astute security-conscious employers look carefully for any gaps in employment. These are not necessarily red flags of trouble. Dates of employment gaps may reflect time needed to find new employment after one job has ended, or it can reflect time off for education, training, relaxation, or healthcare. Such a gap could also reflect jail or prison time or other involvement with the criminal justice system. Careful evaluation should determine why employment gaps have occurred.

In addition to personal and employment references, verification of other information in the application should be undertaken. Such inquiries generally require the approval and sometimes cooperation of the applicant. Managers should always be aware that information derived from commercial services requires careful evaluation. A decision not to hire someone on the basis of a single finding cannot be justified. Commercial services greatly expedite the cost of conducting a comprehensive, relevant, and accurate collection of the applicant's records. However, an employer is able to accomplish the same objectives in collecting information by in-house investigators using available credentials checking manuals, directories, and Internet databases.[8] This type of information includes:

- **Social Security Number (SSN) verification.** Congress never intended SSNs to become national identifiers, but they are. Once issued, an SSN is permanent and is only changed under extraordinary circumstances. The SSN is the best indicator that the applicant is who he or she says he or she is. Guides to understanding the SSN explain the information revealed in the distinctive numbering sequence.[9] SSN verification can indicate the state and approximate year of issuance.

  The importance of SSN as an identifier is beyond question. Therefore, all employers—and individuals—have an obligation to take reasonable efforts to protect the privacy of such numbers. The SSN is vital for both credit checks and criminal records searches. It is important to note, however, that an applicant may be victimized by another person who is currently using or has used that person's SSN for unauthorized purposes. This is called identity theft. This crime could occur without the victim-applicant's knowledge. Hence, problematic information revealed through an SSN check should not be the sole basis, without further confirmation, of eliminating a candidate for employment consideration.
- **Credit history checks.** Credit reports are important as part of a conscientious pre-employment verification. In many—probably most—cases, the employer is not seeking to obtain the applicant's actual or potential debt, which is partially revealed from such reports, but rather wishes to obtain factual directory-type information from an independent source. The major credit bureaus deliver reports rapidly and, due to the automated nature of the business, costs for such reports are low. Prospective employees must specifically authorize the employer to undertake a credit search. A state-

ment of the applicant's rights must be provided and the applicant must authorize the search by signing an appropriate form. Three major credit bureaus are supported by hundreds of independent local and regional credit collection businesses. Both credit bureaus and their customers should be familiar with terms of the Fair Credit Reporting Act (FCRA) as amended. The full text of the FCRA and notices of rights are available on www.ftc.gov/bcp/conline/edcams/fcra/index.html.

- **Military history.** Employers generally are not allowed to ask about the nature of discharge of applicants with military service history. However, asking whether the applicant is registered for Selective Service or was a member of a branch of the armed forces is acceptable. The applicant may obtain a standard form (DD214) detailing his or her personal military service history. It provides information on training specialties, ranks achieved, dates of service, and type of discharge. Locations and dates of foreign service, awards, and distinctions received are cited. Disciplinary information may be revealed. However, offenders have the opportunity to appeal to a military court, subsequent to the time any punishment has been discharged, and request that the court seal the records of any criminal or other charges. In this situation, the potential private sector employer would not be aware of such factors in the applicant's military background.

  Employers may help their applicants obtain a DD214 by providing them with form SF 180, Request for Military Records Form. This may be downloaded from www.nara.gov/regional/mprsf180.html.

- **Motor Vehicle Reports (MVR).** Even if the applicant is not expected to drive a vehicle in the course of his or her intended employment, an MVR check may be wise. The record provides directory-type information and may also reveal other facts about the candidate that could be important in arriving at an employment decision. For example, the MVR may reflect an address different from the one that appears on the application; the employer might wish to determine the reason for this difference. Further, if the applicant must or could potentially drive in the process of employment, the logic of an MVR search at the onset of employment consideration is apparent. MVRs contain directory-type information but also data on height, weight, eye and hair color, date of birth, when the license was issued, violations, often recent convictions, accidents, and other facts. Employers may use the services of a professional records verification service (see below) or contact state motor vehicle agencies directly to obtain such information.

- **Civil records searches.** When filling a management position, an employer may wish to evaluate the civil records of the applicant. Such searches can identify possibly concealed information about the candidate, including other names used, addresses, former employers, and the existence of judgments, liens, bankruptcies, and pending litigation. Although, many types of liens and Uniform Commercial Code filings are available free of charge, a commercial records checking service may expedite the process. In addition, the expertise and time required to retrieve them explains why many larger employers farm out this task. In some states and cities, telephone queries are answered, while such information is increasingly available over the Internet. For a list of state and local sites with free Internet access, visit www.brbpub.com.
- **Criminal history.** Employers in protective services are eager to determine, and rightly so, whether prospective employees have a criminal history that would render them unsuitable for a position of trust. In the past, employers often believed that *any* arrest or conviction for a criminal offense should automatically lead to a rejection of the candidate. However, today many factors have changed employers' ways of thinking about this issue, reflecting the position that certain types of arrests or convictions should not necessarily block a person from obtaining employment for a security position. The key factors are the relevancy of the candidate's offense to the nature of the security work anticipated and the particulars of the offense. Relevant details include the age of the offender at the time of the crime, how long ago it occurred, its severity, and the presence or absence of subsequent illegal behavior since the offense originally took place.

  Criminal records are public documents. However, since states keep their records in different locations and have their own procedures for releasing them, access can be difficult for someone unfamiliar with the process. Most criminal records are retained at the county as well as the state levels. Over 5,000 locations exist within the U.S. where such criminal records may be found. Depending on the locality, such records may be obtained by mail, fax, commercial service, the Internet, or in person. Fees for providing copies of the records and supporting documents often are charged. Usually, search information requires the person's full name, date of birth, years to be searched, and SSN.

  Potential employers are advised to consider relevant felony and misdemeanor records only in evaluating the candidate's total credentials for employment. That is, currently many counties also provide arrest records that did not lead to conviction. For the sake of employment consideration, such information must not be considered. However, sometimes in the process of conducting a background investigation, an employer discovers an open arrest warrant for the applicant, often in a distant jurisdiction. In such circumstances, the employer should contact law enforcement immediately.

The obtaining of criminal history records is highly desirable for comprehensive background investigations of a new candidate. A potential employee may sign a criminal background check release form, including SSN, date of birth, any aliases used, and a driver's license number to expedite the search. This form should be witnessed and notarized. Alternatively, an employer may ask an applicant to obtain a Certificate of Good Conduct from the police department of the communities in which the applicant has lived. This should not be considered a definitive record since the applicant may have committed a crime and been convicted and punished in a different jurisdiction from the one in which the applicant is currently residing and from which it is supplying good conduct documentation.

Employers of security personnel can satisfy their needs for criminal records in communities in which state security guard registration exists and in which the state vets all such applications through the National Crime Information Center (NCIC) database maintained by the Federal Bureau of Investigation. The NCIC collects and organizes records provided to it by state criminal justice agencies. Thus, an employer may have a high level of confidence that the applicant has not been convicted of a crime, making him or her unfit for security employment, as defined by the offenses cited as relevant in the states which require NCIC screening for security personnel. At present, only a few states require such a vetting measure. (For further discussion, see Chapter 10.)

Is it possible for the applicant to possess a security conviction and yet not have it revealed by comprehensive police record checks? Yes. One way is by an error in the record-keeping process, such as the failure of authorities to receive such information from the state or local level. Another way is for the applicant to appeal to a court for a certificate of relief to be granted after discharge from incarceration or termination of parole. This certificate seals the offender's record from public scrutiny. An exception is for certain government jobs in which the applicant must agree to have records of the offense unsealed temporarily for scrutiny by government employment background investigators.

In some employment situations, even a "clean" police record does not necessarily mean that the applicant does not have a past felony conviction. However, a diligent review of the applicant's work history dates should reveal any unexplained time gap that could reflect jail time or other reasons to give the employer pause.

- **Educational records.** Over 4,000 post-secondary academic institutions and programs exist in the U.S. About four out of five such institutions will verify degrees received and years of attendance of past students and graduates. Educational institutions will confirm such requests by phone, mail, and fax, usually without charge. The employer can provide a Request for Educational Verification signed by the applicant for those institutions that do not readily provide such information.

  If the employer desires a transcript of academic records, a release generally is needed, including the name, year graduated, campus attended, and SSN. At a few institutions, e-mail requests by students for transcripts are accepted. The student's signature is required in order to release a transcript to a third party. The student may request that the transcript be mailed directly to the prospective employer, in which case the envelope should be retained and be considered part of the document. Institutions usually charge fees for transcripts.
- **Other records.** Prospective employers may scrutinize other information provided by the applicant. Some examples are vital statistics, workers' compensation records, and licenses. Verification of such data may help raise the confidence level the employer has in the applicant. Such information also sheds light on the applicant that might otherwise not come to the attention of the employer. Professional licenses and certificates may reveal such information as the date of issuance, current status, expiration date, and field of certification. A license and a certification are indications, but not a guarantee, of competence.

  Disciplinary information from candidates usually is not available from the granting authorities. In many cases, a person disciplined by an organization responds by resigning from it. Nonetheless, it is reasonable to document credentials offered by a potential worker prior to offering him or her employment. Candidates might be asked to produce original diplomas or certificates as proof of purported achievements.

The Polygraph

Efforts to determine the honesty of a person go back a long time. Many of the techniques developed to identify honesty were crudely related to observations that psychological stress modifies behavior and physiology. The early Chinese gave suspected liars rice powder and asked them to spit it out.

If the powder was dry, then the person was believed to be dishonest. This was based on the premise that lying affected emotion, which, in turn, had a measurable physiological effect—in this case, deception was believed to lower saliva production.[10] In other cases, truth seekers used methods to detect lies that were hardly reliable or scientific. These included the Ordeal of the Red-Hot Stones and the Ordeal of the Red-Hot Iron.[11] If the suspect survived the walk on the coals or the touch of the iron, he or she was deemed to be innocent of the charges.

Determining what constitutes a lie and what constitutes the truth is not always clear. Deception is a normal part of life. Not surprisingly, it frequently occurs when a person wishes to obtain employment or some other benefit and suppresses information that could damage that opportunity. But in employment circumstances, the nature of the information required tends to be specific. A prospective employer might desire unequivocal responses to certain questions, such as "Are all the facts in your application for employment true?" or "Have you ever stolen any item from a previous employer valued at more than $25?" The answers to such questions, if reasonably obtainable by some objective measurement, can be valuable in an employment decision. This was an expectation of the polygraph examination in the U.S. until the late 1980s.

Polygraph examinations originated in the 19th century, when an Italian criminologist named Cesare Lombroso invented a blood pressure recording device that recorded pulse and blood pressure changes as subjects were being questioned. William Multon Marston, a U.S. lawyer who also studied psychology, publicized the possibilities of using Lombroso's device to aid in distinguishing truth-telling from deception.[12] The first practical use of a polygraph was undertaken by August Vollmer, police chief in Berkeley, California. Vollmer assigned John A. Larson and later Leonarde Keeler to develop the modern device. The instrument collects human physiological responses to various yes or no questions. The device measures and records minute blood pressure changes as questions are posed, and graphically displays changes in breathing, pulse, and electrodermal response.

During a polygraph examination, subjects usually are put at ease and are informed on how the examination will be conducted. Sensors are attached physically to the interviewees and they are asked a series of relevant questions about substantive matters, which are contrasted with irrelevant ones. The polygraph examiner identifies a baseline from physiological responses to irrelevant questions and uses the points of comparison of relevant questions to irrelevant ones to identify possible deception. This is called the relevant/irrelevant technique.

Over most of the 20th century, many law enforcement units, government, and private industry in the U.S. used the polygraph as a means of determining deception in pre-employment vetting, criminal investigation, and for other purposes. But given the relativity of truth itself, is the polygraph reliable for such use? In 1984, a Department of Defense research report summarized 42 studies concerning the polygraph under varying circumstances. The report observed that the polygraph could create two kinds of errors. False negative errors are erroneous decisions that an individual is not deceptive when he or she is actually deceptive. False positive errors are erroneous decisions that a person is being deceptive when he or she is being actually truthful. These are the most frequent types of error. Nonetheless, the report concluded: "Used with prudence, and a full knowledge of its limitations, the polygraph will continue to play a role in our criminal justice system and counterintelligence operations."[13]

A year later, the Office of Technological Assessment (OTA), the research arm of the U.S. Congress, conducted its own review and assessment of the scientific evidence on the validity of polygraph testing at the request of the House Committee on Government Operations.[14] OTA's findings further underscored the broad utility of polygraph examination in numerous applications, but concluded that unanswered questions remain about the validity of polygraph use in some circumstances, such as pre-employment screening.

By the mid-1980s, according to an estimate based on a survey conducted by the American Psychological Association (APA), as many as 5 million polygraph examinations were administered in the private and public sectors, mostly concerning private sector employment.[15] Most observers in the field believed that this estimate was far too high. A Congressional report put the number of polygraph examinations at "over a million" a year—300,000 of them for employment purposes alone in the private sector.[16] About 20,000 federal exams were administered each year. But Congress, facing pressure from unions, passed the Employee Polygraph Protection Act (EPPA) of 1988 (29 U.S.C. 2001 et seq.). This act sought to discourage the widespread use of polygraph testing in non-criminal investigations and it succeeded in that regard. The EPPA prohibited most private employers from using polygraph examinations either for pre-employment screening or during the course of employment to detect any unreported workplace theft. With the passage of this law, use of the polygraph decreased substantially in the private sector. However, federal, state, and local government retained the right to use the polygraph as part of their employment decisions, and some governmental units continue to do so.

Nonetheless, the polygraph was not totally excluded for use in pre-employment circumstances. The act permits polygraph tests to be administered in the private sector, subject to restrictions, to certain prospective employees of security service firms. These include security guards and armored car and alarm monitoring and response employees. Specifically, Section 2006 of the EPPA does not provide blanket exemption on the use of the polygraph for uniformed and plainclothes security personnel *unless* the personnel are being engaged in facilities, materials, and operations concerned with production, transmission, or distribution of electric or nuclear power; public water supply facilities; shipments or storage of radioactive or other toxic waste materials; and public transportation of currency, negotiable securities, precious commodities or instruments, or proprietary information.

The EPPA does not elaborate on what constitutes "precious commodities or instruments," or "proprietary information." One national security guard company claims that all of its new security personnel directly or eventually will be involved in such activities, which necessitates polygraph vetting of all prospective security personnel. However, most other security service providers, large and small, routinely have not required polygraph screening since the passage of EPPA. Like other factors in employment decisions, polygraph examinations have limitations (see Box 3.6). Other measures have grown in use and are discussed in the next section.

Box 3.6. Beating the Polygraph: The Aldrich Ames Case

Use of the polygraph by the military, law enforcement agencies, and national security staffs has provided such success that government received exemption from the polygraph limitations cited in the Employee Polygraph Protection Act (EPPA) of 1988. However, nobody claims that results from a polygraph examination are infallible. Spies have been trained to overcome the skills of an experienced polygraph examiner by using tranquilizers, biofeedback techniques, or pressing toes down at critical points during questioning.

The most serious breach of security in the U.S. intelligence system occurred despite the fact that the convicted traitor, Aldrich Ames, was given routine polygraph tests by the Central Intelligence Agency (CIA) over an 11-year period. Yet Ames was able to identify to the KGB at least 11 clandestine CIA agents, four of whom were executed, plus other information.

How was Ames able to overcome a system that appeared to have worked well in the past? Ames, a veteran CIA case officer, seemed to do nothing more devious than act friendly with his polygraph examiner. Ames first started working with the KGB in 1985 for cash payments that eventually totaled $2.5 million. The year after he began spying, Ames was recorded as deceptive on a polygraph question related to his personal finances. In 1991, deception also was reflected on a question as to whether he had ever worked for the Soviet Union. In both circumstances, Ames was given four days to rest and then retake the polygraph examination, which he then passed. The deception on Ames's 1991 relevant question as to possible spying was not forwarded to the FBI until two years later. Hence, it might be said that Ames did not "beat the polygraph," but that he "beat the system" of evaluating such readings.

Source: D. Waller, "How Ames Fooled the CIA," *Newsweek*, May 9, 1994, p. 24.

Pre-Employment Testing

Applicants for employment may be screened by a variety of standardized objective instruments designed to measure personality, aptitudes, interests, and achievements of potential employees. Test anxiety should be minimized and coaching on how to achieve the desired results should be considered unethical. Different types of test instruments are available to employers, which will be discussed in the following sections.

Psychological Stability. For some positions, prospective employers will wish to be assured that all reasonable efforts have been made to identify and assess relevant psychopathology in the applicant. Such testing is complex and invariably involves collaboration with a psychologist trained in testing methods. The potential employer normally is not interested in classification, etiology (causation), methods of diagnoses, and other facets of abnormal psychology, as interesting as they may be. The employer's goal is to authorize relevant tests that can identify abnormal traits that could lead to dangerous or otherwise unacceptable work performance from such an employee. For example, an employer is likely to be concerned with candidates for employment who evidence mental disorders that would make them risks in stressful circumstances where they would be required to carry weapons.

A psychologist educated in personnel selection and training has many psychological instruments from which to choose as part of a behaviorally oriented screening process. These include the Minnesota Multiphasic Personality Inventory and the Sixteen Personality Factor Questionnaire.

The Minnesota Multiphasic Personality Inventory (MMPI) is a self-report test of 565 true-false questions. The MMPI was developed in the 1940s to assess psychological adjustment problems in mental health settings. It was developed empirically by finding items that separated two groups of individuals: one with known psychiatric problems such as anxiety, depression, and psychosis, and the other, "normals."[17] Uses of the MMPI include pre-employment screening, evaluation for promotion, performance assessment, disability evaluation, and return-to-work evaluation. It is also used to determine whether personality factors are related to job success.

An individual taking the MMPI answers true-false questions, which are assigned *T*-score values on different scales (for example, scale I = hypochondriasis, scale II = depression, and so on). Profiles are then drawn for comparisons to normals; these profiles can be created for highly specific employment circumstances. In 1982, the MMPI was revised, that resulted in the publishing of the MMPI-2 in 1989. The new version includes an updated normative sample for seven regions of the U.S., and is balanced for demographic characteristics.

The value of the MMPI-2 as a screening instrument has been widely recognized. The psychologists J.N. Butcher and S.A. Coelho describe the scales that are valuable in evaluating candidates for law enforcement and protection employment:

> Thus, it appears that common personality problems to be wary of in selection of police and security personnel include applicants who are impulsive, superficial, overactive, manipulative, easily frustrated, and immature. These personality traits are likely to show up in pre-employment screening as elevation on scales 9 (mania), 4 (psychopathic deviate), 3 (hysteria), or 1 (hypochondriasis).[18]

Under the Americans with Disabilities Act, MMPI-2 is considered to be a medical test and, therefore, should be administered only after a conditional job offer has been given to the applicant. MMPI-2 test results should not be placed in an employee's personnel file due to ethical requirements mandating confidentiality of psychological and medical information. They should be filed separately and securely. MMPI-2 results should not be used solely as a basis to deny individuals employment consideration. Proper interpretation—conducted by a trained psychologist—requires a review of an individual's life history and assessment of current functioning.[19]

Another psychological personality test used extensively for security employment in the U.S. is the Sixteen Personality Factor Questionnaire (16PF), currently in its fifth edition. The 16PF has been the subject of more than 4,000 published research articles. Like the MMPI-2, the 16PF requires professional users to have had graduate training in psychological test interpretation. Also like the MMPI-2, the 16PF is considered part of a "screen out" assessment strategy for psychopathological potentials. The state of California currently requires it in the selection of all state police officers. The 16PF, which is composed of 185 items, is faster to take than the MMPI-2. Testing can be self-administered through a reusable booklet or via computer administration.

But such psychological tests are only part of a screening process. J.M. Fabricatore, a psychologist who critiques such psychological instruments, emphasizes the importance of considering testing results as only one factor in an employment decision:

> Every security administrator/selection executive/chief of police would dearly love to have a 10-min. unfakable psychological test that costs $3, can be scored in 30 seconds, be interpreted by a non-professional, is ADA/EEOC invulnerable, and has a predictive validity of 0.99. The difficult and complex truth is that selecting applicants for security/law enforcement assignment is always a human judgment.[20]

The Clear Purpose Test. In addition to personality psychological tests, another type of testing is even more widely used in pre-employment screening for security personnel. Many security employers—large and small—screen all prospective employees with a test instrument that identifies specific personality and behaviorally oriented traits, particularly oriented toward honesty. Unlike the MMPI-2, the 16PF, or others of the genre, these "clear purpose" tests may be administered before an offer of employment is contemplated. A version may be given to the applicant when he or she enters the employment office for the first time.

The term "clear purpose" refers to testing goals that focus on specific qualities such as integrity, propensity of drug or alcohol use, and likelihood of being terminated from employment for cause. A clear purpose test may signal the presence of an undesirable workplace trait. For most of their history, these tests have been called paper-and-pencil instruments, and are often self-administered with a pencil and booklet. But increasingly, the test-taking process involves interface with a computer, which then scores results, providing them instantly to clients. Other test-scoring variations exist, including mailing, faxing, or e-mailing answer sheets to the test publishers for scoring and results.

Because of the importance of employee honesty, employers have been eager to embrace objective pre-employment testing as part of the evaluative process. Tests that focus on integrity traits have grown in importance for entry-level employees in certain industries. These pre-employment screening instruments have achieved substantial support among security-conscious employers for a number of reasons, as shown in Table 3.2. Test publishers have issued two types of measurements, either one of which—or both—may be included in the same test. The first type is referred to as an overt integrity test and measures theft attitudes. This type of test includes questions about frequency and extent of theft and counterproductive activity in general; one's punitiveness and rumination about theft; the perceived ease of theft; and an assessment of one's personal integrity.[21] The second type of test asks applicants to self-report their frequency and amount of theft and other illegal or counterproductive activity.

Other Tests. Another type of test, labeled as "disguised-purpose" or "covert," is linked to normal-range personality devices. These are broader in scope and are not explicitly aimed at theft. The items do identify other desirable features in employment such as dependability, consciousness, social conformity, thrill-seeking, trouble with authority, and hostility.[22]

Still another pre-interview questionnaire is created from a pool of true-false items. The instrument identifies a set of questions that, it is believed, are related to specific aspects of reliable and productive work behavior and can differentiate against a population with contrary responses.[23] These crite-

Table 3.2. Pre-employment Integrity Screening Methods Compared

Screening Methods	Convenience Issues	Main Problems	Main Advantages
Integrity Tests (also called "clear purpose" tests)	Can easily be made part of the usual screening procedure.	Company representative(s) must be trained to appropriately use test scores. Not all integrity tests are thoroughly validated. Should not be confused with clinical personality testing.	Validity evidence exists. Generally non-offensive. No adverse impact (meets EEOC guidelines). May discourage dishonest applicants from applying.
Personnel Interviews	Usually part of hiring procedure and are often time-consuming.	No evidence of validity with theft criteria. Difficult to determine applicant's truthfulness in discussing theft and counterproductive activity. Can lead to charges of bias or discrimination.	Inexpensive (already part of hiring procedures). Structured interviews show more promise than traditional interviews.
Reference Checks	Are often time-consuming.	Little evidence of validity. Most misconduct is undetected. Previous employer reluctant to give negative information.	May increase truthfulness of applicants. Verifies information provided on application forms and resumes.
Criminal Background Checks	Commonly available, yet lengthy turnaround required.	Not all criminals are on record. Likely to exhibit adverse impact. Information obtained must be job-relevant. Many states are introducing restrictive legislation.	Complete, reliable, verifiable data can be obtained (although the process may be burdensome).
Credit Checks	Quick but somewhat costly.	Relevance to theft not clear. May not meet EEOC guidelines.	Obtains information relevant to financial need and fiscal responsibility.

Source: Adapted from J.W. Jones and W. Terris, "Selection Alternatives to Pre-employment Polygraph," *Recruitment Today*, May/June 1989, pp. 24–31.

rion-keyed true-false questions plot the test takers' responses for seven scales, including trustworthiness and safe job performance.

Other areas to consider with regards to pre-employment personality tests include the following:

Testing the Tests. Employers can select from numerous test publishers. These publishers should meet certain criteria so that their tests follow legal and privacy considerations, including:

- **Reliability.** Tests should produce the same results from individuals who are retested. Publishers demonstrate this quality by retesting and determining coefficients of agreement between the original test and the retest.
- **Validity.** Does the test deliver what it is supposed to? Test publishers use a variety of research strategies to establish validity. These include correlation with polygraph test results; correlation with anonymous admissions; the use of a time series to examine aggregate rates of inventory shortage before and after the introduction of a test program; and the comparison of test performance by groups hypothesized to differ in integrity (for example, felons compared to test takers without criminal records).[24]
- **Legality.** The testing instrument must follow federal and state statutes relating to pre-employment screening. If an applicant is denied employment in part because of scores from a pre-employment test and institutes a legal action against the would-be employer, the test publisher must be prepared to demonstrate how its test conforms with such laws.
- **Utility.** Assuming that a test meets the criteria of validity, reliability, and legality, its utility to the workplace must also be established. Major reviews of integrity tests were undertaken by the Office of Technology Assessment (OTA) in 1990[25] and by a panel from the APA[26] in 1991. The OTA report was critical about integrity testing, but neither called for legislative remedies nor suggested alternatives. The APA report supported the validity of such testing, but generally criticized test publishers for their failure to cooperate with independent researchers.

  The OTA report found no studies conducted by independent researchers in which detected theft was used as the criterion. However, a few years after the OTA report was released, H. John Bernardin and Donna K. Cooke, at Florida Atlantic University, studied a group of 111 employees hired by a major retail convenience store chain over a three-year period.[27] The researchers found that a test designed to detect deviant/non-deviant characteristics successfully predicted theft for a group of convenience store employees.

No significant differences on the test emerged as a function of race, gender, or age. Other studies have identified the value of such tests in reducing counterproductive behavior in the workplace.[28]

Job-Related Skills Testing

Employers know from experience that prospective workers differ in terms of personal abilities and behavioral characteristics. Apart from integrity and counterproductive behavior, prospective employers can identify specific minimum levels of skills required for the workplace. For example, security employees who must write concise, coherent reports can be tested on their writing abilities. Those who interact with the public can be evaluated on their interpersonal problem-solving traits. Employers may learn from experience that certain personal characteristics are the marks of high-performing employees in their industry and can obtain standard tests to use as a means of identifying the presence of these desired characteristics. A vocationally oriented psychologist can aid employers in selecting such tests.

Reviewing the File

In an ideal situation, at this point in the vetting process, a candidate for employment might have had a brief interview, completed the application, and have taken a clear purpose pre-employment screening test. Meanwhile, personnel have used the time to verify information contained in the application. This method has saved costly assets of the organization—namely, interview time—from being deployed wastefully in interviewing candidates who do not meet the requirements. An applicant may be scheduled for an interview as the applicant's employment folder is being completed.

The Employment Interview

Due to the time required, this is the most costly aspect of pre-employment screening. Nonetheless, one or more interviews are desirable to evaluate prospects for security employment before making a final decision. Many reasons exist for why this widely used selection procedure is employed. First, written and verbal references are valuable in providing insight into the applicant's fitness for the particular position, but can be incomplete. A personal interview can help to fill out missing information from references. Next, the information collected prior to the interview often contains details that are divergent with details provided by the applicant on the application form. The interview provides an opportunity for these uncertainties to be explained. Finally, the interview process may help the employer to determine the ability of the applicant to explain why she or he is interested in the position and why he or she would or would not perform the job well.

The interview has many flaws, however. The applicant may be coached to answer questions in a certain way in order to please the interviewer, but that fail to reveal the true motivations behind seeking the position. Also, some interviewers have biases that prevent them from objectively identifying the best candidates for the available position. Indeed, computer-based interviews might accomplish the same goals as face-to-face integrity interviews, with greater objectivity and speed as well as less cost.[29] Still, most employers prefer that job prospects be interviewed because the interview is an efficient means of resolving incomplete or confusing data about the application. In addition, the face-to-face interview presents an opportunity to examine a candidate's non-verbal communication skills. Also, some candidates might not accept a position in an organization in which they were not interviewed personally. Finally, in positions of critical importance, the opportunity to raise probing questions and ascertain how the applicant deals with critical issues is afforded by the interview process. Just as the application form—paper or electronic version—must avoid discriminatory-type questions, the face-to-face interview must do the same (see Box 3.7).

Box 3.7. Fair and Unfair Pre-employment Interview Questions

State and federal employment regulations guide what may or may not be asked on applications and during interviews. The employer's labor lawyer or local human rights officials can direct employers to local variations in prohibited questioning. Questions related to race, religion, and ethnic or national origin are prohibited. The following are other general topics of what is fair or unfair to ask:

Unfair Inquiries	**Fair Inquiries**
Age of applicant	If the applicant will be older than a certain age, say 21, at the time the job will begin
Citizenship	The applicant must be able to complete an I-9 form to prove his or her right-to-work in the U.S.
Arrests	Felony and relevant misdemeanor convictions
Marital status or dependent children	If the applicant has any reason why he or she might not be able to work certain hours if the job ordinarily would require it
Military discharge	Request a DD214 as part of the screening process
Medical history	Require medical tests before a final offer of employment is extended
Submit a photograph	May be taken once a job offer has been made

Interviews might be in the form of a single individual session, several sessions over an extended period of time, a panel involving two or more interviewers with the interviewee at the same time, or a combination of these. The trend at the entry level is to conduct one or two brief interviews so that a variety of candidates can be compared. At the management and executive level, the trend is toward multiple contacts over an extended period of time with different persons involved, all of whom eventually pool their observations before a decision is made.

The interview may last from a few minutes to over an hour. The interview process can be construed to be a test. Therefore, interviewers should assure that the questions raised relate to requirements of the job. Typically, interview types are divided into several categories:

- **Structured.** The questions follow a progression and often are prepared in advance. The interviewer may ask every prepared question to every candidate.[30] This is the most frequently used interview style, though aspects of other types may be included. Interviewers take this opportunity to ascertain technical and interpersonal skills. Interviewers are likely to ask a series of open-ended questions that require the applicant to respond at length, rather than with a yes or no answer. An example of an open-ended question is: "Tell me about a problem you encountered in your previous position and how you overcame it."
- **Unstructured.** This type of interview is more free in form and often seems random, with features of an informal conversation. Information is likely to be copiously provided by the candidate, but it may not be adequately job related.
- **Depth.** This style is used for careful evaluation of the candidate's fitness for a particular critical position. It may carefully review contents of the application form and raise hypothetical situations calling for on-the-spot problem solving by the applicant.
- **Panel.** To the applicant, interviews by two or more people at the same time may seem more stressful, but this is not necessarily so. Panel interviews permit several interested persons within management to make a collective decision and to permit the same verbal responses and non-verbal cues to be evaluated by more than one person.
- **Stress.** This type of interview is similar to the structured type except that it introduces difficult and sometimes challenging questions to the applicant to ascertain how he or she might react. In some cases, this interviewing format poses questions for which no plausible answer is available or for which the applicant is highly unlikely to possess the information required in order to properly answer the questions.

Assessing the Candidates

With the completion of the pre-employment screening process, management must now decide which candidates should be offered positions. All the information that has been collected will be assessed. This may be conducted by the employer or by an independent reviewer who has not personally met the candidate.

To determine whether the employer is discriminating against a protected class in its testing or screening process, data should be retained on who passes the process and who fails. If the pass ratio of the percentage in the protected group is less than four-fifths of the majority group, the employer may be creating an adverse impact, according to the law.[31]

Example:
Percent passing from the protected group: 50 percent
Percent passing from the majority group: 75 percent
Ratio = .50/.75 = .667 or 67 percent
Four-fifths = .8 or 80 percent

Therefore, this test does not meet requirements of the "four-fifths rule" and is not in compliance with EEOC guidelines. The employer would be advised to re-evaluate the screening process, determine why some groups are being excluded, and revise the process in accordance with the findings.

Pre-Employment Drug Screening

Many workplaces have a drug-testing policy. It usually requires such testing after a conditional offer of employment has been made. All applicants should be informed in writing of the organization's substance abuse policy and drug screening procedures. In cases in which the potential employer conducts work on behalf of the federal or state government, such a policy is a legal requirement. For example, in Title V of the Omnibus Drug Initiative Act of 1988, including a provision with the short title of the Drug-Free Workplace Act of 1988 (PL 100-609), all businesses contracting with the federal government and all grantees receiving federal financial assistance must certify that they have in place policies directed toward the creation and maintenance of a drug-free workplace.[32] The act does not require pre-employment drug screening. However, many organizations establish this procedure as a policy.

Prospective employees should be requested to sign an informed consent form to permit substance abuse testing. This notice may state that a confirmed positive result may produce a rejection of an employment offer. The form may also state that the failure to consent to the test will result in the application process being incomplete, hence, not leading to a final offer of employment.

Screening test samples are usually analyzed at clinical centers. The usual technique is an Enzyme Multiplied Immunoassay Technique (EMIT), which analyzes the urine sample of the prospective employee. This measures the presence of a drug or drug metabolite by comparing a test sample of urine with a reference solution. The test requires about 90 seconds to produce a reading. A positive reading for the detection of substances tested requires an additional confirmatory test before results are shared with the applicant or a decision about employment should be made by management. A second test, with greater specificity than the initial procedure, is conducted. An example of this is gas chromatography/massspectroscopy (GC/MS). Preferably, the test is conducted on the initial urine sample. Organizations should engage the services of clinical laboratories meeting high professional standards to minimize false positives and to assure test accuracy while minimizing cost of the tests.

The employer may decide that a medical examination for general health determination might also be required after a conditional offer of employment. This may occur when the applicant has had physical disabilities in the past and the employer requires a medical opinion on the potential for recurrence of the problem.

The Final Offer of Employment

Following the collection of all relevant information on the applicant, including drug and medical documentation, a final offer of employment is extended. The next step in the employment process is orientation and training, subjects of great concern to successful security operations management.

SUMMARY

The secret of success in many organizations is to hire and retain the right people. Reliable pre-employment screening is at the core of successful operations management. This is particularly the case for protection programs. Security personnel are selected in much the same way as other workers in an organization; however, higher standards for determining applicants' previous legal, moral, and ethical behavior guide employment decisions.

DISCUSSION AND REVIEW

1. Why is it incumbent upon employers of security personnel to vet prospective employees more thoroughly than in non-security positions?

2. In a civil action, how can a claim of negligent hiring have a better chance of succeeding?
3. What is the risk to security operations managers if they are personally cited in a civil action involving a claim of negligent security?
4. What are the inherent risks of personal references? What are the desirable reasons for retaining such procedures despite the limitations?
5. How have technology and the Internet changed pre-employment screening?
6. Several federal acts protect the privacy of individuals. If an employer in the process of screening an application discovers an open arrest record for an applicant, what is the recommended course of action?
7. Does the Employee Polygraph Protection Act (EPPA) make an exception for certain employers? If so, who are they?
8. Explain the difference between reliability and validity in pre-employment tests.
9. Under what circumstances might a test for psychological stability be highly desirable in employing security personnel?
10. Explain how the "four-fifths rule" serves as a means of identifying employment bias.

ENDNOTES

[1] J.G. Service (January 1988). "Negligent Hiring: A Liability Trap." *Security Management*, p. 65.

[2] J. Chuvala and J.A. Gilmere (April 1992). "Legal Consequences for Negligent Retention, Supervision, and Training of Employees." *Security J.*, 3(2): 87–90.

[3] R. Fischer quoted in ibid.

[4] J.W. Jones and W. Terris (May/June 1989). "Selection alternatives to pre-employment polygraph," *Recruitment Today*, p. 24–31.

[5] S. Rhodes and K. Springen (January 13, 1997). "Economy: Yup: Help Wanted," *Newsweek*, p. 52.

[6] J.D. Werbel and J. Landau (1996). "The Effectiveness of Different Recruitment Sources: A Mediating Variable Analysis," *J. Applied Social Psychology*, Vol. 26, p. 1337.

[7] I.A. Lipman (July 1988). "Personnel Selection in the Private Security Industry: More than a Resume." *Annals, AAPSS*, 498: 83.

[8] *Guide to Background Investigations* (1997). Tulsa, OK: TISI; C.R. Ernst and M. Sankey (1998). Find Public Records Fast. Tempe, AZ: Facts on Demand Press.

[9] S. French (2000). *Who Are You? The Encyclopedia of Personal Indentification.* Mt. Shasta, CA: Intelegence Here, pp. 46-62.

[10] B. Kleinmutz and J. Szucko (1984). "Lie Detection in Ancient and Modern Times: A Call for Contemporary Scientific Study." *American Psychologist*, 39:766–76.

[11] J.K. Barefoot (Ed.) (1994). *The Polygraph Story.* Washington, DC: American Polygraph Association.

[12] D.T. Lykken (1981). *A Tremor in the Blood: Uses and Abuses of the Lie Detector.* New York, NY: McGraw-Hill Book Company.

[13] "The Accuracy and Utility of Polygraph Testing" (1984). Washington, DC: Department of Defense.

[14] *Scientific Validity of Polygraph Testing: A Research Review and Evaluation—A Technical Memorandum* (1983).Washington, DC: U.S. Congress, Office of Technology Assessment.

[15] American Psychological Association (1991). *Questionnaires Used in the Prediction of Trustworthiness in Pre-employment Selection Decisions: An A.P.A. Task Force Report.* Washington, DC: American Psychological Association.

[16] See Note 14, pp. 3 and 25. Also see Hearings on S. 1815. Senate Committee on Labor and Human Resources (Prohibited use of Lie Detectors), 99th Congress, 2B Session, April 23, 1986.

[17] J.N. Butcher and S.A. Coelho (1997). The Minnesota Multiphasic Personality Inventory-II (MMPI-2). *Security J.*, 8:121–24.

[18] Ibid.

[19] G.F. Sumprer (1997). Review of the Minnesota Multiphasic Personality Inventory-2. *Security J.*, 8:125–27.

[20] J.M. Fabricatore (1997). The Sixteen Personality Factor Questionnaire (16PF). *Security J.*, 8:162.

[21] P.R. Sackett and J.E. Wanek (1997). "Integrity Testing: An Overview." *Security J.*, 8:11–18.

[22] Ibid.

[23] G.L. Borofsky (1997). "The Employee Reliability Inventory (ERI)." *Security J.*, 8:55–60; and J.S. Rain (1997). "Review of the Employee Reliability Inventory (ERI)." *Security J.*, 8:61–63.

[24] P.R. Sackett and J.E. Wanek (1997). "Integrity Testing: An Overview." *Security J.*, 8:11–18.

[25] Office of Technology Assessment. (1990). "The Use of Integrity Tests for Pre-employment Screening." Washington, DC: Office of Technology Assessment.

[26] L.R. Goldberg, J.R. Grenier, R.M. Guion, L.B. Sechrest, and H. Wing (1991). *Questionnaires Used in the Prediction of Trustworthiness in Pre-Employment Selection Decision: An APA Task Force Report.* Washington, DC: American Psychological Association.

[27] H.J. Bernardin and D.K. Cooke (1993). "Validity of an Honesty Test in Predicting Theft Among Convenience Store Employees." *Academy of Management J.*, 36 (5): 1097–1108.

[28] R.A. Kuhn (1990). "The Attack on Employer's Rights." *Security J.*, 1: 74–80; G.L. Borofsky, H.J. Klein, and W. Davis (1993). "Pre-employment Screening for Unreliable Work Behaviors: An Opportunity to Work Cooperatively with Human Resource Managers." *Security J.*, 4:185–92.

[29] B.C. Jayne (1997). "The Utility of a Computer Interview to Screen for Positions of Public Safety." *Security J.*, 8:205–208.

[30] D.J. Still (1997). *High Impact Hiring: How to Interview and Select Outstanding Employees.* Dana Point, CA: Management Development Systems, p. 139.

[31] Ibid.

[32] J. Fay (1991). *Drug Testing.* Boston, MA: Butterworth-Heinemann, pp. 23–24.

Additional References

S. Abrams (1977). *A Polygraph Handbook for Attorneys.* Lexington, MA: Lexington Books.

A. Gale (Ed.) (1988). *The Polygraph Test: Lies, Truth and Science.* London: Sage Publications.

D.T. Lykken (October 1974). "Psychology and the Lie Detector Industry." *American Psychologist*, pp. 725–39.

J.B. Miner and M.H. Capps (1998). *How Honesty Testing Works.* Westport, CT: Quorum Books.

Special Issue: Security and Law Enforcement Pre-employment Testing (May 1997). *Security J.*, 8(1–2):1–208.

D.J. Still (1997). *High Impact Hiring: How to Interview and Select Outstanding Employees.* Dane Point, CA: Management Development Systems.

J.F. Vaughan (1999). *Avoiding Liability in Premises Security,* 4th edition. Atlanta,GA: Strafford Publications.

4

TRAINING AND DEVELOPMENT FOR HIGH PERFORMANCE

> Training is everything. The peach was once a bitter almond; cauliflower is nothing but cabbage with a college education.
>
> —Mark Twain, *Pudd'nhead Wilson*

The training of employees and the development of their skills and careers is a critical and time-consuming activity within security operations. Successful training is directly linked to high performance on the job. High standards are more likely to be achieved with a relevant, engaging period of learning. Successful training ensures that the employees meet the short-term needs of the employer, while further development enhances the skills and career paths of workers. It also provides the organization with employees who understand the corporate culture and who are prepared to meet the future needs of the workplace. Numerous approaches regarding training and development have been used in the history of management. Before considering them, the organization needs to establish the training and development resources required and how to teach new employees their tasks.

Security as a vocation is constantly changing. Skills need to be kept up to date, and relevant knowledge must be shared with workers. Yet at the operational level, security is not akin to a skilled craft in which a period of apprenticeship is expected. Usually, security knowledge and procedures needed, say, before an initial patrol or system monitoring assignment, can be taught much more quickly than in an apprenticeship. But security practitioners at all levels must constantly learn more about their endeavors. Training and education thus not only help work become more satisfying for individuals, but also produce direct benefits to the employer as well.

WHY TRAIN ANYHOW?

Training provides employees with the skills and information needed to do the job well. However, additional factors mandate the importance of security training. These are more relevant than in many other vocations. The following are the main reasons why training is important in high-performance security programs:

1. The law requires security training in many states. Pre-assignment or initial training is required in a number of states, and the trend for such state-mandated learning is continuing. According to data collected by Guardsmark, Inc., in 1986, 12 states required pre-assignment training for unarmed security officers. By 2000, that number had grown to 18 (see Box 4.1). Requirements for armed security officers grew from 22 states in 1986 to 33 states in 2000.
2. Many states also require refresher training. The states that mandate refresher training for unarmed security officers have grown from two in 1986, to six in 2000. Meanwhile, refresher training for armed officers was mandated in 24 states in 2000.
3. Many security positions require more training than the legal minimum or a few hours prior to the initial assignment. Security officers who deal extensively with the public, who are armed, who are expected to respond appropriately in emergencies, and who interact with complex technology are examples of those who require more training than the minimum. The length of such training and education will be according to the actual need. For example, in schools where Peace Officer Standards of Training (POST) exist, school secu-

Box 4.1. Security Officers' State Training Requirements (Based on the 50 States Plus the District of Columbia)

	Number/percentage	
Requirements	**1986**	**2000**
Initial/Pre-assignment	12/23%	18/35%
Unarmed Refresher	2/4%	6/12%
Armed Pre-assignment	22/43%	33/65%
Armed Refresher	?	24/47%

Source: Guardsmark, Inc., December 1986 and June 30, 2000.

rity officers usually are required to complete a minimum of 40 hours pre-assignment training.[1] In other cases, training of security officers can last weeks. For example, three weeks of pre-assignment training, in addition to state-mandated training, are required by New York City's Health and Human Resources Administration before security personnel are assigned to homeless shelters. And in Sweden, security officers for a private security services company are trained almost six months before being assigned to certain government-contracted assignments.

4. Training and retraining produce important measurable benefits. Training reduces risk and loss. Research by Liberty Mutual Insurance Company indicates that a mere 2 percent of industrial accidents occur because employees didn't know how to perform the task safely or properly.[2] Only 6 percent of accidents are due to equipment failure. The vast majority of industrial accidents—92 percent—happen because workers failed to perform their tasks properly. Performance deficit may be diminished by training and refresher retraining.
5. Lack of adequate training may be the basis of a successful tort action. The failure to provide adequate training for security personnel can lead to a successful plaintiff's action for negligent security that could be upheld at an appellate level. David A. Maxwell, a lawyer and educator specializing in protective matters, writes: "Employees should be subjected to training prior to assignment. . . When the role at hand involves a potential for harm or injury, the standard for training rises to the risks."[3] Courts have held that when employees are not trained on how or how not to use weapons and an injury results, the employer may be negligent.
6. Requiring security officers to be trained is a reasonable public policy. In the 1950s and 1960s, the average police officer possessed a high school education and was trained in a police academy for a few weeks. Today, police officers in cities and industrialized states often must possess a minimum number of college credits or an awarded academic degree in order to be considered for law enforcement. Meanwhile, academy training averages five to six months for urban police forces.
7. Workers value employers who provide good training. Many workers are attracted to employers who provide training programs and educational benefits. This self-selection enhances the quality of the pool of prospective employees. Once hired and trained, employees are likely to remain longer with such employers, returning to the workplace the benefits of the training and education they have achieved.

The analogy between public police and private security services is awkward and debatable on many points. Yet it is apparent that many similarities also exist between the two. Private security personnel "protect and serve" in much the same way as police officers on patrol do, albeit on private property and for private interests. Therefore, it is reasonable as a public policy that security officers be trained to a level at which they can do their jobs well, while understanding the legal rights and limitations of their roles.

THE TRAINING MANAGER OR OFFICER

High-performance security training programs are headed by training managers or supervisors. Their responsibilities include consulting on training policy and content with senior managers and others involved in establishing training goals. They further seek to ensure that such objectives are achieved. The training manager or supervisor organizes internal training courses and may teach some or all of them. Typically, such a person is closely involved in the instruction of new employees and newly promoted supervisors. The chief trainer is expected to remain informed of the latest materials and technologies to enhance learning of all who come under his or her tutelage. This person also is involved in the evaluation of training programs, though analysis of training results may occur elsewhere in the organization.

The training department itself may be a single person who juggles other responsibilities. In smaller programs, the security manager serves as the chief trainer, but also has other responsibilities. Large departments may have several staff members, mount a variety of programs, and produce their own curricula and training aids. Trainers may be obliged to use whatever space can be assembled or they may use a specially designed and dedicated learning facility, perhaps with distance learning capabilities.

PLANNING TRAINING AND DEVELOPMENT REQUIREMENTS

Security-oriented programs require different types of training for different purposes. In most organizations, entry-level training receives the greatest attention. These programs cannot become static. Key aspects in training programs require review and renewal so that programmatic information remains fresh and pertinent.

Changes in the workplace and society make good training and development important. Many executives think of training as a cost burden. However, the training process should provide precise, desired results for management, with measurable changes in workplace performance. Training methods welcome experimentation in order to improve results and reduce costs. For example, technology-based learning can decrease the cost and enhance learning effectiveness for new workers.[4] Students remain with the program until all critical elements can be tested and passed.

While a large number of vocations require some pre-assignment training, that obligation is particularly clear-cut for security positions. The *Report of the Task Force on Private Security* refers to training as "a vital determinant of job performance" and proposes standards that would be monitored by state boards.[5] Several states have mandated specific requirements for training for security personnel: proprietary and contract workers by law must meet a prescribed curriculum possibly involving pre-assignment training, post-assignment training, and periodic retraining. Security programs are obliged to keep details of individual training achievements for use in potential civil litigation for alleged inadequate training.[6]

In today's Information Age, the need for training and personal development is never-ending. Consequently, effective operations must be on the alert for any new specialized training that is required, including short sessions on new laws, technologies, and strategies.

Training is thus defined as a process of learning specific skills and knowledge required by the employee to carry out an existing job or to complete a new one. It is vocationally oriented. The process is mostly aimed at operational-level workers, both proprietary and contract. Training merges with a separate learning category, development, which seeks to advance the informational, critical, and analytical skills of persons responsible for managerial and executive responsibilities. Development will be discussed later in this chapter. For now, we will look at the initial contact of a new employee with the formal organization.

THE ORIENTATION

Orientation introduces new employees to the history, ethos, objectives, and available resources of the organization. In large organizations, orientation and training programs may be in the hands of dedicated personnel. However, operational-minded managers generally are involved closely with the training process. They are likely to appear at orientation sessions to make new employees feel welcome and to emphasize the personal concerns of the director.

An orientation provides a variety of information types to new employees. The orientation process usually serves as a welcome to new employees from representatives of management and contains the following elements:

- The history and ethical basis of the organization
- What management expects from all employees
- What employees can expect from management
- What the organization does; its trends
- Characteristics of the department where the employee will be assigned

- The reporting structure (an organizational chart may be provided)
- The purpose of security in the organization and the value management places on it
- Security and safety policies at the workplace

The orientation covers certain fundamentals so that all new employees will become familiar with the same organizational requirements and expectations as those who have gone before them. Orientation also covers practical information, such as keys to be assigned; hours to be worked; payroll procedures; the restroom locations and use policy; the alcohol and drugs policy; the firearms policy; the personal telephone use policy; and the policy regarding personal visitors. Some employers request that new employees sign a statement that they have heard and understood policies introduced during the orientation process.

At such an orientation, new workers complete forms required for tax and statutory benefits purposes. The company handbook is distributed and may be reviewed in depth with the new employees by the orientation leader. At the conclusion of the orientation phase, new workers typically feel upbeat about having achieved an offer of employment from the organization and they are now eager to learn specifically how they are expected to perform their work.

Training Content

Prior to beginning the training process, the operating manager will expend considerable energy to identify what information should be transmitted to new employees. No two employers are likely to be identical in their training objectives, protocols, and needs. Content of the training program will emerge from the ethos and needs of particular employers. Still, some subjects related to security services are likely to be found in most programs.

The content of the training material should be related to the job description created for the position being filled. The frequency and importance of various tasks undertaken by the workers should be related to the amount of training time allocated. For example, fires at the workplace are not frequent occurrences, but they are emergencies when they do occur. Therefore, adequate time should be provided for the worker to learn about fire suppression equipment and use.

The following are specific training factors for certain types of security personnel:

Pre-assignment Training. Security personnel require training, subsequent to orientation, before being assigned to a position in an operating program. The number of hours devoted to this phase of training is variable. For example, the State of New York currently requires 8 hours of pre-assignment training followed by 16 hours of basic training within the first 90 days of employment.[7] The Task Force on Private Security in 1976 recommended a minimum of eight hours formal pre-assignment training with a minimum of 32 hours of basic training within three months of assignment. In the Task Force's standards, a maximum of 16 hours could be supervised on-the-job training. Many employers, however, combine pre-assignment training with basic training before the employee is assigned a position. The number of hours that organizations actually train security employees differs widely. Most train for the minimum number of hours required by law; however, some programs have determined that extensive additional training is necessary due to the nature of the demands of such protective employees.

The trend in the U.S. and in other industrial nations is to increase the level of initial training required before personnel are assigned to posts of responsibility. This is because of the growing legal burden to train security personnel so that they are aware of their legal obligations in dealing with the public. Failure to understand such principles could lead to a charge of inadequate security in a tort action. Employers may also increase the amount of training because security employees often must interact with complex systems or because further job-specific training is necessary.

Workers generally require training even if they come to the job with experience in police departments, the military, private security companies, or in proprietary security departments that have provided training of their own. This is because the employer will wish to be assured that permanent employees share the common curriculum of all employed in such positions. An exception to this policy is usually made for temporary security workers who work under close supervision and for active police officers who are working temporarily, that is, moonlighting, in security positions.

The eight-hour pre-assignment training course, proposed by the Task Force for entry-level security personnel, is divided into four segments: orientation; legal powers and limitations; handling emergencies; and general duties, as shown in Table 4.1. The course is to be used in classroom instruction, possibly in conjunction with audio-visual aids, and is concluded with a test to assure that the content has been understood and mastered.

Basic Training. As previously mentioned, many employers combine orientation with pre-assignment training and basic training. However, some employers prefer to provide basic training in modules over time for new security workers. The Task Force on Private Security recommends a minimum of 32 hours of basic training in addition to the pre-assignment phase, as shown in Table 4.2. This training should be completed over a three-month period and may include a maximum of 16 hours on-the-job training.

The growth of training required for security personnel is similar in some ways to the increased training for public law enforcement officers. In

Table 4.1. Private Security Eight-Hour Pre-assignment Training Course

Section I: Orientation

Two hours are spent on the following topics:

- What is security?
- Public relations
- Deportment
- Appearance
- Maintenance and safeguarding of uniforms and/or equipment
- Note-taking/Reporting
- Role of public law enforcement

Section II: Legal Powers and Limitations

Two hours are spent on the following topics:

- Prevention versus apprehension
- Use of force
- Search and seizure
- Arrest powers

Section III: Handling Emergencies

Two hours are spent on the following topics:

- Crimes in progress
- Procedures for bomb threats
- Procedures during fires, explosions, or other emergencies
- Responding to alarms

Section IV: General Duties

Two hours are spent on the following topics:

- Fire prevention and control
- Inspections
- Interviewing techniques
- Patrol procedures
- Safeguarding valuable property
- Safety
- Surveillance

Source: Based on a model originally prepared by the Private Security Advisory Council, included in their Model Private Security Licensing and Regulatory Statute.

Table 4.2. Private Security 32-Hour Basic Training Course

NOTE: A minimum of four and a maximum of 16 classroom hours should be allocated in each of the following sections, and a maximum of 16 hours of supervised on-the-job training should be permissible.

Section I: Prevention/Protection

- Patrolling
- Checking for hazards
- Personnel control
- Identification systems
- Access control
- Fire control systems
- Types of alarms
- Law enforcement/private security relationships

Section II: Enforcement

- Surveillance
- Techniques for searching
- Crime scene searching
- Handling juveniles
- Handling mentally disturbed persons
- Parking and traffic
- Enforcing employee work rules/regulations
- Observations/description
- Preservation of evidence
- Criminal/civil law
- Interviewing techniques

Section III: General Emergency Services

- First aid
- Defensive tactics
- Fire fighting
- Communications
- Crowd control
- Crimes in progress

(continues)

Table 4.2. *(Continued)*

Section IV: Special Problems
• Escort services
• Vandalism
• Arson
• Burglary
• Robbery
• Theft
• Drugs/Alcohol
• Shoplifting
• Sabotage
• Espionage
• Terrorism

Source: *Private Security: Report of the Task Force on Private Security* (1976). Washington, DC: Government Printing Office, p. 103.

1973, the National Advisory Commission on Criminal Justice Standards and Goals recommended 400 hours training for sworn police officers. Since then, most urban police departments require police cadets to successfully pass requirements in a program that is about six months in length. Smaller departments also require academy training, but the time allocated is closer to the 400 hours proposed in 1973. In policing and security, instruction on legal issues, public relations, and new technology add to the quantity of time required in order to be fully effective in these jobs.

Firearms Training. The vast majority of private security personnel are not armed in the course of their employment; however, some are. Security officers who are expected to carry firearms in the course of their employment invariably face additional burdens on training. The Task Force on Private Security recommends that security personnel be required to complete successfully a 24-hour firearms course that includes legal and policy requirements, or submit evidence of competency and proficiency, prior to assignment to a job that requires a firearm. The course of training is divided into classroom and range components, as shown in Table 4.3.

Weapons proficiency requirements should be met on an annual basis if the employee continues to require the weapon as part of his or her duties. State and local requirements for the possession and use of firearms provide specific regulations about firearms retraining and range experience for that

Table 4.3. Proposed Standards for Security Guard Firearms Training

---

Classroom-Based Training

Topic I: Legal and police restraints (3 hours)

- Rights of private security personnel to carry weapons and the power of arrest
- Statutory references
- Policy restraints

Topic II: Firearms safety and care and cleaning of the revolver (2 hours)

- Nomenclature and operation of the weapon
- Performance of cartridge
- Safety practices on duty and at home
- Range rules
- Care and cleaning of the weapon

Topic III: Successful completion of written examination (1 hour)

- At least 20 minutes on the above topics with a minimum passing score of 70%
- Should be designed so that persons with other or prior experience can demonstrate competence in the subject areas.

Range-Based Training

Topic I: Principles of marksmanship (2 hours)

Topic II: Single-action course (8 hours)

- A silhouette target with a distance of 25 yards is used with 30 rounds fired under different circumstances for qualification of which the minimum passing score is 18 hits (60%)

Topic III: Double-action course (8 hours)

- The distance and target are the same, but trainees operate from a crouching position under different circumstances and must score 43 hits out of 72 attempts (60%)

---

particular jurisdiction. Instructors are recommended to be qualified through the National Rifle Association or other comparable qualifications program.

The armored car industry, which requires most service employees to carry firearms, has created its own set of standards. The Training Committee of the National Armored Car Association has proposed an outline of basic firearms training for its employees, as shown in Table 4.4.

Table 4.4. Firearms Training for Armored Car Personnel

- Company and industry policy on use of weapons
- Legal limitations
- Firearms safety
- Care in firearms cleaning
- Basic revolver training
- Combat firing
- Use of gunports
- Use of shotgun
- Qualification and certification

Source: Training Committee of the National Armored Car Association.

Investigative Training. Investigators vary widely according to their assignment type. Usually, a combination of classroom, home study, and on-the-job training is provided. For example, the May Corporation, a major retailing chain, has produced a 60-day training program for undercover operatives. Much of this time is spent observing the performance of highly proficient security investigators. This field experience is supported with study of a proprietary manual. After considerable observation and discussion, the store detective is ready to operate, with the instructor acting as a coach.

Ongoing Training. This refers to flexible, continuous, individualized programs to ensure that private security personnel are kept informed of pertinent developments in the field. This information is provided during roll-call training, by visits from supervisors to security personnel on posts, with mailed or distributed training bulletins, and by other means. This may be provided through bulletins, telephone messages, e-mail, and other means that can inform security personnel of important subjects that are not related to the day-to-day routine. Such means serve to emphasize best practices. Easily forgotten procedures can be reviewed and emphasized with testing.

TRAINING TECHNIQUES

Training may be achieved by using a number of techniques that are adaptable to specific needs. Most of these approaches are suitable for entry-level operational personnel, but they can be successfully used at other employment levels as well.

Classroom

Training individuals classroom-style has many advantages. One instructor can provide training to many learners at the same time. In addition, numerous learning aids (lectures, discussions, films, videotapes, computer aided demonstrations, role playing, and others) can be used during the same training process. For these reasons, classroom training is efficient, uniform, and economical. Furthermore, this method promotes bonding among new employees who come to know each other as fellow learners in a supportive setting. Classroom instruction is valuable for all tiers of learning.

At the management level of training, the case history method is sometimes used. This introduces a hypothetical or an actual unresolved organizational problem, with the management options then considered. Often, groups of managers work in teams to compete with each other in identifying optimal solutions.

Case histories have been an integral part of training and education for over a century. Christopher Columbus Langdell introduced the case history method to students at Harvard Law School in the 1880s. This method replaced the previous technique of law school professors slowly reading their notes to students, who would copy them and later commit their points to memory. The case history method extended from law schools to other educational venues, including graduate education in business and management.[8] Participants in case history learning experiences absorb considerable information about a situation and are asked to analyze the material and present their own recommendations to help the organization resolve its identified problem. This helps participants enhance their understanding of specific circumstances while at the same time hone their decision-making skills. In less challenging circumstances, hypothetical case histories are used to replace extended real-life examples for edification and discussion. Leaders often pose questions for students such as "How could the developments in this situation have been avoided?" or "What other circumstances does this circumstance remind you of?" or "What are the hidden losses in this case scenario?"

The disadvantage of the classroom method is that it may be too passive. Learners are present physically and may appear to participate in the learning process, but actually may tune out or not comprehend the necessary information. Further, classroom instructors vary considerably according to their effectiveness. While most are enthusiastic about helping students learn, some cover material inadequately, serving as harmful role models for new employees. Additionally, classroom training is not hands-on training. Lecturers may discuss procedures and share illustrative or personal experiences to make points clearer; however, if the learner needs to interact with technology or others, he or she may need direct exposure to

these circumstances, guided by the instructive presence of an experienced worker. This process is described next.

On-the-Job (OTJ) Training

This type of training occurs while employees perform actual job-related activities. They are neither in a classroom nor in another type of learning facility, though these may be incorporated in the process. The learner is in an actual work location guided by an experienced, competent, and reliable worker who serves as a supportive instructor. For example, novice store detectives learn the craft of making successful apprehensions by observing and shadowing experienced detectives. The detective trainees do not make apprehensions, but observe them, participate in interrogations, and may witness the signing of apprehension reports. In time, the instructor will determine whether the learner is prepared to conduct his or her own apprehension, and at this point the new employee will be encouraged to do so under supervision. If the apprehension is conducted generally as desired, it will signify that the OTJ training process has achieved its objectives and that the store detective will begin to work under less direct supervision.

Managers are generally positive about OTJ training because the learning process is direct and relevant. Instructors may describe the desired performance and then watch someone learning the new skills. Errors can be identified and corrected immediately and on the spot. Desired behavior can be supported by immediate praise, re-enforcing the desired actions. The disadvantage of OTJ training is that it is costly. While the instructor and learner are conducting useful work during the training process, the quantity of work achieved is no more than what a single experienced worker alone would perform, and often less. Furthermore, OTJ training depends on the availability of motivated instructors who will nurture the learning of others. Not all of even the most highly qualified workers are able to meet these requirements.

Computer-Aided Instruction

Computer-based learning has had a substantial influence on the training process in recent years. Unquestionably, this dynamic process is continuing and will aid learning endeavors by increasing interactivity. Program enrichment and dramatic special effects make the material memorable.

The subject material is logically created, presented in programmed format, often enlivened with animation and audio materials, and designed to elicit responses to assure comprehension of the material. When the learner answers a question correctly, the program acknowledges it—often with personal references that may be part of the programming: "Good work, Joseph!" "That's your fifth correct answer, Tina!" If the learner presents an incorrect answer to a question, the program will suggest "try again" and await the correct answer to provide the expected praise. In some cases, the learner will be referred back to earlier materials in the sequence where the critical information was initially presented. By this looping back process, the learner will review and presumably understand the desired instructional material before being able to proceed further with the lesson. Unlike conventional learning, where some students will have full comprehension and others "passable" but incomplete learning, computer-aided instruction allows full achievement by all participants. Further, the training manager will have details on what questions in the testing sequence a particular student initially had trouble with. These wrong answers may suggest the need for additional training by the learner. However, they may also indicate that the teaching material was not sufficiently clear or compelling for learners to understand more quickly what was to be comprehended.

The disadvantages of computer-aided instruction relate to complexity and cost. The learning process is related to specific facts, policies, and procedures. The time and effort required to create the learning segment of this type of instruction is extensive, though computer techniques have reduced the costs in the past and promise increasing future efficiencies. Still, the programming activity often has a designed rigidity that makes the student less able to deal with unexpected occurrences, so frequently part of security experience. That's why computer-aided learning is rarely the sole way trainers have employees learn new material. And although computer learning gives all participants a full understanding of the key points in the subject material, retention of such information will differ among learners.

Computer-aided instruction may supplement other modes of training. Such instruction may be delivered at workstations as part of instructor-led courses, at commercial learning centers, within a corporate or security department training center, or wherever the learner has access to a computer with a modem. Some programs may be self-paced and delivered via laptops. Programs are available for different levels of complexity—from entry-level basic patrol instruction to advanced systems and data security (see Box 4.2).

Box 4.2. Outline of an Advanced Security Programming Course: Implementing Windows NT Security

---

The following program provides an example of a course to be learned over five days. Hands-on exercises provide the learner with practical experience. These include observing a brute force attack on NTFS files, using tools to gain unauthorized administrator privileges, and controlling access to files using privileges and permissions.

Box 4.2. *(Continued)*

Introduction to Windows NT Security

- An Overview of Security Objectives
- Developing a Windows NT Security Policy
- Trusted Computing Base (TCB)

Encryption and Authentication

- Encryption Techniques
- Authentication Methods

Windows NT C2 Certification

- Microsoft's Security Commitment
- Practical Implications of C2 Security

Administering Windows NT Security

- The NT Security Sub-system
- NT Security in Action
- Planning Domains
- Managing Accounts and Groups
- Password Vulnerabilities

Object Security

- Overview of Access Control Lists (ACLs)
- The Registry
- The NT File System (NTFS)

Auditing Windows NT

- Mechanics of Auditing
- Common Auditing Scenarios

Network Security

- Protecting Shared Resources
- Identifying and Preventing Attacks
- Securing Networked NT

The Evolution of NT Security

Source: Learning Tree International

Audio-Visual (AV) Materials

Security management facilities may possess rich resources of films and tapes for training and educational purposes. AV resources can enliven classroom learning and may also be incorporated into computer-aided instruction. AV resources are valued because they can draw upon dramatic situations, actual images from past crimes or incidents, and the use of corporate officials or professional actors to convey desired messages. AV materials use voice, action, and special effects just like any Hollywood production.

AV materials have an initial high cost for acquisition, but they can be used and reused for future training iterations. Some AV programs for security use include a manual for instructors and student handouts. AV material can be exciting and informative when well produced, although a few productions fail to achieve their desired goals and are boring, while other films can quickly become dated and obsolete. The trainer will preview such materials and select those that best meet the needs of the workplace. Often site-specific AV materials can be developed on videotape or digital format at moderate cost.

Demonstrations

Often following the acquisition of new products, systems, and software, employees must be trained in their use. Such training may be provided with demonstrations of the actual materials acquired and intended to be used in the workplace. (Occasionally, such products may be "demo-ed" prior to their purchase to obtain critical feedback from workers.) Demonstrations may be incorporated into OTJ training and classroom training as well. For new products and services that are acquired, initial demonstrations may be given by vendors, although some vendors have computer-based instructions to aid in the process. Demonstrations are important because they represent the most efficient means by which new corporate investments in hardware or software can be disseminated to persons most likely to use them on a daily basis.

T Groups (Sensitivity Training)

These highly participatory learning methods are intended to improve trainees' skills in working with other people. This is achieved by increasing their ability to understand how others react to behaviors encountered in the workplace. T (for training) groups are small sessions that may take place without a leader for much of the time. The group will be given a task, and members will create working relationships with each other to achieve the goal. T group members eventually may see the subtleties of their verbal and non-verbal communications with each other. A manager may use such a technique to explore feelings of sexism, racism, and other issues that might hamper workers from interacting productively and harmoniously with each other. For example, if concern for harassment in the workplace exists, T groups can evaluate the nature of how people feel differently about remarks and behaviors that might seem like normal communication or joking to some, but which are offensive to others. Such a process generally leads to greater sensitivity on the part of those who might be the cause of, bystanders to, or recipients of such behavior.

Role Playing

A leader or coach may describe a particular situation to a group and then ask members of the group to improvise the scenario. For example, the trainer may wish to instruct a new group of retail security personnel on how to apprehend a shoplifter. One volunteer would play the shoplifter's role and others would act as security agents intent on apprehending the offender. The remaining learners would observe and critique the process. The unrehearsed acting by the participants provides an opportunity for the leader or coach to stop the action at different points and ask trainees what they think is right or wrong about action as it is occurring. The use of unstructured drama in role-playing can substantially aid trainees in understanding procedural and behavioral lessons that need to be learned. For example, a security manager who is responsible for healthcare or social service facilities may determine that T groups are valuable in helping trainees understand the nature of their patients' or clients' emotional status, which could be fragile or dysfunctional. Role playing scenarios are mini-dramas that are usually enjoyable for instructors and participants alike, leading to a generally stimulating learning experience.

Other Techniques

Numerous other types of training also exist, but are less frequently found in security training programs. These include apprenticeship or extensive OTJ training; behavior modeling methods that provide immediate feedback; and skill analysis, in which relevant workplace behavior is assessed for weaknesses.

FIREARMS TRAINING

The need for firearms training is generally codified by state and local regulations. In 1999, 32 states and the District of Columbia required initial training for armed security personnel. Of these, 24 also required refresher training.

Generally, specific training by a certified firearms instructor is required prior to registration. In some cases, states that do not require initial training have requirements on a county or city level.

Managers of security programs and operators of security service businesses have depended less on armed employees than in earlier years. The *Rand Report*, released in 1970, stated that 49 percent of private security personnel carried firearms. By the year 2000, that percentage was well under 10 percent. Further, at the time of the Rand research, only 19 percent of private security personnel carrying firearms had job-related training on use of the firearms they carried.

The decline in the arming of private security personnel is mostly related to the increased liability from accidental or unintentional injury from the possession of firearms. Most security directors have determined that they can design security programs that meet their objectives without arming security personnel. Nonetheless, numerous examples exist where security officers probably should be armed. In such cases, responsibility for assuring that armed security personnel are initially trained and maintain their annual training requirements is a responsibility of both the individual security officer and his or her direct employer. Software programs can keep track of completed training and provide reminders when annual refresher firearms training should be scheduled.

The *Report of the Task Force on Private Security* proposed a standard of a 24-hour firearms course—or its equivalent—prior to assignment to a position requiring a firearm. The proposed standard further required re-qualification annually as long as the security officer continued duties requiring use of a firearm. In New York state for example, security guards carrying firearms require an additional 47-hours of initial training and eight hours of annual training in addition to the basic training required by the Security Guard Act of 1992, as amended.

ONGOING TRAINING

Most high-performing security programs require formal training on an annual basis for all security personnel. This training maintains the basic skills, knowledge, and judgment employees need to perform their assigned duties. Such training helps keep performance levels high as the re-learning counteracts the tendency of skills—once learned but not used—to decline in their reliability. It additionally is presented to keep employees informed on issues such as changes in corporate directions, legal aspects, criminal and loss-related patterns and trends, and technological advancements.

Ongoing training can be presented using the techniques described earlier in this chapter. For routine emphasis, personnel may also receive brief training before the start of a shift. This is widely incorporated into law enforcement programs. The same technique can be logically applied to circumstances in which a large number of security officers congregate before leaving for their posts. Lectures, handouts, and brief audiovisual presentations, for example, can emphasize the importance of a point in the training agenda. For security officers who are few in numbers or who work at dispersed locations, training can be provided by phone-in messages, through contact with supervisors, and from printed material left for or mailed to such employees.

Ongoing training is flexible and draws upon the resources most pertinent to the application of the workplace. This type of training is needed for several important reasons. First, existing skills become obsolete and need refocusing. Further, employees need to learn more so that they can absorb other tasks within the organization during times of need. This is referred to as cross-training. In addition, retaining is often needed.

SECURITY TRAINING FOR NON-SECURITY PERSONNEL

In many situations, the number of security personnel budgeted will not be adequate to achieve the level of staff coverage desired by the director of security. Even if this is not the case, the strategy in many organizations is to involve all employees in the protective process. Therefore, the security manager or the security training manager will enlighten non-protective workers in the rationale and methods of workplace security and safety issues (see Box 4.3). In large proprietary organizations, security managers routinely speak about relevant protective issues at orientations for new classes of trainees.

TRAINING FOR TRAINERS AND SUPERVISORS

In recent decades, attention has been devoted to training the trainers as well as aiding supervisors and managers to be more effective in their positions. Such efforts improve the quality and performance of the services provided. Specific training-for-trainers courses are available at trade and educational institutions throughout North America. While not specifically related to training for protective purposes, the brief courses expose trainers to principles that can be applicable to loss prevention needs. One benefit of the program is that trainers have the opportunity to be exposed to new techniques and strategies by attending such training sessions and conferences.

DEVELOPMENT AND EDUCATION FOR MANAGERS AND EXECUTIVES

Managers and executives also must grow in their positions. Because of the knowledge-based content of managers' work, the term "development" is

Box 4.3. Training Non-Security Personnel

Ongoing security training need not be just for protection personnel. Many security directors train non-security workers on security aspects of their job assignments, in effect making them part of the security process. At Manhattan East Suites Hotel (MESH), a chain of 10 suite hotel properties in the New York City area, the director of security trains all employees who have public contact in security and safety principles. "It's a challenge to schedule the training times since the segments must be done when the employees' regular jobs will not be interfered," notes William J. McShane, director of loss prevention of MESH.

McShane divides the training into seven modules, each 10 to 15 minutes in length. The sequence of the modules to be learned is not considered vital. When employees complete all the segments in the security and safety training program for MESH, they receive a certificate of completion. The training program stresses aspects of guest safety, protection of personal property, emergency procedures, and legal liability.

Source: *Security Letter*, November 1, 1995, Part I, p. 4.

---

used for the ongoing cognitive and skills growth of managers and executives. Development is fostered by education, which can be provided in a variety of settings. Attendance and participation at programs provided by specific security-related and general management organizations can serve personal developmental needs. Examples of specific programs include those offered at symposia, conferences, and exhibits of the American Society for Industrial Security, the International Security Conferences, and meetings held by other industry-specific groups that support security training and education. General management organizations that offer courses include the American Management Association, the Conference Board, and numerous academic institutions found throughout North America and abroad with programs and courses in management and organization. Additionally, graduate and undergraduate institutions provide continuing education courses in security management and business administration, criminal justice, and industrial/occupational psychology.

Distance learning is a recently developed teaching setup in which the student is not present in a classroom with the instructor, but learns via a closed-cable television link. The term also refers to self-paced learning that may be monitored by an educational institution. Distance learning helps students come in contact with instructors and resources not available in the learner's immediate area. It further permits self-motivated individuals to study where they are and when they can. However, most academicians believe at present that, while distance learning has much value, its value has not met the test of time like conventional group learning.

Many employers encourage self-paced development and education by providing partial or full tuition reimbursement for employees who attend institutions of higher learning on a part-time basis. This encourages employees to pursue degree programs that will enrich their careers. Workers may also learn by taking management-sponsored correspondence courses on subjects of relevance to the workplace (see Box 4.4).

Certification is a process whereby an individual is awarded a designation after having demonstrated competence through education, pertinent experience, and independent verification, usually through testing. Among security practitioners, two certifications have stood the test of time. The Certified Protection Professional (CPP) designation is awarded by the Professional Certification Board of the American Society for Industrial Security. The CPP program began in 1979, and over 3,000 individuals around the world meet CPP requirements. Re-certification requires proof of continuing education and growth and must be verified every three years. The other main certification in the protection field is the Certified Fraud Examiner (CFE) indication issued by the Association of Certified Fraud Examiners. The CFE program started in 1988, and numbers over 2,000 practitioners, often with auditing and investigative responsibilities. Re-certification for the CFE also is required on a triennial basis.

Employers generally view apt certification of managers positively, since the process provides an independent indication of relevant capability. Senior managers often encourage managers to obtain professional certification, and to keep it once earned.

Box 4.4. Correspondence Courses

Security training normally occurs during conferences, symposia, meetings, and courses. But learning can also occur on an ongoing basis through lessons delivered by mail or by the Web.

One of the time-tested ways to learn is through correspondence courses. In such courses, learners can proceed at their own pace and may select courses that are of greatest interest or importance to them. Allied Security, a national security services firm based in King of Prussia, Pennsylvania, offers a home-study course for all employees. The 11-volume self-study program includes topics such as report writing, appearance and professionalism, and conflict management. Completion of each volume is verified by tests, which are graded at the training facility. Tangible incentives are awarded to learners who achieve a certain level of points on the quizzes.

MEASURING EFFECTIVENESS

Security training is too important not to be validated and evaluated. Yet all forms of training cannot be completely and objectively or quantitatively assessed. This is a difficult issue for quantitatively oriented managers: it is not possible to prove that by providing a certain number of hours of instruction on a particular topic that a measurable effect in workplace performance can be determined. Nor is it always possible to see that this effort will have a definable economic impact. Still, certain fundamental questions can be raised, as suggested by Leslie Rae[9]:

> Has the training satisfied its objective?
>
> Has the training satisfied the needs of the clients or workplace?
>
> Are people operating differently at the end of, and as a result of, the training?
>
> Did the training contribute directly to this behavior?
>
> Is the learning achieved being used in the real work situation?
>
> Has the learning contributed to the production of a more effective and efficient worker?
>
> Has the training contributed to a more effective and efficient (hence, more cost-effective) organization?

The first three questions relate to the nature of training itself, while the last four are concerned with the effect of training on the work. According to Rae, questions arise in validation and evaluation that are subject to assessment. These include:

- **Content of training.** Is it relevant? Is it up to date?
- **Method of training.** Were the methods used the most appropriate ones for the subject? Were the methods used most appropriate for the learning styles of the participants?
- **Amount of learning.** What was the material of the course? Was it new to the learner or merely a rehash of information previously learned? Was it useful, although not new to the learner, as confirmatory or revision material?
- **Trainer skills.** Did the trainer have the necessary skills to present the material in a way that encouraged learning? Did the trainer have a positive attitude about learning?
- **Length and pace of the training.** Was the learning of the essential material of appropriate length and pace? Were some aspects too extensively covered? Were others provided insufficient time to adequately learn?
- **Objectives.** Did the training achieve its desired objectives? Did the learner have the opportunity to try to satisfy personal objectives?

- **Omissions.** Were any essential points omitted from the learning process? Was material included that was not essential to the learning?
- **Learning transfer.** How much of the learning is likely to be put into action on return to work? If it is to be a limited amount or none at all, why is this so? What factors will deter or assist in the transfer of learning?
- **Accommodations.** If the training facility is within the control of the trainer, was it relevant to the type of training provided? Was the learning in a facility suitable to the occasion? Were adequate refreshments and comforts available?
- **Relevance.** Was the course/seminar/conference/workshop the most appropriate means of presenting the learning activity undertaken?

After a period of time, learners may be questioned about their reflections on the material they had learned. Such issues as these may be assessed:

- **Application of learning.** What aspects of your work now include elements that are a direct result of the learning event? What new aspects of work have you introduced as a result of your learning? What aspects of your previous work have you replaced or modified as a result of your learning?
- **Efficiency.** How much more efficient or effective are you as a result of the training? Why or why not?
- **Retrospective analysis.** With the passage of time and attempts to apply learning, are there any changes you would wish to make to your outcome answers?
- **Ongoing evaluation.** The assessment of learning can be conducted by providing questionnaires to learners asking them to rate their experiences. For example, learners can be asked to evaluate the relevance of their training on a scale. One scale frequently used is the Likert Scale, which has seven levels, where 1 represents the lowest score, 4 represents neutral or the midpoint, and 7 represents the highest score. Gradations between the extremes and the midpoints allow variations of feelings to be identified. Such questionnaires can be scored quickly by management, though interpretation will take longer. Many managers prefer to evaluate training shortly after it is completed by a group and then again at some point in the future when the lasting effects of the training may be assessed.

  Questionnaires may also include open-ended questions about the content and process of the learning experience. Examples of such questions include: "What did you most like about the training you received?" "What did you least like?" "Do you believe others in your position should receive this training program in the future?"

In addition to questionnaires, managers may wish to establish control groups in which the performance of those who undergo training is contrasted with the performance of those in a control group. Clearly, control group research is not desirable for situations in which it is necessary to train all workers about critical skills or knowledge.

Another way to evaluate training is through direct observation, in which training evaluators or supervisors report on the performance of workers following the completion of training. (Presumably, these evaluators would have had experience observing these employees prior to the training process and are thus in a position to note performance differences.) Training or Human Resources personnel may also conduct in-depth interviews with trained workers to assess what they achieved during the learning process. Questions tend to be specific, asking workers what is different about the workplace processes since the training concluded. Additionally, trained individuals may wish to keep journal accounts of the ways in which their training has affected their behavior and performance.

Training assessors collect such questionnaires, research data, interview notes, and journal accounts in an attempt to evaluate the effects of training and determine the value such training has to the employer. The training evaluator also is likely to speak informally over time to those who have been trained and to those who supervise or manage those who have been trained and obtain useful subjective information on how the training has been beneficial or has failed to meet its objectives.

SUMMARY

With the complexity and diversity of security management tasks today, adequate initial training for workers is expected. This may be supplemented by ongoing in-service training for operational workers. Managers and executives also require development through education at conferences, seminars, and academic courses. Such training sometimes seems costly to managers who do not understand its benefits. Training should be planned to achieve or exceed the objectives of the employer. A variety of training methods can be considered to meet the requirements of cost-effective learning. Workplace training requires validation and review as well as long-term evaluation and follow-up.

DISCUSSION AND REVIEW

1. What are the salient arguments for training new security personnel?
2. How has the Information Age affected content and delivery of training programs?

3. What appears to be the main emphasis to the eight-hour pre-assignment training course proposed by the Task Force on Private Security? What is the reasoning behind pre-assignment training?
4. Define ongoing training and compare it with in-service training.
5. What are the strengths of the case history method for training? What are its weaknesses?
6. Under what circumstances would management provide T group in-service training?
7. What are the inherent limitations in measuring the effectiveness of training? What reasonable measures might a manager take to collect data to help assess a recently completed training program?

ENDNOTES

[1] J.B. Hylton (April 1998). "Is Security Training Getting Short Shrift in Schools?" *Security Management*, p. 102.

[2] M. Fletcher (October 1995). "Encouraging Safety Not Always Easy," *Business Insurance*, p. 3.

[3] D.A. Maxwell (1993). *Private Security Law: Case Studies.* Boston, MA: Butterworth-Heinemann, p. 15.

[4] L. Thornburg (January 1988). "Investment in Training Technology Yields Good Results," *HR Magazine*, p. 37.

[5] National Advisory Committee on Criminal Justice Standards and Goals (1976). *Report of the Task Force on Private Security.* Washington, DC: U.S. Department of Justice, p. 87.

[6] Plaintiffs' lawyers frequently include "negligence in training" in civil actions when it is alleged that a tort to the plaintiff would not have occurred had the security employees directly involved been trained better.

[7] The New York State Guard Act of 1992, as amended in 1994, provides a state registry of security officers including their current license status. Potential security guard employees are fingerprinted by licensees or other designated persons or entities. The fingerprint cards originally were screened by the Federal Bureau of Investigation's National Crime Information Center. The law was later amended to limit the criminal records check to the state database. Other states continue to use the FBI database.

[8] The case history method in professional schools presents the student with detailed and accurate facts about a situation as the basis of classroom discussion. A possible security-oriented case history: A department faces reorganization due to a merger. How should the new structure operate?

[9] L. Rae (1986). *How to Measure Training Effectiveness.* New York, NY: Nichols Publishing Company, p. 4.

Additional References

M. Adams (February 1999). "Training Employees as Partners." *HR Magazine*, pp. 65–70.

S.M. Brown and C.J. Seidner (Eds.) (1988). *Evaluating Corporate Training: Models and Issues*. Boston, MA: Kluwer Academic Publishers.

J.D. Calder and D.D. Sipes (April 1992). "Crime, Security, and Premises Liability: Toward Precision in Security Expert Testimony." *Security J.*, 3(2):66–82.

J.D. Facteau (1995). "The Influence of General Perceptions of the Training Environment on Pretraining Motivation and Perceived Training Transfer." *J. of Management*, 21(1):1–25.

I.L. Goldstein (1989). *Training and Development in Organizations*. San Francisco, CA: Jossey-Bass Publishers.

International Foundation for Protection Officers (1998). *Protection Officer Training Manual*. 6th edition. Boston, MA: Butterworth-Heinemann. (An Instructor's Guide for the 6th edition is also available.)

J.P. Leeds (October 1994). Legal Concerns in the Use of Psychological Screening Tests. *Security J.*, 5(4):212–16.

L.G. Nicholson (1997). *Instructor Development Training: A Guide for Security and Law Enforcement*. Boston, MA: Butterworth-Heinemann.

R.J. Phannenstill and F.S. Horvath (July 1991). "A Comparison of Computerized Interviewing of Job Applicants with a Personal Security Interview." *Security J.*, 2(3):172–79.

M.A. Quiñones (1995). "Pretraining Context Effects: Training Assignment as Feedback." *J. of Applied Psychology*, 80(2):226–38.

5

SUPPORTING AND MOTIVATING SUPERVISORS AND STAFF

> There are few jobs more difficult but at the same time more interesting than that of supervising people.
>
> —William R. Van Dersal

Executives and managers work through other people. No matter how talented senior and middle management may be, results are achieved through the combined efforts of the larger organization working together. A key component to operational success of the organization is its supervisors. Supervisors are also called first-line managers—the lowest level of the managerial hierarchy, but management nonetheless. Supervisors differ from middle and senior managers in having larger directing and controlling responsibilities and less to do with planning, organizing, and staffing.

The titles of those subordinate to supervisors differ according to the nature of the employer. Operational-level security staffers may receive titles specific to their position, such as security guard or officer; agent; investigator; store detective; alarm console operator; firefighter; or documents classifier. A few organizations use vaguer titles, such as associate or team-member. Generically, these individuals may be termed "staff" (not to be confused with headquarters senior staff) or just "workers." Senior management often seeks to find a dignified and enhancing term to refer to workers. For this chapter, first-level managers will be referred to as supervisors and the individuals they oversee will be referred to as staff.

SUPPORTING SUPERVISORS AND STAFF

For the work to be achieved and the desired results attained, management must not only provide ample workers capable of achieving or surpassing goals, they must also provide different types of support so that these staff members can thrive. The best planning and team of co-workers will come to naught if those persons do not have what they need. In a sense, the failure of the staff in such circumstances is a failure of management. The requirement of routine supplies, vehicles, and forms is clear enough. Less obvious are the procedural support commitments that are found in higher performing organizations. The first section of this chapter, therefore, considers these specific needs for this cadre of workers. Nobody, for example, would blame an autoworker on an assembly line for not doing his or her fair share if that person didn't possess critical parts or tools needed to do the job. The same analogy holds with security employees, though in most cases what these individuals require are not nuts, bolts, and side panels, but mostly intangibles that are central to getting the job done. Providing the staff with various support items and resources is not merely a desirable action for first-line and middle management, it is a duty.

To achieve desired results, management plans conscientiously so that resources are ready before they are likely to be needed by supervisors or staff. But before that occurs, it is necessary to find the right person for the right slot. That is the role of placement.

Placement

Fitting people and their jobs together is the first step in successfully bringing newly trained workers into the job stream. Management has determined that openings are available for particular shifts, days, and levels of experience. The Human Resources and training officer plan the assignment so that newly trained workers can be placed in positions to work without delay. Prior to the assignments, the supervisor is consulted on facts involving new staff workers assigned to his or her unit. At this point, the supervisor may review the training files and pertinent information about the new worker. He or she may conduct a brief interview to ascertain that the nature of the job and the characteristics of the worker are compatible. This is no time for uncertainty. For example, if the new employee begins by working the night shift, is he or she truly prepared to accept that assignment for a minimum period of time? This is when the supervisor determines whether any reasons exist as to why the placement would not be successful. In addition to scheduling issues, the supervisor will review the particular nature of the position to assure that the fit for the worker to the position is appropriate. This includes an analysis of:

- The specific nature of the tasks to be performed
- The knowledge required for the site and the particular duties
- The equipment or supplies the security worker will require
- Any particular physical requirements, such as lengthy standing or walking, or the possibility of having to endure substantial temperature changes during a tour
- Any unusual requirements relative to the site and the position

Supervisors are likely to take for granted the physical circumstances of the job and the particularities of the personalities involved. This attitude could be counterproductive to staffers' success. Therefore, the supervisor needs to consider carefully any means by which the new worker might not succeed in this placement and ascertain how further information or additional support might help mitigate any possible assignment difficulties early on.

Providing new employees with a personal welcome is practiced at some workplaces. At Southwest Airlines, the new-hire orientation program includes signage: "New Hire Celebration: You, Southwest, and Success." Thus, Southwest Airlines provides an opportunity for new workers to be welcomed by their peers in a relaxed, friendly atmosphere, focused on success.[1]

Within the context of supervision, several principles guide the relationship between supervisor and the supervised. The principles need to be learned when a staffer is promoted to supervisor or when an outside person is hired into a primary or secondary management position:

1. Staffers must understand exactly what is expected of them. Much of the information shared between the supervisor and the new worker will have been covered during training. However, the supervisor provides her or his own second orientation as a specific introduction to the job.[2] This process is highly meaningful as the staffer is now about to start working and knows that pleasing his or her supervisor will now become a priority.

During this second orientation, the supervisor likely reviews topics of utmost importance at the time the worker begins the assignment. A number of issues are routinely covered at such times:

- **A few words about ethics and fair working conditions.** Security people work with the trust of others. The trade and professional organizations have codes of ethics, binding the members who belong to such groups. But ethical issues do not stop there. At the job site, the nature of the ethical commitment of the organization itself and of its entire staff should be shared with the new worker. The supervisor may communicate informally a message: "You're fortunate to join the leading manufacture of smart widgets in the industry. Since the day we were founded, all of us here have been committed to serving our customers with the best products and services. Our commitment to fair, honest, and honorable dealings equally involves vendors and employees. Security is important in our success and growth." The new staffer may be referred to the organization's ethical statement and information on the workplace ethics and ethos. This statement may be printed in the employee handbook or be posted at different locations in the workplace. By taking a moment to call the worker's attention to these statements, their importance is bound to be enhanced in the staffer's estimation.

The supervisor may point to the statement of principles involving the dignity and fair play for all employees. If an employee has a complaint and it cannot or is not handled adequately by the supervisor, the new staffer should learn that higher management will be available to hear the dispute or issue. If the general manager has an open-door policy, this fact should be mentioned as well. These comments are relevant for proprietary organizations. When employees for the outside contractors have unresolved workplace complaints, mechanisms for dealing with them should be structured and communicated to those involved.

- **What the organization does.** After a period of orientation and training, it might be assumed that the new worker has a clear idea of what the organization actually does. This is not usually a safe assumption. The supervisor should therefore discuss with the new employee the nature of the organization: what it does; what the strategy appears to be; what the organization is particularly proud of; and what current challenges and difficulties the organization faces. Having established the ethical nature of the workplace, the supervisor should make it clear what the organization stands for and what impediments are in the way of the goal from being achieved.

  The new staffer may think that he or she knows what the organization's reason for being is, and probably does to some extent. But relating what takes place at the job site and putting it into a larger framework of the organization's total goals and strategy is useful and valuable in the early interactions between the supervisor and new staff members.
- **What the job involves.** The training received by the new worker is valuable, providing general, legal, operational, and emergency information to the new security worker. But on the job site itself,

the nature of the tasks to be performed by the worker need to be reviewed. This is the time for the supervisor to make sure instructions are understood. This is also a time to make clear the extent to which the supervisor will be available for assistance, the ability to contact other co-workers on the job sites for any assistance, and steps to take in an emergency situation.

Initially, the new worker is likely to assist the efforts of an experienced person who acts as an on-the-job trainer and facilitator for the new staffer. The supervisor usually introduces the two parties to each other and observes the nature of the interaction. If the on-the-job trainer has not trained a new worker before, the supervisor will stay in touch with both on a more frequent basis than would otherwise be the case. Before leaving the new worker with the experienced security staffer, the supervisor may again review important aspects of the job. The supervisor also will emphasize possible pitfalls and ask whether the new worker understands what has been said and whether he or she has any other questions.

- **Where to get what's needed for the job.** The new staffer will need access to a supervisor and co-workers in order to obtain general information. Specialized information and supplies and services may be needed for the position as well. The worker is on post and may need instant help, replacement parts for equipment that fails, or routine supplies that are unexpectedly exhausted.
- **How the quality of work is to be measured.** Workers in production and service positions are interested in learning what the quality of work is to be expected. This is particularly the case when trainers have stressed how the organization places emphasis on superior results. Security officers may have the quality of their work assessed primarily by direct observation and frequent interaction with supervisors. However, other methods are also available, including measurements and electronic recordings of data collected by the officer; observations and informal and formal reports by other workers; observations and reports by senior managers interacting with security personnel; and questionnaires and comments from the public.

Alarm monitoring operators may have quality of their work measured by the speed, manner, and accuracy with which an alarm condition is responded to. Central monitoring operations often have quality control standards that will be subject to supervisory or management review.

Investigators usually are evaluated on the accuracy, completeness, and insightful evaluation of the investigative reports they complete. Such reports may be strong enough to use successfully in legal actions.

Armored car personnel may be measured by the care by which they note the deliveries and pickups from the various locations on their routes. Attention to security procedures will be another factor in determining quality of job performance. For example, does the guard vary routine as much as possible? Are firearms maintained in a safe and responsible manner? Does someone remain locked within the armored vehicle at all times, if company policy requires this measure?

Security technicians are judged by their ability to complete a task fully, without the need for subsequent service calls. For example, a particular alarm system installer might take a few minutes longer than a peer to install a system. But that technician may complete the job with higher quality and instruct users on applications well so that further service calls will be less frequent.

The staffer may be informed that quality is a constant issue to be discussed with the supervisor. However, near the end of the probationary period and during formal reviews, the results of such findings will be discussed specifically.

- **How the quantity of work is to be measured.** Quality matters; however, *quantity* does as well. In security work, the nature of tasks performed by security personnel is increasingly measurable by security systems that track completed tasks. Quantity of tasks completed by personnel is an important and usual basis of evaluating workplace performance.

  Security officers may be measured by the frequency of security rounds to posts they must cover. Also, the number of specific recorded services may be a basis of evaluation. For example, security officers who conduct escorts, key runs, or vehicle checks may be recorded by data entry systems. Over time, these will serve as a basis for comparison between security guards working during similar time shifts. Such reports may also be aggregated to document total services performed by the security unit over a period of time.

  Alarm monitors may be evaluated by the number of customer interactions during a work period. To be sure, a slow number of alarm conditions may not reflect badly on a monitoring operative if that is the reality. In such cases, however, management will wish to evaluate the circumstance to determine whether an alternative way of managing alarms could be arranged to achieve optimal use of personnel, if an alarm monitor is not busy enough. In handling telephone communications with system customers, management may set a desired goal of maximum average customer contact time, for example, two minutes. Signals to the operator and supervisor can indicate when that point has been reached.

Armored car personnel may be monitored by the number of runs conducted by the group in the course of a shift. Security technicians may be judged by the number of installations and service calls completed during a period of work. Depending on the nature of the tasks, the assessment may be on a daily, weekly, or monthly basis. Adjustments can be made for time off or other circumstances that otherwise would make comparison meaningless.

In many cases, the quantity of work completed by security service workers is not comparable to that of other employees where measurable units of production or sales can be counted. That does not imply, however, that supervisors and their managers are not concerned about measuring the quality and quantity of individual personal efforts. Supervisors and their managers constantly search for fair, logical means by which work can be measured and assessed.

- **Relevant resources.** The supervisor will make the new staffer feel more prepared to deal with the tasks at hand by providing resources directly related to the organization and to tasks at the specific job site. Here are some important examples:
  - **Procedure manual and post orders.** In many security positions, a manual will outline the workplace tasks and expectations. Additionally, specific post orders describe what is required of the worker for the specific workstation. The supervisor will review these site-specific requirements, making sure the material has been thoroughly covered.
  - **Job description.** The new worker has been selected because he or she is capable of performing the required task and has been trained to do so. The job performance should remain accessible for the worker. However, it should be made clear that job descriptions are not immutable documents, but rather change with the times as the nature of the job is modified to fit different circumstances.
  - **Rules and regulations.** Requirements of the workplace need to be reviewed early in the relationship between the new staffer and the supervisor, even if the training provided covers many or all of these points. If the facility does not permit smoking, that should be mentioned. The policy for personal use of telephones, facsimile machines, and Internet resources should be discussed. Visitors are usually not permitted on the job except for exceptional circumstances. Many operational security staff are expected to arrive a few minutes before the beginning of their shift or scheduled responsibilities. They

may also remain for a few minutes after their work is over. However, presence at the workplace when one is not scheduled to be at the job is not permitted in many security operations. If the workplace requires security personnel to possess and carry firearms or weapons of any types, specific regulations will pertain to them. Otherwise, employees who bring weapons to the job without authorization are subject to sanctions, including dismissal.

- **Property.** The worker needs to know what property she or he will be assigned and how that property should be cared for. Property may include keys, two-way radios, cellular phones, laptop computers, data collection devices, manuals, vehicles, uniforms, weapons, and other items. The employee's role in protecting electronic assets, including programs, hardware, and output, should be made clear. Password protection issues should be covered.
- **Hazardous materials.** The worker needs to know what his or her rights are concerning any chemicals or material safety issues that could affect them on the job. Specific training on the observations of and response to any hazards must be provided before staffers are responsible for such materials.
- **Organization chart.** This helps the worker understand the personnel structure and reporting relationships at the place of employment.
- **History and CEO's welcome.** A leaflet, booklet, or video of the organization's history may help make the workplace more alive and relevant to the new staffer. It may also make the new staffer feel more connected with the breadth of operations of the employer. Sometimes, in large organizations, a video from the CEO welcoming new workers is presented.
- **Career tracks.** Most new employees have cloudy visions about how they can grow in the job and where their initial positions can lead them. The supervisor or manager should make clear what the realistic career growth opportunities are and how individuals will have an opportunity to take advantage of them when opportunities arise.
- **Compensation.** Human Resources will have established the pay level for the new employee. Therefore, this topic is not one the supervisor usually needs to dwell upon. However, frequency of pay reviews and their significance over the first two or three years of employment may be discussed with the employee. This discussion should not indicate that pay increases will be automatic, unless this matter is a workplace policy. How compensation is delivered physically or electroni-

cally to workers should also be explained. The location of the compensation office or its telephone number should appear in the literature provided to the new staffer.

- **Benefits.** Technical questions on benefits are best referred to the human resources department. However, supervisors may provide some general advice, if asked, on the different types of benefits when a choice is required. Supervisors should never put themselves in a position of forcefully arguing for one type of benefit option over another, but rather answer questions so that new staffers are fully apprized of the opportunities management has made available. A new worker may be grateful years later if a supervisor encourages her or him to participate in the employer's 401(k) self-contributory retirement or other similar program.
- **Creature comforts.** New workers need to know where the restrooms are on each floor, and where the cafeteria or canteen is located and the policies related to its use.

2. Staffers need general guidance to do their jobs properly. Despite superior efforts of employment selection, training, and orientation, circumstances may occur within the scope of employment in which the worker will need further assistance. This is the task of the supervisor.[3] Several types of such guidance may be needed:

- **Routine job support.** The supervisor's presence supports and guides security personnel in dealing with circumstances where the vision of a more experienced person could be of benefit. Frequently, these situations involve public contacts by security personnel in which the supervisor later may critique the performance of the worker. Often, the supervisor may be present to take responsibility for managing the situation on the spot. Supervisors also provide guidance to the worker on organizing tasks, writing reports, and dealing with unexpected events.
- **Public relations skills.** Much of the activity of security workers concerns the public. Therefore, the enhancement of interpersonal skills is a concern of the supervisor and middle managers. The successful supervisor provides support to enable the staffer to understand his or her emotional predisposition in dealing with the public.
- **Information support.** The job environment changes constantly. Risks change. Procedures are altered. Personnel come and go. Workers expect to be informed by supervisors of the routine but significant news and information pertinent to the job. During contacts with staffers, supervisors are expected to share factual infor-

mation that can help the worker understand the current circumstances at the job site.

- **Specialty information.** When a new procedure, tool, device, database, or system is introduced to the workplace, the staff person may require assistance beyond training. At such times the worker turns to the supervisor for specialized assistance in mastering the new resource. In this sense, the supervisor continues the training by providing field support for workers.

3. Staffers deserve to be recognized for good work. Kenneth Blanchard and Spencer Johnson have provided a simple but compelling method of emphasizing the positive. Supervisors and managers are expected to recognize good performance they observe among staffers to help their behavior become better still.[4] Supervisors know that prompt recognition of good work reinforces it, increasingly the likelihood that the desired behavior will become learned and be expressed as a natural behavior. Supervisors are the logical authority figures to observe and comment upon superior and commendable behavior that they observe. Normally, the supervisor will provide the direct positive feedback to sustain desirable performance. However, on some occasions, the supervisor may arrange for her or his manager to deliver the praise.

Supervisors have a number of means available of delivering the positive feedback that most workers crave. For example:

- **Verbal feedback.** The simplest and most direct means of recognizing good work is by a verbal message. The message should be direct and comprehensible. The Blanchard-Johnson thesis suggests that people should know up front that the supervisor is going to let them know how they are doing. People are praised immediately on good work (see Box 5.1). The praise is specific. The manager says how good he or she feels about what the worker did and how these actions help the organization. The manager stops for a moment of silence to let the person being praised "feel" how good the manager or supervisor feels. Further encouragement is offered. Finally, the manager shakes hands or touches people in a way to make it clear that the manager supports the staffer's endeavors in the organization. For example, the manager might say: "That last visitor was irate when he arrived. You listened sympathetically to what was bothering him. Then you provided a helpful suggestion. In the end, he left in a better frame of mind. That's an ideal way to handle such circumstances." Speaking directly, specifically, and objectively to workers about their exemplary behavior stimulates the acknowledged behavior to be remembered, and possibly further improved upon.

Box 5.1. ***The One Minute Manager*** **on Praising Good Work**

Two wise authors, Kenneth Blanchard and Spencer Johnson, have provided valuable guidance on how people who manage others may achieve goals productively and humanely. In *The One Minute Manager,* the authors describe a mythical extraordinary manager who achieves success with such brilliance that he leads a stress-free high-achievement orchestration of his subordinates' performances. Much of the book describes the simple premises that lead to this achievement. The mythical marvelous manager believes in catching an employee "doing something *right,*" reinforcing this desirable behavior:

1. Tell people up front that you are going to let them know how they are doing.
2. Praise people immediately.
3. Tell people what they did right; be specific.
4. Tell people how good you feel about what they did right, and how it helps the organization and the people who work there.
5. Stop for a moment of silence to let them "*feel*" how good you feel.
6. Encourage them to do more of the same.
7. Shake hands or touch people in a way that makes it clear that you support their success in the organization.

Source: K. Blanchard and S. Johnson (1983). *The One Minute Manager.* New York, NY: Berkley Books, p. 44.

- **Formal written praise.** A letter or memorandum that summarizes superior behavior can also reinforce desired behavior. The difference is that such communication represents a higher level of thought and effort: the tribute requires discernment and concern. The written praise also can become a permanent part of the worker's record.

  Some supervisors pause about bestowing workers with written praise because they fear that the laudatory remarks could return to haunt the writer. For example, an employee whose work was once praiseworthy might use letters in defense when that employee is disciplined for cause. However, managers who are consistent can use written communications to complement workers as well as to correct behavior (see Chapter 7). The practice of formalizing approval of superior behavior in written form often produces a sense of pride and a commitment to further good efforts that extend beyond verbal honorable mention.

Formal written approbation is also appropriate when a director or manager at headquarters wishes to applaud superior behavior of those working at distant sites. The process of formal approval from headquarters has a powerful impact on individual performance, sustaining it in ways words alone could not achieve (see Box 5.2). Writing costs little and means much.

- **Provide certificates.** A supervisor might provide a certificate of good service, signed by the supervisor and manager and presented thankfully to workers, as a tangible example of recognition.
- **Promotion.** The supervisor may use good work as a basis of promoting the worker to a higher classification.
- **Better scheduling or conditions.** To reward the deserving, the supervisor may be able to find preferred hours, days off, locations, or conditions for the workers.

Box 5.2. A Letter from J. Edgar Hoover

For 47 years, John Edgar Hoover (1895–1972) not only led the Federal Bureau of Investigation, he was the FBI itself personified. Hoover was in his post too long and used his power to achieve control by attempting to destroy or silence those with whom he did not agree. In the years following his death, his reputation has been sullied by many writers.

Nonetheless, Hoover should be regarded as a brilliant manager whose efforts turned a poorly supervised group of politically appointed investigators into a highly efficient and ethically salutary organization. Hundreds of security directors in public and private organizations were trained during the Hoover years and were influenced from the controlling reaches of Hoover's Pennsylvania Avenue office suite.

Hoover trained special agents in charge of FBI field offices to observe and report to the central office details of superior field performance by individual special agents. Days after a significant event, special agents would find in their mail a letter personally addressed from FBI headquarters. With trepidation they would open it. Inside would be a brief recitation of the special agent's recent actions and would end with the observation that this behavior was in the "highest traditions of the Bureau," signed in ink by the director himself. Some special agents had received a number of such citations by the time their careers ended. The letters from Hoover were powerful motivators and served as lasting treasured mementos to the special agents who received them. (The opposite was also true. Hoover would criticize disappointing behavior of special agents with the same precise language used to praise.)

- **Award merchandise.** Supervisors may award gifts of merchandise to personnel under some programs. Professional Security Bureau (PSB), a security services firm based in Nutley, New Jersey, provides booklets of coupons worth points toward merchandise for meritorious security to their personnel. The booklets are given by PSB to their clients as well as PSB supervisors and managers. Security officers use the coupons to redeem merchandise: sports clothing, travel bags, pens, clocks, pocket calculators, and numerous other items. The on-site PSB supervisor and client both can reward the security officer tangibly for superior performance.
- **Provide a night out.** In Great Britain, many service businesses, including security firms, provide recognition for superior performance by awarding the worker with an evening's entertainment and dinner for two on the town. These are popularly called "bennies." Scaled-down bennies may be concert or theater tickets without food.
- **Cash or bonus.** Almost all workers like to be told they have done a good job with cash.

This list is not complete, and is limited only by imagination. Some supervisors will find other unusual ways of recognizing their employees, which we will discuss late in this chapter. The tangible and intangible recognition that a supervisor might provide all carry with them degrees of desire and significance. Surely, excessive use of any reward system can cause it to become unproductive, leading eventually to its collapse. The astute supervisor focuses upon desired behavior to support it with a reward that is appropriate to the circumstance. Such tangible or intangible rewards should not disturb the web of workplace personal relationships by unfairly rewarding some persons and ignoring others.

Ideally, such awards and benefits are deserved and are seen by other workers who are not recipients, nonetheless, as a cause of general congratulations to the persons being recognized. Good work is recognized publicly so that the recognition and rewards can have a quantum effect on the immediate recipients and all other workers in the unit. Other programs like Employee of the Month and Office of the Month can reward superior achievements.

4. Staffers deserve constructive criticism for poor performance. Desired behavior is achieved more effectively by emphasizing positive traits rather than trying to suppress negative ones. Still, poor performance occurs in the workplace. At such times, the supervisor must provide constructive criticism so that such behavior can change. In most cases, the worker is aware of the substandard performance and awaits reaction from the supervisor. If such a reprove is not forthcoming, the worker may lose respect for the supervisor and also feel permitted to engage in a continuance of such activities.

Unlike the public reward for good behavior, poor performance deserves prompt, constructive, *private* reproach. Again, the Blanchard-Johnson thesis has a strategy for responding to undesired workplace performance (see Box 5.3). The following steps ensue:

i. The supervisor summarizes specific unsatisfactory behavior and asks if the facts are essentially correct. If the worker starts to explain or otherwise not answer the question immediately, the supervisor

Box 5.3. *The One Minute Manager* on Correcting Poor Work

According to the Blanchard-Johnson thesis, successful managers leave subordinates alone to do their work undisturbed. However, sometimes work is not satisfactory and a reprimand is necessary. The following presents the essence of the Blanchard-Johnson human relations strategy:

1. Tell people beforehand that you are going to let them know how they are doing and in no uncertain terms.

The first half of the reprimand:

2. Reprimand people immediately.
3. Tell people what they did wrong; be specific.
4. Tell people how you feel about what they did wrong, and in no uncertain terms.
5. Stop for a few seconds of uncomfortable silence to let them *feel* how you feel.

The second half of the reprimand:

6. Shake hands, or touch them in a way that lets them know you are honestly on their side.
7. Remind them how much you value them.
8. Reaffirm that you think well of them but not of their performance in this situation.
9. Realize that when the reprimand is over, it's over.

Source: K. Blanchard and S. Johnson (1983). *The One Minute Manager*. New York, NY: Berkley Books, p. 59.

cuts off the digression until the worker agrees that the facts are essentially right.

ii. Next, the supervisor briefly states that the *behavior* was unsatisfactory. The supervisor avoids criticizing the worker personally, but rather makes it clear that the particular acts were unacceptable. The supervisor attempts to avoid circumstances in which the worker vaguely blames others for the situation. The burden for the correct behavior is on the worker alone in most cases. (In some situations, additional training may be needed to correct the cause of the substandard performance; other measures may also be indicated.)

iii. The worker acknowledges the poor performance that has been described.

iv. The supervisor asserts that she or he has confidence in the worker's abilities, and a reference might be made briefly to positive aspects of the worker's performance. The supervisor indicates that the interaction has come to an end by lightly touching the worker on the shoulder or elbow. The worker returns to his or her duties. The process is over.

5. Staffers should have opportunities to show that they can accept greater responsibilities. Many high-performance workers are not happy with their status quo. They want to improve themselves. Astute supervisors realize that the needs of such self-improvement can be channeled productively when the employee feels that he or she has the supervisor's support in advancing in the workplace.

Promotion might be an eventual option for the worker seeking greater responsibilities. In the meantime, the worker may be given opportunities to demonstrate a competency in new circumstances. The worker may be assigned a special project, perhaps one that the supervisor does not have time for. Yet such an assignment requires burdens on the supervisor: the original worker may require some time realignment from his or her previous schedule; work normally done by the worker may be performed at less than its usual level; and other workers may sense that the special assignment reflects undue favoritism and create new problems for the supervisor in the process. However, such assignments produce insights also, making the extra burdens on the supervisor worthwhile. The worker who is willing to accept increased or more challenging responsibilities deserves the chance.

In many organizations, supervisors find ways of distributing special assignments and responsibilities among many staffers, indeed, among everyone who seeks additional opportunity. In all cases when such workers have demonstrated their abilities, such achievements should be recorded in their personnel file.

SAFETY AT WORK: THE RESPONSIBILITY OF SUPERVISORS

The American workplace is generally safe, or safer than it was in past generations. In the 19th and early 20th centuries, for example, security guards were the knights of ownership, protecting private property from striking workers. In this type of assignment alone, scores of security workers were seriously injured in attempts to fulfill their duties. Similarly, numerous thieves, vandals, and strikers were killed or injured by zealous private security personnel.

Even now, security staffers have a higher risk than most employees of serious injury and death.[5] Many security personnel have been victims of personal assaults at one or more points in their careers. In a few cases, injuries have reached the level of felonious assault. About 20 to 40 security guards lose their lives each year in the course of their employment.

Not all security positions are equally dangerous. The night watchman in a warehouse or museum needs to fear the risk of an accident more than the possibility of felonious physical injury. The investigator who works primarily with the telephone and computer and the central alarm station monitor operator are at greater risk of falling than from being attacked. Nonetheless, according to the National Institute for Occupational Safety and Health (NIOSH), security personnel rank fourth in terms of rates of workplace homicides when adjusted for number of workers, as shown in Table 5.1.

Table 5.1. Workplace Homicides to Employees in High-Risk Industries in the U.S. (1980–89 and 1990–92)

	1980–89		1990–92*	
Industry	**Number**	**Rate**	**Number**	**Rate**
Taxicab services	287	26.9	138	41.4
Liquor stores	115	8.0	30	7.5
Gas stations	304	5.6	68	4.8
Security services	152	5.0	86	7.0
Justice/public order	640	3.4	137	2.2
Grocery stores	806	3.2	330	3.8
Jewelry stores	56	3.2	26	4.7
Hotels/motels	153	1.5	33	0.8
Eating/drinking places	734	1.5	262	1.5

Note: Rates are per 100,000 workers per year.
*Data for New York City and Connecticut were not available for 1992.
Source: "Violence in the Workplace" (June 1996). Cincinnati, OH: National Institute for Occupational Safety and Health, *Current Intelligence Bulletin* 57.

Physical assaults to security personnel are much more numerous than fatalities. Here again, security personnel are disproportionately victimized relative to most other vocations. According to a study based on the National Crime Victimization Survey (1992–96), security personnel ranked fourth in occupational victimization for nonfatal violent incidents, adjusted for the number of workers, as shown in Table 5.2.

Workplace violence is a matter of concern for security supervisors and managers in many industries. Those at greatest risk work in armored car courier services; banking; food service; healthcare (especially emergency rooms); retailing (when making apprehensions); and at places of public assembly, such as theaters and concerts. Any location that has on its premises substantial cash and other liquid assets—check cashing businesses, cashiers, and jewelry businesses—creates risks for security and non-security employees alike.

Security of workers is a management responsibility, extending from planning through training and supervision. This principle does not absolve, certainly, an individual security worker who abandons common sense and flaunts rules and regulations to take a risk that results in harm. Nonetheless, the supervisor bears inordinate responsibility in dissuading workers from actions that result in unacceptable risk-taking behavior. Further, the supervisor has direct responsibility in deciding whether a particular situation is too

Table 5.2. Occupations of Victims from Non-fatal Workplace Violence (1992–96)

Occupation	Average Number	Rate
Police	234,200	306.0
Corrections	58,300	217.8
Taxi drivers	16,100	183.8
Security	71,100	117.3
Bar workers	26,400	91.3
Mental health	50,300	79.5
Gas station attendants	15,500	79.1
Convenience/liquor stores	61,600	68.4
Junior high teachers	47,300	57.4
Bus drivers	70,200	45.0

Rate is per 1,000 workers. Calculated from National Crime Victimization Surveys for 1992–96. Source: G. Warchol (July 1998) "Workplace Violence, 1992–96," Washington, DC: Bureau of Justice Statistics, Report No. 168634.

dangerous for trained security personnel. In such cases, the supervisor should opt instead to keep security personnel away from the danger until the risk can be reduced to an acceptable level.

Similarly, when security staffers are in an environment in which workplace safety is poor, the supervisor holds direct responsibility on behalf of management in deciding the extent to which risk may be reasonably assumed. Chemical plants, healthcare facilities, and nuclear process industries are a few of the workplaces in which hazardous materials exist. Such facilities can be well managed and pose little risk for security personnel, yet wherever hazardous materials or conditions exist, supervisors have an obligation to assure that their people are not exposed to undue risks. In addition to hazardous chemicals, such factors as lighting, workplace design, safety devices, structural features, and even ventilation, when harmful, can present an unacceptable risk for employees.[6]

All employees have a right to know about the hazardous materials located within their workplaces. They have the right to obtain safety data sheets, which should be provided within five days of any request by an employee. While security personnel normally will not personally handle hazardous materials at such facilities, should the possibility exist, they need to be properly trained in emergency response procedures relative to such substances.

WHY BE A SUPERVISOR, ANYWAY?

The previous section indicated that "security supervisors are the direct link between the management of a security organization and the security officers who carry out the duties associated with the security function."[7] The security supervisor has to be willing to accept responsibilities, and must have demonstrated competence in technical, human, and conceptual skills. Often, the inducements to become a supervisor do not appear to be great, and many competent and valuable security staffers decide not to accept the promotion to supervisor when offered one. Yet some compelling reasons exist for why operations-level workers may accept the challenge of a promotion:

- Achievement is reflected in the new position, buttressing one's self-image
- Authority and control come with the position
- Creativity is encouraged: the supervisor is a resourceful problem-solver
- Opportunity to higher positions such as middle management may be made possible
- Pay and other benefits usually, but not invariably, are better

DUTIES OF EMPLOYEES TO SUPERVISORS AND THE WORKPLACE

Up to this point in the chapter, obligations of the supervisor to his or her staff have been discussed. However, security employees themselves have obligations to their supervisors, management, and, by extension, to the management of the employing organization. This extends to contract employees and consultants who provide services. Such duties are not unreasonable burdens, but rather reflect the nature of the particular employment and its requirements. These are bound to change according to the workplace and its characteristics. Many of them will be described in the employee manual. As an example, the following are 12 rules and regulations established by a major casino for its employees, including security personnel. Note that rules 9 and 10 are specific to a gaming environment. The list may stimulate the reader to consider the context for other occupational settings:

1. Employees will conduct themselves in a manner promoting good public relations and goodwill.
2. Use of profanity, rude behavior, and lack of consideration for customers or other staff may be considered grounds for disciplinary action or termination of employment.
3. The use of force by company employees is forbidden, except as a last resort to protect the life of a customer, fellow employee, or oneself. In such a situation, only the minimum force necessary is acceptable. Protection of property IS NOT considered grounds for use of force. Persons violating this policy may be prosecuted criminally or civilly or both.
4. The use or possession of alcohol, narcotic drugs, or any type of weapon while on company property is strictly forbidden.
5. Family members will not be hired as co-workers.
6. Sexual harassment—or any act that may be considered as sexual harassment—is strictly forbidden. Any reports of such conduct will be fully investigated.
7. No one other than authorized employees or official visitors are to be given access to any office or storage areas at any time.
8. Cash collection policies and requirements for recording transactions must be followed exactly as written procedures indicate. Failure to do either may result in termination of employment.
9. Employees will not play games, be a partner in games with customers, or finish a game for a customer or one which a customer has abandoned.
10. All refunds or settlements of customer disputes will be conducted as prescribed and recorded on the proper form.

11. Employees will not store or hold packages or valuables for customers.
12. Employees will not discuss company business, policies, or practices with any person not authorized access to such information.

This example of one organization's general rules and regulations reflects concerns about general civility, the drug-free work environment policy, sexual harassment, and policies and procedures that, if not followed, can promote internal and external crime. In this case, employees were asked to sign a statement that they had read and understood these rules and regulations. Further, employment terms stated in the employee manual are also to be considered conditions of employment and violation could lead to termination. The new employee signs and dates such a statement. The signature is witnessed by a supervisor or Human Resources manager.

Rules and regulations of other security departments commonly stress punctuality, care of uniform, personal hygiene (showering before reporting to duty), and respect of the employer's property.

MOTIVATING SUPERVISORS AND STAFF

Supervisors and managers at all levels are responsible for achieving and maintaining high levels of workplace productivity. As part of this objective, management is expected to maximize use of time. Personal time management techniques can be useful for management at any level. But apart from using one's time well, a separate conceptual category—motivation—is addressed to help supervisors and managers understand this topic.

TIME MANAGEMENT FOR SUPERVISORS AND MANAGERS

The first chapter of this book discussed the seminal contributions of Frederick W. Taylor in establishing the time study method for improving workplace productivity. This concept is applicable in process work in which employees are responsible for measurable throughput. But the tasks of first- and second-level management require time for planning, organization, direction, and problem solving. The need for time devoted to planning is hard for many managers to find. Although nobody has more than 168 hours per week, some individuals are able to accomplish more than others. Certain time management techniques can help people accomplish more.

Business school professor Henry Mintzberg wrote: "Free time is made, not found, in the manager's job; it is forced into the schedule. Hoping to leave some time open for contemplation or general planning is tantamount to hoping that the pressures of the job will go away."[8] Many books have been written on time management, and numerous others are doubtlessly awaiting their day. They have influenced millions. Nonetheless, time management is a topic not readily adaptable for quantitative research; therefore, it is not always clear which methods work best. Close observation of successful managers identifies attributes that lead to achievement and greater productivity. Superior time management is part of that achievement path. The following are some widely used techniques to increase managerial productivity.

The ABC Technique

Not all tasks are equally important. A manager can face a seemingly endless series of demands on his or her time. Yet concentrating on what's most important, making these tasks the priority, and working on them until they are completed will enable the most important work to be completed before less important items.

The ABC technique directs users to divide all work-related matters to be handled into three categories: A, vitally important; B, of nominal importance; and C, unimportant relative to A and B. The manager learns to identify what is A, B, or C quickly upon being presented with a task or opportunity. Greatest effort is devoted to A; B is next; and what time is left goes to C.

The Pareto Principle

Popularly called the 80/20 rule or Pareto's Law, this is another control scheme like the one above that identifies workplace tasks in order of their importance. The principle is named for Vilfredo Pareto, an Italian economist.[9] In 1908, Pareto proposed that the distribution of wealth and income in a population was mathematically predictable. In time, the principle expanded to fit other circumstances.

While the ABC analysis denotes three categories of analysis, the Pareto technique focuses on the inverse relationship between the percentage of items in each of a set of subclasses and the importance of such subclasses. The principle is best illustrated by thinking about customers or clients in an organization. Not all are equal. Through the use of conventional analysis, some customers or clients will be particularly important. The Pareto principle posits that 80 percent of an organization's sales are due to only 20 percent of the customers, and vice versa.

The 80/20 ratio is not meant to be taken literally; but the point holds across numerous applications that a small number of issues or customers—

external or internal—are more important than others to the operation. The principle has limitations and dangers. It encourages proponents to focus on the important 20 percent in the organization. That certainly seems wise, but at some point various components of the important 20 percent may disappear for reasons over which the service supplier has no control. And components of the less profitable or urgent 80 percent might become more important. Supervisors and managers must consider that all customers, clients, and tasks are important, but that the important differential over time can be helpful in prioritizing what to do first, if competing demands for service are received at about the same time.

Time Analysis Management

Why is it that some people are more effective when they appear to work no harder or longer than others in the same position? Perhaps they have learned and practice the previous principles. To determine whether excessive emphasis is placed on matters that are not true priorities in a manager's agenda, time analysis may help.[10]

One key to time analysis is to obtain an accurate record of how one's time is used. Time diaries may be kept for that purpose. A manager who keeps track of a typical week and then analyzes how his or her time is used may discover that too much time is devoted to tasks of minor importance. This leaves important obligations on which the manager is judged with less time than desirable to complete or to develop fully. In the process of keeping a time diary, the diarist quantifies time used for various activities and determines whether this pattern is satisfactory and justifiable. Often, the diarist is surprised to learn that counterproductive habits have crowded out valuable hours that should be allocated to priority work. This could bring the awareness that changes time-use priorities.

Delegate Everything Delegable

Supervisors and other managers have subordinates to whom work may be delegated. Directing and controlling activities generally are not amenable to being delegated.[11] However, numerous other activities that a manager might ordinarily take on can be delegated. Routine memos and drafts of letters; reminder calls; organization of forthcoming meetings; normal requests for supplies; and related matters may be delegated to subordinates. This provides the supervisor and manager with more time for planning and for routine communication with workers, customers, and others.

Using Technology for Greater Efficiency

The highly productive manager constantly searches for ways to use time more effectively. Technology dramatically helps meet that objective. Personal computers loaded with time-saving software, laptops, palm computers, cellular phones, beepers, facsimile machines, and portable dictating systems have greatly enhanced the means by which managers have become more effective in achieving their goals. A device or system that saves a manager as little as an hour or two a week is highly valuable over an extended period and is well worth the capital investment required.

Clean Desk or Messy?

The vision of the manager with the desk completely empty of papers—except for the matter being worked on at that moment—seems powerfully etched in the minds of many CEO watchers. Such clean-desk managers are considered on top of things, cool, and fully focused on the decision immediately facing them. But are these persons really more efficient or effective? No strong evidence supporting the clean desk or messy viewpoint exists, though many high management achievers seem to have little paper on their desks. Others achieve much—with piles upon piles of papers, records, books, and communications messages scattered everywhere.

Motivation Matters

Managers seek to elicit peak performance from subordinates. According to entrepreneur and writer Andrew S. Grove, high output from subordinates can be achieved in two ways: from training and motivation.[12] Both can improve performance, yet training an individual with a low degree of motivation will produce far fewer benefits than training those who possess a high degree of motivation will. Figure 5.1 graphs the connection between motivation and training. Motivation research identifies the attitudes, habits, and motives that trigger desired behavior. A few theories have served to underpin the thinking of contemporary management and have even affected workplace performance.

Theory X and Theory Y

This famous management construct proposed by Douglas McGregor encompasses the complex and dynamic relationship between personality

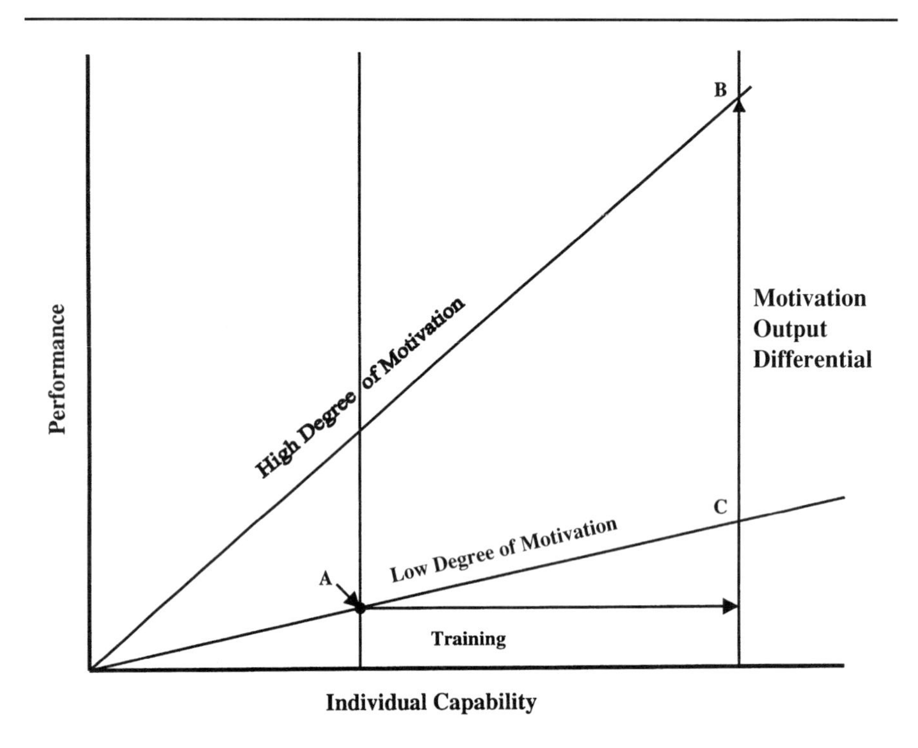

Managers have two ways to improve performance: training—or retraining—and motivation. Point A indicates the beginning of training, which serves to increase performance. Persons with high motivation increase capacity even further, eventually reaching point B instead of point C.

Source: Andrew S. Grove (1985). *High Output Management*, New York: Vintage Books.

Figure 5.1. Linking Performance to Degrees of Motivation and Training

and motivation.[13] A Theory X manager reflects authoritarian leadership favoring centralized decision-making. Theory X is often thought of as the view of old-line management. Workers are to be supervised closely. Theory X managers tend to hold the following views:

1. Most people do not like to work and will avoid it whenever possible.
2. The average person prefers to avoid responsibility and would rather be told what to do than have to make a decision.
3. The most effective managers use punishment, or the threat thereof, to achieve their goals.

[21] D. Eden (1990). *Pygmalion in Management: Productivity as a Self-Fulfilling Prophecy.* Lexington, MA: Lexington Books, p. 1.

[22] M. Magnet (July 6, 1987). "The Money Society." *Fortune*, p. 31.

[23] R. Rosenthal (1985). "From Unconscious Experimenter Bias to Teacher Expectancy Effects." In *Teacher Expectancies.* J.B. Dusek (Ed). Hillsdale, NJ: L. Earlbaum Associates.

[24] R. Rosenthal and L. Jacobson (1968). *Pygmalion in the Classroom: Teacher Expectation and Pupils' Intellectual Development.* New York, NY: Holt, Rinehart & Winston.

[25] J.S. Livingston (September–October 1988). "Pygmalion and Management." *Harvard Business Review,* 66(5):122.

[26] T.H. Fitzgerald (1979). "Why Motivation Theory Doesn't Work." In *Harvard Business Review on Human Relations.* New York, NY: Harper & Row, p. 277.

Additional References

M. Bloom (February 1999). "Performance Effects of Pay Dispersion on Individuals and Organizations." *Academy of Management J.*, 42(1):25–40.

R.E. Boyatzis (1982). *The Competent Manager.* New York, NY: John Wiley & Sons.

A. Fisher (September 30, 1996). "Stop Whining." *Fortune*, pp. 206–08.

M.B. McCaskey (November–December 1979). "The Hidden Messages Managers Send." *Harvard Business Review,* 57(6):135–48.

K. Tyler (May 1999). "Take New Employee Orientation off the Back Burner." *HR Magazine*, 43(6):49–57.

6

APPRAISING AND PROMOTING PEOPLE IN SECURITY PROGRAMS

> The secret of success in business of all kinds . . . is a liberal division of profits among the men who make them, and the wider distribution the better.
>
> —Andrew Carnegie

People are an organization's most important assets. Managing them successfully means providing opportunities for growth, including promotion to greater responsibilities. Promotions in formal organizations normally occur after work has been appraised over time as exceeding minimum expectations. Appraisal is the process of evaluating individual performance on the job and assessing it relative to goals and objectives. Appraisal not only is a measure of individual performance, but also identifies potential for future performance and capability. When employees can be evaluated and compared with reliability and fairness, the workplace *and* workers gain from the process.

THE DIFFICULTIES OF PERFORMANCE APPRAISAL

The organizational process of personal appraisal fits into the context of other aspects of management development: forecasting, recruiting, training, compensation and conditions of service, deployment, and management review or audit. Yet many security operations do not provide formal appraisal of workers. This may be because the process is time-consuming, demands judgment, and requires confronting individuals in what can be awkward situations. Some employers feel that statements made at such times can lead to litigation if the individual is terminated subsequent to an unfavorable evaluation. Litigation is more likely if the employee is terminated shortly after a favorable appraisal has been issued.

Such reservations about formal appraisal processes are understandable. However, appraisals' advantages outweigh their disadvantages. A well-conceived and conscientiously operated appraisal program can lead to motivation and growth of individuals, and produce higher performance. With such a process, justifiable criteria must be identified as the basis of promotions. No manager can be guaranteed that certain procedures will prevent the employer from being sued for failing to promote someone. However, the existence of a well-conceived employee appraisal system serves as a deflective shield against spurious civil litigation from employees passed-over for promotion. Beyond the uncommon likelihood of such litigation is the larger issue that all employees will observe that management has sought to establish reasonable, though imperfect, standards of promotion. Instead of encouraging litigation, such programs mitigate it.

Rapidly growing organizations sometimes feel they cannot spare time for appraisals. Yet this process helps identify individuals within the organization who are capable of assuming new responsibilities. In addition, appraisals are important in determining merit increases, special training, and layoffs.

WHO SHOULD BE APPRAISED AND WHEN?

In well-structured organizations, all employees deserve appraisal. Due to the time-consuming nature of the process, appraisals are generally conducted on an annual basis. However, some organizations will conduct appraisals on a semi-annual or more frequent basis. Organizations should schedule appraisals at times that will interfere least with critical activity. In retail organizations, for example, it would not be logical to schedule appraisals in November and December, as workers are busy with the holiday season. However, January and February are good months for retailers to schedule appraisals. Other workplaces will have different operating rhythms that will indicate when the most logical time for formal appraisal should be.

Many organizations schedule appraisals at least one year after the individual has begun a position and at each annual anniversary. However, as this chapter observes, employers frequently appraise workers during probationary periods, which are often particular to that organization or industry and should be mentioned specifically in the employee's manual. The annual review might occur approximately six months to a year after the probationary appraisal has been successfully completed by the new hire, and on an annual basis subsequently.

APPRAISAL FOR ALL LEVELS AND BY ALL LEVELS

How the appraisal will be designed and who may do the appraising differs from organization to organization and within an organization. The appraisal process should be flexible enough so that it produces the best returns for the time required. The following are a number of methods of appraising workers:

- **Top-down.** Appraisals are traditionally considered top-down; that is, a supervisor appraises his or her subordinates. In a hierarchical organization, this will be the expected and usual method of evaluation, and perhaps the only one normally scheduled. The advantage of top-down appraisals is that the more experienced supervisor understands the needs of the workplace clearly and is the best judge of how the subordinates have achieved workplace standards over the previous appraisal period. Also the appraiser knows how that performance can be raised to higher level. The disadvantage of this process is that it is a reflection of an autocratic style of management, especially when other forms of appraisal are not taken into consideration.
- **Bottom-up.** In this circumstance, subordinates evaluate their supervisors. The results of the appraisal document are received by a Human Resources manager, who analyzes the results and shares them with the supervisors involved. The advantage is that this process helps reveal strengths and weaknesses to the supervisor in a way that might otherwise not be discovered by upper management or the individual supervisor. Often, what the supervisor believes is a personal strength—for example, a string of delightful witticisms during the workday—is regarded quite differently by those who are targeted for such remarks on a regular basis. The reverse may also be the case. A weakness that the supervisor believes he or she possesses may be interpreted differently by subordinates, enabling the supervisor to reassess his or her management traits. Many managers find it hard to accept criticisms from subordinates and may ignore their appraisals.[1]
- **Peer review.** This is a situation in which peers evaluate each other. Typically, the results of the questionnaire used in such a process are seen only by a Human Resources manager, who then distills and shares the information with the persons involved. Peer reviews also help identify to management strengths and weaknesses of team members. The drawback is that such a process makes many participants uncomfortable. The process forces coworkers to raise unpleasant issues that possibly could be traced back to them and lead to disharmony.

- **Customer or client reviews.** Often, contract workers are part of the work environment for extended periods, sometimes for years. These individuals should be assessed annually by the contractor who assigns them to the work location. In the event the worker is a sole contractor on an extended assignment, that person may be reviewed much the same way proprietary employees are. Reviews by customers or clients of contract personnel provide the contractor with tangible evidence of worker qualities. They are the persons most in a position to evaluate performance under daily circumstances. In situations where contract workers are employed for extended service to the organization, the appraisal should involve collaboration between both the contractor and management of the contractee. Similarly, security service employees may be appraised by their "customers" within or outside of the organization. A security department within an organization serves the organization as a whole, and individuals who provide those services may be spot-checked periodically by a simplified evaluative document. Generally, senior managers do not opt for this type of evaluation unless criticisms have been raised and need to be substantiated or unless a new program requires evaluation.

WHAT TYPES OF EVALUATION DO WORKERS PREFER?

People being appraised are seldom questioned as to which types of performance appraisal they prefer. Some managers believe that such a preference is irrelevant, while other researchers of appraisal instruments conclude that such an inquiry may be "valuable."[2] In one study, 52 full-time registered nurses were asked how they preferred to be rated, by whom, and for what reasons. The nurses displayed a marked preference for specific methods of appraisal. They also preferred performance appraisal that had certain objectives, such as determining promotion or an adjustment in compensation. By contrast, they were less positive about appraisals that compared themselves to others, that did not include scales, and that were completed by subordinates. (These techniques are discussed later in this chapter.)

WHAT NEEDS TO BE EVALUATED?

If performance appraisal is important, what is evaluated must be of significance to the employer. This issue requires thought; not every trait of an employee should be subject to the appraisal process. Relevancy to the organization's goals is a basis for evaluative activity. The following are examples of goals and skills that may be subject to performance assessment at different levels of the workplace:

1. **Success performing functionally assigned tasks.** The employer may identify a series of activities routinely performed by the worker and specifically linked to performance standards. For this type of top-down appraisal, management would have identified the specific tasks required for the position and would have related them to standards the worker may be expected to fulfill regularly. These tasks flow from the job description originally written for the individual, spelling out in greater detail the nature of the duties undertaken and their appropriate standards. (Examples of this process are found in the following section.)
2. **Trait analysis.** This determines how a worker performs in a specific activity, such as clarity in dealing with the public, efficiency, and reliability. Raters are asked to appraise workers on different scales. Certain scales use such words as "outstanding" (top 2 percent); "excellent" or "superior" (top 20 percent); "above average"; "average"; and "below average" or "needs improvement" to rate the worker. Trait analysis tends to be focused on narrow qualities considered important to management for specific positions. Supervisors who review specific traits of subordinates often have difficulties in providing unfavorable assessments. This characteristic limits, but does not negate, the use of such measurement.
3. **Critical incidents methods.** Performance appraisers using the critical incident process note specific positive and negative actions taken by the worker during the evaluation period of complex actions vital to the job description. Such measurements have been identified previously by management as significant with regards to job function.[3] An example would be the technique by which a protective employee handled an untoward event that resulted in the completion of an incident report.
4. **Behavioral measurement.** One type of behaviorally oriented evaluation is the Behaviorally Anchored Rating Scale (BARS). This scale identifies a number of possible actions by workers and then assesses performance based on a scale from very desirable to very undesirable.[4] BARS graphically rates behavior with specific behavioral descriptions using a numerical scale. Since its introduction in 1963, BARS and its numerous variants have been widely used to evaluate the performance of law enforcement and, to a lesser degree, private security personnel.
5. **Mixed-Standard Scales (MSS).** Another method of performance review, used in policing and security, are Mixed-Standard Scales

(MSS). Such scales describe high, medium, and low performance and force raters to make a choice:

- **High performance:** Takes numerous steps in the patrol area both to prevent and to control crime; educates citizens in prevention techniques; has comprehensive knowledge of preventive equipment.
- **Average performance:** Makes some efforts to emphasize crime prevention in patrol district and has an adequate knowledge of preventive equipment.
- **Low performance:** Has little or no contact with district citizens to inform them of methods of improving their property for crime prevention.

What each standard represents to the rater is not always obvious. The rater indicates only that the worker's performance is "better than," "as good as," or "worse than" the behavior described.[5] However, many appraisal-instrument developers believe that MSS decrease rater leniency. In assessing "crime prevention" qualities of patrol officers, for example, items were included that identified specific dimensions of performance.[6] These included judgment, communications skills, job knowledge, demeanor, tolerance, cooperation, and human relations skills (see Box 6.1).

Box 6.1. A Mixed-Standard Scale (MSS) for Patrol Performance

The following 13 items are used to assess performance in different facets of a patrol officer's job. The appraiser is asked whether an item is an accurate description of the patrol officer's typical performance in that area of work. If so, the appraiser places (0) in the space provided for the worker. If the patrol officer's typical performance is *better* than the item description, then a (+) is placed in the space. If the patrol officer's typical performance is *worse* than the item description, then a (–) is placed in the space. Appraisers rank officers on code sheets that can later be analyzed.

1. Behavior sometimes shows the effects of a stressful situation, but it does not tend to interfere with the performance of duties.
2. Looks neat most of the time, although uniform occasionally reflects a busy schedule.
3. Reports are good, but occasionally need elaboration or clarification. Sometimes has difficulty communicating.
4. Takes numerous steps in patrol area both to prevent and to control crime; educates citizens in prevention techniques and has comprehensive knowledge of preventive equipment.

Box 6.1. *(Continued)*

5. Has little or no contact with citizens to inform them of methods of improving their property for crime prevention.
6. Performance reflects the proper judgment necessary to anticipate, select, and perform the appropriate behaviors in almost all circumstances.
7. Is quite emphatic about the types of people he or she can and cannot work with. Has difficulty getting along with many officers.
8. Shows maximum effort and enthusiasm almost all the time and in almost all circumstances.
9. Carries out assignments and responsibilities with satisfactory standards of performance. Rarely cuts corners or bends the rules.
10. Behavior with others is insightful and skillful, often preventing as well as ending conflicts.
11. Performance must be closely supervised, or it may slip to less-than-adequate standards. Behavior is often designed to find shortcuts in duties.
12. Appearance displays a careless attitude toward the job and the impression conveyed to the public.
13. Works adequately with most people, but has difficulty with some types of personalities. Although willing to break in new personnel, would prefer not to.

Source: H.J. Bernardin, L. Eliott, and J.J. Carlyle (1980). "A Critical Assessment of Mixed Standard Rating Scales," *Proceedings of the Academy of Management*, Athens, GA: Academy of Management, pp. 308–12.

USING A FORMAL APPRAISAL DOCUMENT

A formal employee performance evaluation compares the employee's performance to a set of standards. A series of ratings used by one formal system allows reviewers the opportunity to provide specific assessments for each task, as shown in Figure 6.1. The form illustrated, used by a security program employing over 200 security officers, begins with the noting of directory-type information:

- Section I clearly indicates to the employee that a performance evaluation is part of the expectation and notifies the officer of the extent of the evaluation period.
- Section II identifies the tasks and standards deemed critical by management. During the training process, the new employee

would have become aware of these tasks and the nature of the standards expected by management to be met. This would serve as the guideline for the worker during the time allocated for the evaluation. The employee signs the list of functionally assigned tasks, often in the presence of the supervisor. In a sense, this is a contract between the employee and employer and it is an explicit understanding that the worker is to be judged predominantly on the functionally assigned tasks. Figure 6.1 only provides space for five functionally assigned tasks and standards. These should be the critical broad workplace achievements that management expects from the employee at that particular title, level, and unit. These can be expanded according to the requirements of the position. Management then prepares a master list of tasks to be performed by the category of security personnel, as shown in Table 6.1. Standards of satisfactory performance accompany the master list and are entered into the employee performance evaluation form.

- Section III is completed when the evaluation period has reached an end and the supervisor makes ratings appropriate to the worker's actual performance compared to the standards. The supervisor will use comments and examples to justify the ratings. Failure to include these can lead to the sense that the supervisor has made a decision without proper reference to actual performance on the part of the employee.
- Section IV provides an overall rating after individual tasks are considered. The overall rating takes into consideration the totality of the employee's work performance during the previous period of time. Again, comments and examples are needed to justify the overall rating.
- Section V is where the supervisor provides his or her recommendation for the employee in a probationary period.
- Section VI provides space for specific plans for improvement discussed in the evaluation interview. This section identifies an area or areas in which employee performance improvement is required. Further, the specific means by which such improvement may be achieved is identified. Frequently, the behavior to be corrected can be altered by nothing more than additional personal effort following a discussion with the supervisor. In other cases, additional training may be needed to achieve the desired performance.
- Section VII provides the employee with an opportunity to add written comments to the evaluation. Such comments may be made on a separate sheet of paper and attached to the appraisal evaluation form if necessary. The date of the evaluation interview is also indicated.

SECTION I EMPLOYEE INFORMATION

Employee's Name	Soc. Sec. No.	Evaluation Period: From To
Title and Level	Title Code No.	Functional Title
Responsibility Center Name	Division	Section/Unit

Employee's Status: (Check one) ☐ Permanent ☐ Non-Competitive ☐ Probationary ☐ Provisional ☐ Other (Explain) ____________

SECTION II ASSIGNMENT OF TASKS AND STANDARDS (Complete at beginning of evaluation period)

Supervisor's Name (Please type or print)	Signature	Date
Reviewer's Name	Signature	Date

"I have received a copy of the tasks and standards below."

____________________ ____________
Employee's Signature Date

SECTION III EMPLOYEE'S PERFORMANCE COMPARED TO STANDARDS (Complete at end of evaluation period) Use comments and examples to justify ratings.

RATINGS (Check one)

Master List Task No.	FUNCTIONALLY ASSIGNED TASKS	STANDARDS	Employee's Performance Compared to Standards	Outstanding	Superior	Satisfactory	Conditional	Unsatisfactory	Unratable
1									
2									
3									
4									
5									

Figure 6.1. Non-Managerial Employee Performance Evaluation Form

SECTION IV OVERALL RATING (Complete after rating individual tasks). The Overall Rating is derived from the general tendency indicated by rating the individual tasks, taking into consideration the importance or priority of tasks, the overall work performance on occasional projects or minor tasks which affect the goals of the unit, and the employee's compliance with bureau rules and policies, including attendance and lateness, provided that they have a direct impact on the tasks performed and the achievement of the unit's goals.

A. Overall Rating: (Check one)	☐ Outstanding	☐ Superior	☐ Satisfactory
	☐ Conditional	☐ Unsatisfactory	☐ Unratable

B. Comments and Examples to Justify Overall Rating (Attach additional sheet if required)

SECTION V RECOMMENDATION FOR EMPLOYEE IN PROBATIONARY PERIOD (Check one)

☐ Retention ☐ Dismissal ☐ Demotion (for promotee only) ☐ Extension of Probation

SECTION VI SPECIFIC PLANS FOR IMPROVEMENT DISCUSSED IN EVALUTION INTERVIEW (Attach additional sheet if required)

A. Plans to Improve Employee's Performance

B. Plans to Change Conditions Affecting Employee' Performance

C. Description of Any Training Needed

SECTION VII SIGNIFICANT COMMENTS MADE BY EMPLOYEE DURING EVALUATION INTERVIEW (Attach additional sheet if required)

Date of Evaluation Interview: ____________

SECTION VIII SIGNATURES

Supervisor's Name (Please type or print)	Signature	Date
Reviewer's Name (Please type or print)	Signature	Date

Employee's Statement: My signature below indicates only that I have received a copy of an evaluatory statement on this date and does not necessarily indicate my agreement with the contents of the statement.

______________________ Employee's Signature

______________________ Date

Figure 6.1. *(Continued)*

Table 6.1 . Examples of Master List Tasks and Standards for Security Officers

Master List Task Number	Standards
1. Patrols designated areas of public buildings, reports in uniform, makes rounds, checks that the public and staff are following rules and regulations to prevent crime, vandalism, disturbances, and are safeguarding life and property.	Reports for patrol in uniform and at the designated time. Observation is acute and comprehensive. Makes complete rounds. Follows rules and regulations closely.
2. Reprimands and ejects loiterers and disorderly persons by checking restrooms, stairways, halls, other areas, and advising persons to leave. Uses persuasion to obtain results. Follows rules and regulations to remove unauthorized persons and quiet disturbances.	Areas checked frequently. Persons asked to leave correctly, firmly, and courteously. Force used appropriately and only when necessary. Appropriate rules and regulations followed closely.
3. Guards department's personnel and property. Restricts persons from entering unauthorized areas. Notifies supervisor of damaged and inoperative equipment. Patrols efficiently to prevent personal injury or property damage.	Unauthorized persons barred from restricted areas. Supervisor notified promptly and clearly of damaged and inoperative equipment. Patrols effectively. Removes disorderly persons promptly and properly.

- Section VIII concludes the evaluation process with the supervisor's signature. The employee generally signs the statement at this point indicating that he or she does not necessarily agree with the contents of the statement, but acknowledges that the document is complete. At a later date, a reviewer will add a signature indicating that the employee performance evaluation form has been reviewed by the next level of management. Alternatively, the reviewer may be an independent Human Resources officer. Generally, reviewers are directly superior to the supervisor completing the evaluation form.

JOB PERFORMANCE RATING

In some organizations, job performance evaluations are required during the probationary period, which usually lasts three to six months. A monthly performance rating evaluating new security officers is presented in Figure 6.2.

Security Officer Job Performance Evaluation No.____

NAME: ____________________ TEL. NO: ____________________

UNIT: ____________________ Pager No.: ____________________

DATE HIRED: ____________________ ASSIGNMENT: ____________________

REVIEW DATE: __________ SOCIAL SECURITY: ____________________ SHIFT: __________

ATTENDANCE RECORD			MONTHLY PERFORMANCE RATING & EVALUATION (Space for additional comments, assignments, on back)	
Date & Time	Absent? Tardy?	Cause and/or Comment	Appearance	Evaluation of job performance: reliability, log sheets, ability to learn, initiative, judgment, courtesy, accuracy. etc.

Adapted from: E.T. Guy, J.J. Merrigan, Jr., and J.A. Wanat (1981). *Forms for Safety and Security Management*, Boston, MA: Butterworth-Heinemann.

Figure 6.2. Security Officer Job Performance Evaluation Form

Following the conclusion of a successful probationary period, performance ratings and evaluations can be conducted at the same frequency as with all other employees in that job category. Use of the form acts as an incentive both for the supervisor to note performance on a daily basis and for the subordinate to be aware that the probationary period entails written performance verification on topics important to the work function.

Evaluation of contract employees may be aided by the use of a form designed for this purpose, as shown in Figure 6.3. This document may be

Officer Performance Evaluation Form

In order to assure the effectiveness of Security Service, the Security Department would appreciate your appraisal of officer performance

Officer's Name:__________ Unit: __________ Date: __________

Please complete the following and return to: __________.

Superior	Good	Needs Improvement		
____	____	____	__	Appearance
____	____	____	__	Courteousness
____	____	____	__	Co-operation
____	____	____	__	Effectiveness
____	____	____	__	Job Knowledge
____	____	____	__	Ability to communicate
____	____	____	__	Attitude

Please list recommendations that would improve our security service. Return this form to the undersigned.

____

____

____

____

____

____

____

Security Director

Adapted from: E.T. Guy, J.J. Merrigan, Jr., and J.A. Wanat (1981). *Forms for Safety and Security Management*. Boston, MA: Butterworth-Heinemann.

Figure 6.3. Officer Performance Evaluation Form

completed by the client on a monthly or semi-annual basis and forwarded to the contract company as part of ongoing evaluation. The "Remarks" section provides an opportunity for management of the contract firm to take appropriate action. Of course, any substantive unsatisfactory job trait or behavioral quality should be transmitted to the security services contact supervisor quickly to result in corrective attention. A telephone call to the contractor as soon as such unacceptable conduct is observed may suffice.

Peer reviews may also be considered as part of the strategy to determine job performance, cooperation, and teamwork characteristics. A Colleague Confidential Evaluation form is used for co-workers to evaluate each other's performance, as shown in Figure 6.4. The evaluation director distributes and later discretely collects completed forms from evaluators. The evaluation director looks for strengths, weaknesses, and cooperative patterns in workplace performance where close teamwork is required. These forms are then assessed and transmitted to the workers being evaluated. Organizations tend to use peer-review when mutual cooperation is essential and cannot be evaluated fully by a supervisor due to the complex nature of the work performed by the teammates. The evaluation director must be discrete when sharing results so as not to reveal the identities of the evaluators who have prepared the evaluations, as this could cause friction in the workplace. These documents should be stored separately from other records in the worker's employment file.

THE NEED FOR APPRAISAL DOCUMENTATION

Employees sometimes sue their employers for failure to be promoted, for disciplinary action, or for termination. In such actions, written performance appraisals are likely to become critical evidence. In appraising subordinates, supervisors have to find a balance between encouraging desirable behavior while also identifying non-productive behavior. A pattern of substandard performance could lead to termination that conceivably could be the basis of a civil action by an employee. Many supervisors emphasize the positive and ignore the negative in subordinates. However, such relevant negative features could become worse subsequent to their initial identification in the written performance appraisal. Therefore, the supervisor needs to document any negative behavior, potentially serious enough to be the basis of discipline or discharge, at the earliest opportunity.

Regrettably, many evaluators see only the positive in those whom they evaluate (see Box 6.2). Failing to be observant of a worker's shortcomings is just as harmful as being excessively lenient or strict in judging workers as a group. Other evaluators who judge most or all of their subordinates as "average" may lack discernment and judgment expected of those with supervisory responsibilities. The appraiser needs to include specific, detailed observations, with the time and date noted, of notable workplace

Colleague Confidential Evaluation

Evaluatee: ______________________________

Department: ______________________________

Evaluator: ______________________________

Your comments will be reviewed by the Evaluation Director and will be anonymous to the Evaluatee in Evaluation and Development Summary.

Comments On Overall Performance

- Please describe the Evaluatee's strengths and development areas:
- Pay particular attention to the criteria listed below:

Work Production/Professional Skills:	Teamwork/Cooperation:
• Flexibility/Versatility • Communication Skills (i.e., relaying clear and accurate verbal and written work-related information) • Professional Demeanor	• Contribution to group • Providing assistance to all group members • Answering phones • Sharing firm knowledge/resources

Comment on greatest strengths (provide examples):

Comment on areas in need of further development. Indicate suggestions for developing these areas:

Evaluator Signature: ______________________ Date: ______________

This form is prepared by managers and distributed to team for workers' mutual appraisals. The confidentially completed form is returned to management where it is assessed and used as the basis of appraisal of the evaluatee.

Figure 6.4. Colleague Confidential Evaluation Form

performance. This rigorous method of identifying both desirable and undesirable performance characteristics supports the overall objectives of operations management.

Box 6.2. Rating Workers Objectively in Appraisals

The process of using written appraisals can be counterproductive if the rater is not objective about the person being judged. Raters can improve the accuracy of their ratings by recognizing the following factors that subvert evaluations:

1. **The halo effect.** The tendency of an evaluator to rate a person good on all characteristics based on an experience or knowledge involving only one dimension.
2. **Leniency tendency.** A tendency toward evaluating all persons as outstanding or above average and to provide inflated ratings rather than true assessments of performance.
3. **Strictness tendency.** The opposite of the leniency tendency, this is a bias toward rating all persons at the low end of the quality scale and a tendency to be overly demanding or critical.
4. **Average tendency.** This is the tendency to evaluate every person as average regardless of major differences in performance.

Source: *Effective Phrases for Performance Appraisals* (1994). Perrysburg, OH: Neal Publications.

OTHER WRITTEN APPRAISAL TECHNIQUES

Performance evaluation may involve extended documentation. Prior to evaluation, the standards should have been made clear to everyone involved. They should be realistic, objective, and comprehensible to all security workers in the job category. However, more than one rating category can and should be used. The following are some options:

- **Ranking.** In this type of measurement, the supervisor is asked to simply rank all subordinate workers in the group on a numerical basis from best to worst. This may create a bias against the most recently hired and, therefore, less experienced worker.
- **Paired comparison.** In this circumstance, the supervisor compares subordinates to specific tasks ranking them in order from best to worst according to each criterion. In this situation, the overall highest ranking can be determined from an aggregate of different factors. A negative feature of this process is that those most recently hired—regardless of their level of training—are likely to perform less well than experienced work unit members.
- **Narrative form.** In this rating method, the supervisor writes a discursive passage on each worker to summarize individual strengths

and weaknesses. This helps to provide a human dimension to the worker's performance during the evaluation period.

- **Forced-choice method.** In this measurement, the supervisor selects from a set of statements involving the subordinates in the work unit. The rater selects two items from a group of four descriptive items, one that emphasizes the most characteristic of the worker and another that emphasizes the least characteristic. Perhaps 10 or 12 sets of four characteristics are presented and then subsequently analyzed for each worker. The report is tedious to create and requires construction by a technical specialist, but it provides a portrait of the worker unattainable by other written appraisal techniques. One problem with this measurement is that the sets may be widely different in terms of their significance, and the appraiser may not understand which employee characteristics are deemed most significant.[7] For example:

  1. A. Problems need not be stated in detail for him or her.
     B. Double checks work others do for him or her.
  2. A. Does more than his or her share of the work.
     B. Works to improve his or her main weaknesses.

THE APPRAISAL INTERVIEW

The preceding section discussed written, documented appraisal forms. These are prepared prior to sharing the results with the employee or have been used as part of the evaluation process. If and when transmitted to a subordinate worker, the results must be communicated with tact and clarity. The supervisor who conducts the appraisal interview needs training on how to conduct such interviews. Initially, many supervisors are uncomfortable with the prospect of appearing to judge a subordinate, perhaps saying or doing something that will be counterproductive and could create an unpleasant environment within the workplace. This usually occurs subsequent to an unfavorable rating. The cooperative basis between the supervisor and subordinate could be tainted by the experience and a previously friendly environment may become soured.

These are fears many supervisors possess when they approach appraisal interviews for the first time. In the days and weeks prior to a formal appraisal interview, stress among those involved is common, and is not necessarily harmful. Yet for the vast majority of appraisal interviews, the process is positive and ultimately even enjoyable for both parties. High-performing workers leave the interview enthusiastic to do better work. Substandard workers realize they have a sympathetic supervisor and another opportunity to do better; often, they too feel relieved by the process. The following are guidelines for letting workers know how they are doing:[8]

- Select a quiet, comfortable, and appropriate location for the interview, such as the supervisor's office, if private, or a conference room.
- Plan to avoid interruptions. The process has been on the minds of the interviewees for weeks. They deserve the supervisor's undivided attention during the interview. The supervisor also should allow extra time for the subordinate to discuss workplace-related matters. Such discussions, however, should not serve as a general sounding board for excessive gripes that are far afield of the main purpose of the meeting.
- Put the person at ease. Humor and informality can help reduce tension before substantive issues are considered.
- Conduct the interview in a positive manner. Even if the subordinate requires further training for a skill not fully mastered, the supervisor should emphasize positive job accomplishments, while not minimizing any failings that require correction.
- Review the ratings by category. The categories for security workers can include decision making, dependability, development, interpersonal skills, leadership, learning ability, management ability, motivation, personal qualities, professionalism, quality consciousness, and report writing (see Box 6.3). Not all qualities can or should be covered in the time allocated for such interviews. The review should limit ratings to those considered most central to the tasks of the workplace.
- Keep the interview performance-oriented. The supervisor may wish to avoid the accusatory use of "you" in speaking with the worker and instead emphasize the way certain tasks were completed and their quality.
- Encourage the subordinate to talk. Often, supervisors and subordinates have little opportunity to discuss job performance in a quiet environment. At the interview, the subordinate has an opportunity to state whatever he or she thinks is pertinent to the review. Following the interview, the supervisor may wish to take notes of such statements.
- Respond to objections, problems, and disagreements. The supervisor should calmly accept criticism from the subordinate. Appropriate objections, problems, and disagreements should be resolved by the supervisor within a reasonable period of time after the meeting. (Often simply relating a vexing incident to the supervisor relieves the stress felt by the subordinate.)

Box 6.3. Qualities Evaluated in Performance Appraisals

Managers may need to evaluate particular qualities among security workers. Those qualities differ according to job title and level, and often overlap. The following are some categories by which workplace achievement or deficiency can be measured:

Accuracy
Achievement
Administration
Analytical skills
Communicative skills
Competency
Computer skills
Cost management skills
Creativity
Decision-making skills
Dependability
Development
Evaluation skills
Goals and objectives
Improvement
Initiative
Interpersonal skills
Judgment
Knowledge
Leadership
Learning ability
Loyalty and dedication
Management ability
Mental capacity and application
Motivation
Organization
Performance qualities (general)
Performance qualities (specific)
Personal qualities
Planning skills
Potential
Problem-solving skills
Productivity
Professionalism
Quality
Report writing skills
Resourcefulness
Responsibility
Selling skills
Stress
Supervisory skills
Tact and diplomacy
Team skills
Time management skills
Versatility

Source: *Effective Phrases for Performance Appraisals* (1994). Perrysburg, OH: Neal Publications.

- Concentrate on facts. The truth must guide the supervisor in such appraisal interviews at all times.
- Be a coach, not a judge. The objective is to improve performance, not to pass judgment that, although possibly accurate, will not lead to employee improvement.
- End the interview on a positive and supportive note.

The interview process might seem tedious from the above description. Yet the interview itself, properly planned and executed, takes only a few minutes. At its conclusion, the supervisor may note critical results from the interview and plan any follow-up actions indicated by the information provided by the subordinate. The observer is likely to use this occasion to identify plans to ensure employee growth. After a series of appraisal interviews, the supervisor may analyze teamwork performance and identify technical or behavioral issues requiring improvement.

With the conclusion of the appraisal interview process, the supervisor transmits reports to his or her supervisor for review. This process informs middle management about the advancements being made by operational employees and also by the supervisor responsible for success within a unit. Senior management uses this method, among others, to evaluate the capacities of work units and their managers.

The appraisal interview is a confrontation between a superior person in the management hierarchy and a subordinate one. Like any confrontation, the process can be difficult. Yet the lack of such an encounter can leave superior performance unacknowledged. Equally, unsatisfactory performance may remain uncorrected. Unsatisfactory workplace performance that goes unevaluated and uncorrected is not likely to improve on its own. The appraisal process—from planning through written reports to oral interviews—is an obligation of high-performing management. The process has too many potential benefits relative to risks to be ignored.

ASSESSING PERFORMANCE AMONG DIFFERENT EMPLOYMENT LEVELS

Much of what has been discussed so far is written with the supervisor and subordinate at operations level in mind. These points also are applicable among members of middle and upper management throughout the organization. Assessment for these persons is directed at the different nature of work performed by employees with different levels of responsibilities.

The workplace demands activity that may be divided into three categories: conceptual (analyzing, directing, planning), human relations, and technical skills and operational services. While the proportion of these components differ according to the level of employment, all three are present for all employees, as shown in Figure 6.5.

REVIEWING MANAGEMENT STRATEGY

Middle and upper managers deserve appraisal just as the operational staff and their supervisors do. This section concentrates on the nature of manage-

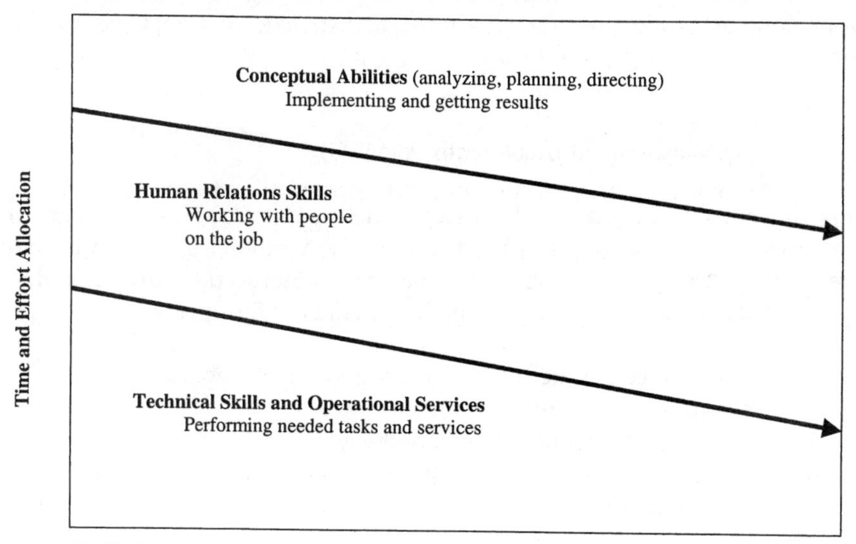

All work within organizations contains a combination of three types of abilities shown above. The proportions of each can vary widely according to the job position and employer. Performance evaluation can be directed to each of these categories weighted according to individual employment responsibilities.

A trained entry-level employee on the job for the first time, without a supervisor or co-worker nearby, may be called on to make an on-the-spot decision of importance to the organization. This certainly is conceivable for security employees with extensive public contact. For this reason, even this operational neophyte must possess some conceptual skills to tackle such a situation.

Nonetheless, the nature of management moves from being less operational to more conceptual depending on the worker's position on the workplace hierarchy. Therefore, middle and upper managers tend to be judged more on analytical, planning, and implementing skills than on technical and operational abilities. Note also that human relations skills remain a significant concern throughout the workplace, regardless of one's position.

Figure 6.5. Performance Evaluation Related to Different Levels of Workplace Responsibilities

ment plans and how they are created and evaluated. The nature of management review focuses more on programs and their success over the previous work period than on human relations or technical service skills. All components are significant. Any one of these aspects cannot be significantly deficient for the manager to retain standing.

A few strategies have had wide influence on organizations for getting work done and adjusting to changing circumstances. The next section discusses some of these techniques.

Management by Objectives (MBO)

This concept was introduced by Peter Drucker and Douglas McGregor in the 1950s and was widely accepted in management circles about a decade later.[9] The success of the MBO model has been embraced by private industry, institutions, and the government. At the core are four principles:

1. The employee and her or his superior jointly set goals.
2. The employee endeavors to meet the goals.
3. Performance is evaluated against the goals.
4. The employee and superior jointly set new goals for the next measurement period.

MBO is a cooperative process, yet for it to succeed, both the goal and the means of achieving that goal must be understood by all parties concerned.

MBO was created with profit-making organizations in mind. It soon became apparent, however, that substantial applicability of MBO to non-profit organizations and government existed. In order to use the model, the organization must affirmatively answer the following questions:[10]

- Does the organization have a mission to perform? Is there a valid reason for it to exist?
- Does management have assets (money, people, a plant, and equipment) entrusted to it?
- Is management accountable to some persons or authority for a return on the assets?
- Can priorities be established for accomplishing the mission?
- Can the operation be planned?
- Does management believe it must manage effectively even when the organization is a non-profit one?
- Can accountabilities of key personnel be pinpointed?
- Can the efforts of all key personnel be coordinated into a whole?
- Can necessary controls and feedback be established?
- Is it possible to evaluate the performance of key personnel?
- Is a system of positive and negative rewards possible?
- Are the main functions of a manager (planning, organizing, directing, evaluating) the same regardless of the type of organization?
- Is management receptive to improving methods of operations?

These questions are geared towards nonprofit organizations, although the same qualities can appear in any bureaucracy. Those who plan operations strategically often choose to design systems in which management has authority but not responsibility. Managers or entire work units normally have authority to conduct activities, and may take credit for any success. But in the event that the desired goals are not achieved, the same individuals or entities can argue that they are not responsible for the results. They may claim they did not have control over all circumstances and resources. Who does? Persons who work in such an environment are "never wrong" because they always take credit for success but shun responsibility for any failure (see Box 6.4). This kind of situation is the target of MBO strategy. MBO makes individual managers or groups conceptually responsible for defining objectives that can be independently verified and justified.

Box 6.4. Linking Management to Authority *and* Responsibility

---

One of the circumstances in which management under-performs occurs when a manager or a group of managers have authority but not responsibility. This is the situation that MBO strategy seeks to make less possible. MBO links authority and the right to take credit for success directly to others, but holds the same manager responsible in case of failure. It might seem astonishing that organizations can have active units of authority without responsibility, yet it has been known to occur. The following are just a few examples:

- **Organizational design.** The Port Authority of New York and New Jersey is a giant bureaucratic conglomerate that owns bridges, tunnels, airports, and office buildings in the New York City area. The Port Authority employs thousands and has revenues in the billions. But who is responsible when things go wrong? Technically, the governors of New York and New Jersey have the responsibility. But it is hard to pin responsibility on two persons in different states who often have divergent and competing interests and priorities. In such situations, if results are unsatisfactory, no single authority answers for them.
- **Programmatic design.** Consider a security program in which the director has the authority to hire or fire personnel, to contract or terminate a contract of a security service provider, and to take other relevant actions. Yet in some circumstances, such a manager may claim to take little or no responsibility unless results are tied to the job. For example, such a manager could claim that insufficient resources, too little time, and changing circumstances were responsible for unsatisfactory results.

MBO brings managers, supervisors, and workers together to share authority for setting objectives and holds them responsible if those goals are not met. MBO can be structured for a particular work group or for a larger organization. The MBO team allocates and distributes personnel, identifies what resources are available, and determines what is needed. This increases the opportunity for involvement by all personnel, provides specialized resources as needed, and uses varying management concepts as appropriate to achieve results. In the end, responsibility and authority, success and failure, rewards and penalties are all linked.

Examples of MBO in Security Applications

In the following sections, we will look at various examples of MBO in action.

Example 1. In this example, the organization is a regional distributor of brand-name gasoline and owns refining, storage, and transportation facilities as well as several filling stations, many with convenience stores attached. Additionally, the company has franchise relationships with many independent operators. The parent company has designed and installed an advanced proprietary central monitoring system and has connected its offices, processing, and retail facilities to it. Soon after installation, two needs emerge concerning the new system.

Objective: More efficient use of proprietary alarm monitoring resources.

Goal No. 1: To reduce the number of false alarms received by the department's proprietary alarm system by 20 percent over a six-month period without losing quality of response.

Program:

1. Re-educate all internal users about the use of the alarm system, focusing on factors that frequently cause false alarms.
2. Designate an employee to review all false alarms within one day of occurrence and determine what actions could be taken to reduce such alarms.
3. Institute measures to verify alarms before calling police, for example, by using closed circuit television with interactive audio capacity to verify a possible alarm condition after receiving the initial alarm signal.
4. Provide management with written reports of alarm activity as they occur. The reports compare incidents on a year-to-date basis to document the trend.

Goal No. 2: Expand use of the alarm system to the organization's customers and franchisees.

Program:

1. Prepare literature on cooperative use of the proprietary system and distribute to all prospects emphasizing technological advancement. Mail a second time two months later to non-respondents.
2. Schedule an open house so prospects can visit the alarm monitoring station and learn of its capacities. Demonstrate its surveillance and two-way communications features.
3. Designate a program manager to call prospects subsequently to determine interest. The program manager should work from a carefully prepared script and should emphasize benefits to being protected within the company's alarm system, though the commitment is optional.
4. Train alarm installers to deal with the company's franchisee's special requirements and concerns.
5. Facilitate insurance savings for participants who contract for the service by providing details to the franchisee's insurance broker or carrier.

Example 2. In the second example, a large urban medical center operates several of its own parking garages combined with nearby open parking lots. Due to the nature of the 24-hour service provided by the facility, a pattern of thefts from cars, vandalism, and occasional car theft has emerged over the previous year, and the problem is increasing. Management is concerned that employees and visitors who use the parking facility are becoming fearful. Victimized employees and visitors have pressured the hospital administration and security director. Perimeter control and other measures are needed to reduce incidents.

Objective: Reduce theft from and of vehicles. Reduce vandalism.

Goal No. 1: Reduce chances of unauthorized access to parking areas.

Program:

1. Seal openings to all the parking garages so that users may not enter indirectly. (A redundant pedestrian back entrance to one multi-story parking lot should be sealed permanently.)
2. Provide additional surveillance at the zone where cars enter and exit. (An office for the parking garage manager on duty should be relocated with windows open to the entrance.)
3. Improve lighting throughout. (Newer lights also cut energy costs.)

4. Install covert and overt closed circuit television, which collects images of all vehicles and drivers arriving and departing from the facility. The videotape recording of the traffic should be located in the office of the security department in the main hospital building. Signs should be posted stating "These Premises under 24-Hour Video Surveillance."
5. During hours of little activity (10 P.M. to 7 A.M.) the garage door should be closed and opened only when traffic is present. Whenever the doors are open, a security person or garage attendant must be near and visible at the entrance in addition to the nearby visible parking garage cashier.

Goal No. 2: Reduce chances of theft from automobiles and of automobiles themselves.

Program:

1. Provide a scooter to patrol the lots frequently and randomly.
2. Install signs to remind drivers not to leave valuables exposed in their vehicles. If customers make their cars tempting to thieves, provide extra surveillance and inform the customers when possible to be more careful.
3. Establish separate parking areas for regular medical center staff and visitors.

Goal No. 3: Reduce fear level of nighttime and early morning patrons of the garages and lots.

Program:

1. Offer an escort service from the building entrance to vehicles for those arriving or departing between 11 P.M. and 7 A.M.
2. Install an alarm and communications system at all levels of the parking garage and lot. If someone presses the emergency button, an audible alarm and a flashing strobe light should be illuminated. An attendant should have two-way communication with the person requiring assistance.
3. Improve housekeeping in the parking garage and lot. Walls should be painted white and unused utility vehicles previously stored in the lot should be removed.
4. Conduct a brief pre- and post-implementation survey of users to determine success.

The MBO technique has many advantages. One criticism of MBO, however, is that it has not been subjected to rigorous analytical standards to determine its success. This is true in security applications as well as other management applications using MBO. Yet this criticism could be raised with many other personnel tools currently in use as well. Further, one review estimates that MBO successes are about five times greater than its failures.[11] However, the benefits of MBO may decline over time, requiring a fresh analysis of situational circumstances that have changed months or years after the original objectives were achieved.

Failure to achieve the desired goals from MBO does not necessarily mean total failure. Planners and implementers in such situations have endeavored through numerous means to achieve success. Their lack of success may eliminate the value of some measures that have been tried and may suggest other measures that could take their place or augment them.

Critical Incident Review

Another way by which managers may be evaluated is through a case review of a significant incident or launched program that occurred over the previous six months or year. The manager for this program might expect to have all aspects of the incident carefully reviewed, leading to an eventual statement or report of findings. In this sense, the critical incident review is like an internal audit of an incident or program.

The reviewers in such a process frequently are assigned from outside the immediate chain of command. They may or may not have any personal knowledge of the incident or program, but possess broad experience and analytical skills sufficient to enable a fair and comprehensive review of the facts. Critical incident reviews are often preferred by the government for program evaluation. The reviewers may be investigators, internal auditors, managers from other operations, or professional consultants.

Problem-Solving Ability

A majority of persons who work in security come in contact with the public as part of their assignments. Security exists partly to provide at strategic locations competent, trained individuals who can deal with exceptional circumstances on the spot. This prompt action prevents a simple issue from becoming a major problem. Management expects security workers to be problem solvers, not deniers. However, security personnel should not be expected to receive and resolve untoward events for which they have not been trained. Security personnel, unlike Human Resources personnel who usually work from their offices, may be found throughout the organization's facilities, dealing with incidents that can be resolved quickly. They are placed strategically about the operations. This enables security practitioners to resolve low-priority incidents so that they will not escalate to significant issues.

Senior management assesses and rewards subordinate personnel who assume responsibility for a potential problem and find a solution. Management observers also seek to assess such behavior and use it as a basis for commendation or promotion. Other individuals may initially tend to ignore the problem, then deny its consequence, and, learning otherwise, temporarily think that others should be blamed, as shown in Figure 6.6. Instead, they should be conditioned to pass through these emotional checkpoints and maturely assume responsibility, which leads toward a solution of the problem presented. Other individuals accept responsibility and readily seek to resolve problematic issues.

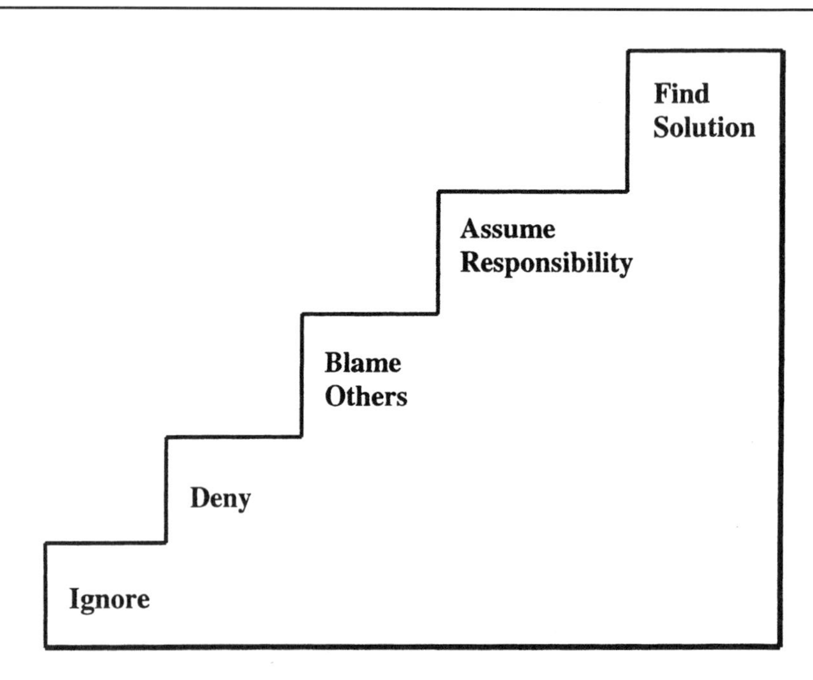

Not all problems presented to security personnel should be solved by them. In many cases, the problem-presenter should be encouraged to solve the problem himself or herself. Other issues are reasonably related to the concerns of security services and the burden for responding belongs to security. When this is the case, security personnel should push through any psychological barriers to resist acting, and assume responsibility for a solution.

A.S.Grove (1985). *High Output Management.* New York, NY: Vintage Books, p. 194

Figure 6.6. From Ignoring to Resolving Problems: A Process

Field Review

The term "field review" refers to programs located at distant facilities. It frequently represents a senior manager's critique of an operational manager's program elsewhere. At such times, the senior manager usually arrives at the distant location, conducts a broad survey of relevant factors, and later reports to local, regional, and senior management of the circumstances found during the scheduled field review (see Box 6.5). The field report evaluates local security management and programs for their strengths or weaknesses.

Box 6.5. Avon Calling: Visits for a Field Review

Field reviews of ongoing security programs give headquarters' management an opportunity to assess and respond to local problems and opportunities. At Avon Products, Inc., Robert F. Littlejohn, vice president for global security, visits some 60 Avon locations each year to conduct field reviews. Littlejohn's procedures for conducting the field review are meant to enhance cooperation and good rapport between global headquarters and the local operating business. Generally, a routine field visit of an operating facility can be conducted in one or two days, depending on the complexity of the facility. The final analysis and report requires hours to compile.

The following are guidelines used by senior managers on their periodic field visits:

1. **Plan ahead.** Schedule the visit weeks in advance so that the occasion will not be disruptive to local operations. (Obviously, if a pressing need exists, the review should be expedited.)
2. **Meet with the general manager at the beginning of the fact-finding visit.** Explain that the nature of the visit is largely consultative and supportive. But also state that if anything of a critical nature for action is discovered during the visit, this will be shared with the general manger by the security evaluator before leaving. Another advantage of meeting with the general manager before beginning the inspection and evaluation is that the evaluator has the chance to learn about any local concerns.
3. **Endeavor to have quantitative data ready to evaluate.** Expect to have incident reports ready for review. Photocopy or download records for later analysis.
4. **Allow time for local security managers to share concerns.** Endeavor to meet significant new employees who have joined the organization since the previous visit.

Box 6.5. *(Continued)*

---

5. **Inspect any new security system that has been installed since the previous visit.**
6. **Concentrate on the most important security problem concerning management at that location.** Additionally, review security matters of concern to headquarters and share what the plans are to respond to them.
7. **Meet with the general manager again before departing.** Share what has been observed and discuss possible problems. If they are of a minor nature, they may be resolved on the spot and not appear in the final written report. If they are major, begin dealing with a solution immediately, following up later until the issue is resolved.

Final reports contain an executive summary, findings, and recommendations for actions. This plan details what work is needed to be done, who is responsible, and when the implementations are to be completed. Global headquarters follows up to make sure that its tasks and those of the local operating business are both completed in a timely fashion.

PERFORMANCE REVIEWS FOR SENIOR MANAGEMENT

Appraisals need not be limited to operational staff personnel and lower and middle management. All employees in an organization committed to systematic evaluation of individual performance may be involved in the review process. This includes senior staff officers and the chief executive officer (CEO). In large complex organizations, the CEO may request that the senior officer for Human Resources conduct the evaluation of all senior staff officers, or the CEO might assume this responsibility. In addition, an outside consultant may be retained for the process.

Performance evaluations of the CEO and senior staff officers are less likely than appraisal further down the management hierarchy ladder. This is because senior staff officers are under scrutiny by a variety of outside sources. Principal among these is the board of directors, trustees, or an equivalent governing body. Review of performance of a different sort also takes place among officers of publicly held companies by stock analysts. Such analysts may follow the company and regularly ask penetrating questions about performance goals and results and then publish their findings and conclusions. Further, senior officials are subject to judgment by the media and shareholders, particularly when an incident has occurred that raises doubts about the capability of the team in the executive suite.

THE LIMITATIONS OF APPRAISALS

A case has been made that appraisals of individuals at all levels can help serve the needs of a dynamic organization. Because of the criticality of loss prevention in many organizations, performance appraisals may be more likely to be an integral part of the management strategy in this department than in others.

Appraisals do have their limitations. Bias in judgment of one individual by another may be difficult to eliminate, particularly in the minds of a subordinate receiving a critical evaluation. A more searching dissatisfaction with the appraisal system has been raised by psychologist Harry Levinson, who argues that job descriptions must be behavior- as well as results-oriented.[12] Levinson suggests employers create dynamic job descriptions in which behavior for different positions—particularly management—is identified in advance. Examples of behavior-oriented questions to be answered in preparing the dynamic job description are:

1. How does this job require the incumbent to handle his or her aggression and attacking capacity?
2. How does this job require the incumbent to manage affection, the need to love and be loved? Is the person required to be a socially friendly leader of a close-knit work group?
3. How does this job require the incumbent to manage dependency needs? Will the individual be able to lean on others who have skill and competencies, or will he or she have to operate alone?
4. What ego ideal demands does this job fulfill? If one does the job well, what is the gratification to be gained? Money? Recognition? Eligibility for promotion? The feelings of great personal achievement?

THE PROMOTION PROCESS

Promotions are substantial changes in a job, normally requiring additional responsibility and entailing greater authority for which increased income, perquisites, and status are provided. Persons considering the prospect of being promoted concentrate on these external benefits. Such benefits serve as important inducements that make promotions attractive for many. Employers also see promotions as a means to sustain vital programs. Promoting someone offers hope for improvement in the sector for which the newly promoted person will be responsible. Consequently, to the employer, the act of promoting someone involves the weighing of numerous factors before the decision is made. For this reason, organizations consider promotions carefully.

Advancement That Stops Short of Promotion

Frequently, management cannot promote someone at a particular time to a new position for various reason. The new position may not be funded at the time and the supervisor or middle manager needs to await budget approval to cover the prospective promotion. Or the management may require more time in order to make a decision because two or more candidates are in competition for the same position and a clear superior choice has not emerged. Additionally, the organization may be facing a change in senior management or in a policy direction that temporarily blocks promotions and other decisions down the chain of command. Still other reasons can delay an otherwise normal procedure.

Despite these conditions, the supervisor and middle manager have several options available to them so that they can provide greater recognition and challenge to their workers. Supportive managers endeavor to encourage subordinates, though the measures available fall short of a traditional promotion. The following are a number of ways in which supervisors can recognize and challenge their workers:

- **New responsibilities.** The supervisor or manager may provide new duties for persons who are exceeding requirements but for whom a promotion at the time is not possible. Sometimes, the assignment of duties to promising candidates can help identify the best performers in contention for the promotion.
- **Desirable perquisites.** If the individual cannot be promoted but deserves recognition, perquisites may be provided in way available to management. These include better scheduling, preferred postings and vacation time, and special training.
- **Special recognition.** Ego gratifying measures may be within the unit's capacity to fund easily and can satisfy the employee by recognizing his or her achievements. This can involve providing the security officer or investigator with a new title—for example, adding "senior" or "executive" to the title. Providing calling cards with the individual's name and position, and changing the person's uniform or office location can also serve as welcome perquisites.
- **New duties now; promotion later.** Pending the increased funding, the individual can effectively assume the new position without officially assuming the new job. When the budget line is approved, the position becomes available to the person to whom it was promised.

WHAT'S WRONG WITH PROMOTION?

Promotions work to advance the needs of the organization. However, several factors make promotions difficult decisions for management.

The Peter Principle

In 1969, Lawrence J. Peter and Raymond Hull introduced a facetious concept of occupational incompetence, named for the principal author. Their proposition stimulated debate almost immediately and has remained controversial since then. The Peter Principle states that in a hierarchy, every employee tends to rise to his or her level of incompetence, or the cream rises until it sours.[13] This notion is best considered a jocular characterization of management ascension proposing that people eventually reach a level they should not occupy. It observes that persons in positions of authority eventually reach their level of incompetence and are not likely to be promoted in the future.

The appeal of this notion is that it correctly reflects the experience of many managers who unflatteringly conclude that their supervisors have been promoted from a level of competence to one of incompetence. This is reflected in Peter's Corollary, which states that in time, every position tends to be occupied by an employee who is incompetent to carry out its duties. The assumption is that work is accomplished only because many persons have yet to attain their level of incompetence. The key to one's health and happiness on the job is not to accept the "Final Promotion." Peter suggests the key is the condition of "Creative Incompetence" in which the individual produces superior work while avoiding being promoted. But how can one ever be sure exactly what that level is?

The Peter Principle and its analogues have been resilient topics in management discussions because they reflect the fact that at least some promotions are failures. Most promotions require individuals to grow with their new positions and possibly make lifestyle changes, such as working different hours and physically relocating. These are stress-provoking circumstances and some people who are promoted later conclude that the promotion was not to their satisfaction. Despite the effort that management has expended to select and promote the right candidate, the process has failed. The best strategy is to permit the individual to resume his or her previous post as gracefully as possible. Surely, the individual has performed optimally in this role previously and can do so again in the future. Numerous security programs over the years have had such experiences and have made the necessary accommodations to people who chose self-demotion.

WHY PROMOTIONS ARE IMPORTANT

Apart from benefiting the individual who is promoted, promotions are good for the work unit and the larger organization in several ways:

1. **The skills are needed to achieve goals.** This is the most obvious reason for promotion: the job has been budgeted for good reason and the work it represents needs to be done.

2. **The person promoted may take the program in a new, desirable direction.** Organizations are constantly changing. Sometimes this is a result of internal forces, such as changes in management strategy; other times it is a result of external factors, such as modifications in the marketplace and advances in technology that affect how the job needs to be done. A newly promoted person may add value in translating these forces to the workplace relative to the previous manager.
3. **Promotions permit individual growth.** Employees at all levels need to feel as if they are growing, although not all employees want challenges beyond those presented in their current positions.
4. **Promotions reward good work.** Unlike one-time bonuses or lesser recognition of competence and promise, promotions are enduring and set a platform to encourage further good work.
5. **Promotions inform all employees that career advancement is possible.** When one person is promoted, other workers are encouraged that greater opportunity awaits them if they are qualified.

SUMMARY

Appraising workers' efforts at all levels is difficult. The resistance to appraisals takes many forms: dislike of judging others; fear of harming workplace comity; and concern that litigation will be filed from dissatisfied workers at some point. Yet the returns are worth the effort. Properly managed appraisal programs can document and correct unproductive behavior before it becomes unchangeable or causes harm. On the positive side, appraisals identify and encourage superior performance in the workplace and help develop talent. There are numerous ways of assessing and ranking employees, all of which must be measured against the needs of the workplace. Promotions help the organization reward those best able to maintain the desired standards of productivity and to bring fresh ideas into the organization in order to move it forward.

DISCUSSION AND REVIEW

1. Discuss specific reasons why personnel appraisals serve the interests of the organization.
2. How can top-down appraisals be designed to make judgments that are less likely to be biased by autocratic behavior?
3. How do appraisal guidelines differ for employees during the probationary period?

4. How can the halo effect be mitigated among managers who must assess subordinates?
5. Why should an evaluator look forward to an appraisal interview?
6. If work can be divided generally into three categories—technical/service, human relations, and conceptual—how do personnel appraisals change relative to these characteristics?
7. What accounts for the popularity and resiliency of MBO?
8. Why should organizations want to promote employees?

ENDNOTES

[1] L. Reibstein (June 13, 1998). "Firms Ask Workers to Rate Their Bosses," *Wall Street J.*, p. 15.

[2] J.L. Jordan and D.B. Nasis (June 1992). "Preference for Performance Appraisal Based on Method Used, Type of Rater, and Purpose of Evaluation." *Psychological Reports*, 70:963–69.

[3] S.J. Carroll and C.E. Schneier (1982). *Performance Appraisal and Review Systems.* Glenview, IL: Scott, Foresman and Company; E.R. Moulder (1995). *Performance Appraisal and Compensation Programs in Local Government.* Washington, DC: ICMA.

[4] D.P. Schwab et al. (1975). "Behaviorally Anchored Rating Scales: A Review of the Literature." *Personnel Psychology*, 28:549–62.

[5] H.J. Bernardin and R.W. Beatty (1984). *Performance Appraisal, Assessing Human Behavior at Work.* Kent Human Resources Management Series. Boston, MA: Kent Publishing Company.

[6] H.J. Bernardin, L. Eliott, and J.J. Carlyle (1980). "A Critical Assessment of Mixed Standard Rating Scales," *Proceedings of the Academy of Management*, pp. 308–12.

[7] R.M. Guion (1986). "Personnel Evaluation." In *Performance Assessment: Methods & Applications.* R.A. Berk, Ed. Baltimore, MD: Johns Hopkins University Press, p. 365.

[8] Effective Phrases for Performance Appraisals (1994). Perrysburg, OH: Neal Publications.

[9] P. Drucker (1954). *The Practice Side of Management.* New York, NY: John Wiley & Sons; D. McGregor (1960). *The Human Side of Management.* New York, NY: McGraw-Hill.

[10] D.D. McConkey (January 1973). "Applying MBO to Nonprofit Organization." *S.A.M. Advanced Management J.*, p. 12.

[11] J.M. Kondrasuk (1981). "Studies in MBO Effectiveness." *Academy of Management Review*, 6:419–30.

[12] H. Levinson (July–August, 1970). "Management by Whose Objectives?" *Harvard Business Review*, 48(4): 125–34.

[13] L.J. Peter and R. Hull (1969). *The Peter Principle.* New York, NY: William Morrow & Co.

Additional References

C.D. Fisher, L.F. Schoenfeldt, and J.B. Shaw (1993). *Human Resource Management*. 2nd Edition. Boston, MA: Houghton Mifflin.

G. Hoover, A. Campbell, and P.J. Spain (Eds.) (1994). *Hoover's Handbook Of American Business 1995*. Austin, TX: Reference Press.

P. Krass (Ed.) (1998). *The Book of Leadership Wisdom*. New York, NY: John Wiley & Sons, p. 375.

PART 2

SPECIAL ISSUES IN SECURITY MANAGEMENT

7

DISCIPLINE AND DISCHARGE

There is occasions and causes why and wherefore in all things.

—Shakespeare, *King Henry V*

The previous chapters discussed management's role in selecting, training, and supporting employees successfully. However, some employees do not meet the expectations of management, despite appropriate interventions by their supervisors. At such times, supervisors and middle managers must resort to more strenuous measures in order to promote satisfactory work performance.

When the performance of employees does not meet minimum expectations, and a few words by the supervisor have not achieved their objective, a disciplinary procedure may be considered. The word discipline is derived from the Latin *discipere,* meaning to grasp or comprehend, and from *discipulus*, meaning pupil. The term a "disciplined worker" may be used to refer to an employee who is reliable and completes required tasks. Yet the word in contemporary use often is equated with punish (from the Latin *punire*, denoting penalty and pain). However, in a workplace context, the two words should be regarded as distinct concepts. Supervisors use—or should use—discipline not to penalize subordinates, but rather to improve their behavior in order to meet objectives of the workplace. As Henri Fayol observed in his seminal book *General and Industrial Management*, poor discipline is the result of poor leadership. Good discipline occurs when workers and managers know and respect the rules governing activities in the organization.[1]

This chapter discusses why discipline is necessary, how operating programs use it, and what its pitfalls are. It also discusses the ultimate breakdown in the employee/employer relationship, namely discharge or removal from employment.

WHY SOME EMPLOYEES FAIL TO ACHIEVE DESIRED STANDARDS

In a well-planned and functioning workplace, most workers meet the minimally acceptable standards most of the time. But what of those who do not? It is useless for security managers to launch into a disciplinary mode before considering the reasons for poor performance. In fact, many reasons exist for why subordinates do not perform at a satisfactory level. These possibilities may be divided roughly into two categories according to their significance and credibility, as follows.

Explanations that may satisfactorily explain poor performance:

1. **A process critical to work malfunctioned.** For example, utilities and support mechanisms for security operations sometimes fail. If a worker monitors alarm signals and the computer crashes, making it impossible for the worker to respond to the alarm in a timely fashion, the alarm console operator cannot be held responsible for substandard performance.
2. **Contradictory orders are given by another supervisor.** The skills and job understanding of supervisors should be interchangeable. Yet if the primary supervisor sets the worker on a particular task, leaves the scene, and another supervisor preempts that original order, the worker should not be held accountable for not respecting the requests of the initial supervisor.
3. **The task requested is illegal.**
4. **The task required is immoral or unethical.**
5. **The request is unsafe or dangerous.** In such cases, the supervisor is responsible for the subordinate and should not have placed the worker in a position where safety is an issue.
6. **The worker has an acute health or personal problem.** Supervisors tend to be lenient when an otherwise well-performing worker has an acute health problem or a personal emergency. But the nature of much security work requires regularity and reliability, and frequent performance exceptions are disruptive to its goals.
7. **The worker does not have the capacity to do the job.** This suggests a failure in selection and training. While this is a possible explanation for poor performance or behavior, the situation is unlikely to occur in carefully managed operations. If it does occur, discharge of the employee is indicated. Another option is to assign the worker to a different type of position where he or she may have the capacity to perform satisfactorily.

*Explanations usually **not** satisfactory to explain poor performance:*

1. **Insufficient supplies or materials are available for the worker.** The absence of objects or substances routinely required at the worksite

usually should not be an excuse for substandard worker performance. Workers generally ought to be able to recognize a situation in which needed materials are low and when reordering them is appropriate. For example, security officers may be expected to complete incident reports in a timely fashion on a prescribed form. The lack of availability of such forms should not be used by a security officer as a reason for not completing a report. Security officers are expected to be flexible in such situations. In this case, the details of the report could be written on plain paper, if necessary, and attached to the correct form when it becomes available. When something critical is required for worker performance but is not available due to a failure by management, workers should not be penalized.

2. **The employee has been improperly trained.** The corrective in such cases is to reassess the training process, if the objection seems reasonable, and re-train the individual. However, training that was carefully planned, adequately taught, and certifiably completed by testing to assure comprehension works against this as an explanation for poor performance.
3. **A co-worker prevents the employee from completing a task.** Employees are responsible for their specified duties. Saying that a co-worker prevented the task from being completed is not a tenable excuse. Exceptions exist, for example, if one worker's behavior against another was harassing or flagrantly offensive and the worker could have notified management of this circumstance but did not.
4. **There is insufficient time to complete the task.** This explanation would not be valid if work is assigned with an accurate understanding of how much time is required to complete it. The time required is generally measured against the performance of established workers so that a reasonable work objective can be met.
5. **The worker does not like to do a particular task.**
6. **The employee dislikes the supervisor or vice versa.**

THE PSYCHOLOGICAL BASIS OF NON-COMPLIANCE

While some reasonable explanations for unsatisfactory workplace behavior are situational, others are psychological. That is, the worker has a conflict with the methods prescribed by the supervisor or the workplace itself. Unconsciously, the worker may resist authority seen in the embodiment of the supervisor. Failure to respond according to training and directions may be an act of rebellion that reflects deeper unresolved psychological conflicts on the worker's part. In extreme cases, the behavior may mask an adjustment disorder.[2]

Consider the situation of a new uniformed security officer. The individual has been provided with a complete set of clothing, carefully selected by the employer. During training sessions, emphasis on the use and care of the uniform are stressed. Other uniformed security officers seen as role models by the novice during training are properly dressed. Nonetheless, on occasion, security officers may not be dressed according to regulations. This failure to be properly dressed can be explained by the security officer as a situational exception, a lapse in judgment. This may be possible and excusable. However, a repeated pattern may be interpreted as a sign of resistance—even hostility—to managerial requirements.

The supervisor's role at such times is not one of psychoanalyst for the errant subordinate. Workers who do not meet the standards of quality and behavior generally achieved by their co-workers require discipline. To fail to do so makes the employer a conspirator to poor performance.

WHY SOME SUPERVISORS DO NOT DISCIPLINE WELL

A few tasks in the workplace are difficult for novice supervisors. The previous chapter mentioned appraisal interviews as a possible awkward task for supervisors. Providing adequate discipline for under-performing workers is another often onerous task for novice supervisors. But unlike appraisal interviews, which generally emphasize positive features of performance, disciplinary contact between the supervisor and subordinate is different.

The disciplinary process is awkward for both parties. Supervisors in particular may rationalize their inaction with regards to disciplining fellow workers (see Box 7.1). Yet like the appraisal interviews, if disciplinary measures are not taken, work performance could deteriorate. A lack of consistent disciplinary action by a supervisor in the workplace can be the basis of arbitrators' or civil court judgment against the employer. Poor performance without the presence of corrective action trains the worker that such behavior is permissible. The poor performance then is established as a fixed pattern.

Box. 7.1. Why Supervisors Fail to Discipline

A supervisor asks two workers to perform an unpleasant task, but one that is included in their job description. A few minutes later, the supervisor notes the workers taking a break for coffee. It is not their break time and the employees did not have permission to take a break. The supervisor observes the situation and leaves the room without making a comment to them. The workers never perform the requested task and nothing is ever mentioned again by the supervisor.

Box. 7.1. *(Continued)*

Why do some supervisors fail to address employees' refusal to perform an assigned task? Edward L. Harrison, a professor of management at the University of South Alabama, surveyed supervisors from several industrial organizations participating in management development seminars. Based on their responses from his questionnaire, Harrison came up with 14 main reasons why managers fail to discipline workers:

Percentage	Reason for Failure to Discipline
42.9%	The supervisor had failed to document earlier actions so that no record existed on which to base disciplinary action.
40.4%	The supervisor believed that he or she would receive little or no support from higher management for the disciplinary action.
29.2%	The supervisor was uncertain of the facts underlying the situation requiring disciplinary action.
20.9%	Failure by the supervisor to discipline employees in the past for a certain infraction caused the supervisor to forego current disciplinary action in order to appear consistent.
20.5%	The supervisor wanted to be seen as a "good guy."
19.5%	The employee involved was a close friend of the supervisor.
14.8%	Job demands and conditions made it inconvenient for the supervisor to discipline the employee.
13.9%	The supervisor was uncertain of provisions in the labor agreement pertaining to the situation involved.
13.5%	The supervisor provoked the employee infraction.
13.3%	The employee problem was one that should be dealt with through the employee assistance program rather than through a disciplinary penalty.
11.9%	The supervisor was concerned that a disciplinary penalty might result in a charge of racial or sexual discrimination.
10.9%	The supervisor did not want to draw negative attention to his or her own operation.
6.9%	The supervisor was reluctant to penalize the employee because the employee was a union officer.
5.9%	The supervisor did not want to spend time on the grievance that might result from a disciplinary penalty.

Source: E.L. Harrison (April 1985). "Why Supervisors Fail to Discipline." *Supervisory Management*, 30(4): 18–22.

In effect, the supervisor who fails to appropriately correct the worker hurts several parties. The deficient worker becomes complacent and may feel encouraged to flout other rules. Other workers observe the substandard co-worker and wonder why he or she is permitted to deviate from the organizational policy. The quality of their work, too, may decline. The organization then becomes affected by an insidious decline in quality performance. Desired production goals or qualities of service are not met as planned. Finally, the supervisor realizes that by failing to correct subordinates, his or her job is not being performed as it should be. All of this is moot if the supervisor identifies errors in performance in a timely fashion, corrects them discretely and with dignity for the worker, and restores the employee to desired productive levels or behavior.

HUMAN RELATIONS–ORIENTED MANAGERS

In the lore of the workplace, the bosses of yesteryear are remembered as Simon Lagrees or Ebineezer Scrooges. The characters are harsh, brutish, and cruel. Lagree remains unrepentant until the end of *Uncle Tom's Cabin*, though Scrooge experiences a personal epiphany resulting in a happy ending for *A Christmas Carol*. While the fictional portrait of past employers may have been stereotyped, it is beyond debate that the contemporary workplace is a kinder and gentler place for employees than in past generations, even concerning disciplinary matters.

Abusive bosses are no longer accepted in the workplace. Yet not all managers have the sensitivity of trained counselors. Why should they? Managers have strengths and weaknesses like everyone else. Still, the reality is that contemporary managers are better attuned to and more tolerant of moods and feelings of workers than in the past. This is due to increased education, workplace training, changing workplace ethos, a growing understanding of psychological dynamics, changing employment laws, and the risk of litigation or forced arbitration from complaints about supervisors from subordinates.

PROGRESSIVE DISCIPLINE TO SAVE WEAK WORKERS

Management expends a great deal of effort on recruiting, selecting, training, and supporting employees. Considerable investment in the employee has been made by the time he or she begins productive work. Further, the employer wants and needs workers to succeed. In such circumstances, the desirable goal is to turn any substandard behavior into acceptable or superior behavior with the least amount of stress and strain. Among some supervisors, the temptation to discharge an errant worker for solid reason often is strong and may be justifiable. However, management has a financial incentive in endeavoring to improve substandard performance in lieu of dismissal if at all practicable. Therefore, a deficiency in behavior should not lead directly to a disproportionate response on the part of the supervisor. The response must be balanced, impartial, and appropriate to the circumstance.

Corrective or progressive discipline is an increasingly used disciplinary procedure by unionized and non-unionized employers alike. In unionized workplaces, such actions often are mandatory and included in collective bargaining agreements. These agreements usually state that discipline should be corrective in nature, rather than punitive, but they do not say that discipline must start with a letter of warning and be increased after every subsequent behavior. Supervisors may initiate formal disciplinary action through the issuance of a letter of warning or suspension if an employee's actions do not improve after discussion. This process forces the manager who supervises the under-performing individual to make decisions and be held accountable. Resolution to problems is meant to be achieved expeditiously.

The stepwise or progressive disciplinary procedure, shown in Figure 7.1, is used in many formal organizations and is meant to enhance levels of trust, communication, and dispute resolution. The sequence of events is as follows:

1. **Infractions at work or poor performance.** First, a potential disciplinary offense or behavior must come to the attention of a supervisor. Typically, these include chronic absenteeism, not being present at a post without leave, the performing of an unsafe act, poor work performance, and failure to follow instructions. When confirmed, the supervisor evaluates the infraction further before confronting the errant employee.
2. **Supervisor investigates any significant predisposing factors.** The supervisor determines if any significant predisposing factors might explain why the worker committed the infraction or performed poorly. Assuming the worker was tardy or absent from work (poor performance), was someone sick at home? Or were there extenuating factors such as inclement weather? Despite being tardy, did the worker call the supervisor to warn of his or her late arrival or absence so that scheduling adjustments could be made? Has the offense occurred previously? With what frequency? After gathering all related facts, the supervisor is now ready to discuss the performance deficiency with the worker.
3. **Supervisor warns the worker.** In a calm, non-confrontational manner, the supervisor speaks briefly and quietly to the subordinate about the undesirable behavior. A supervisor may wish to employ a

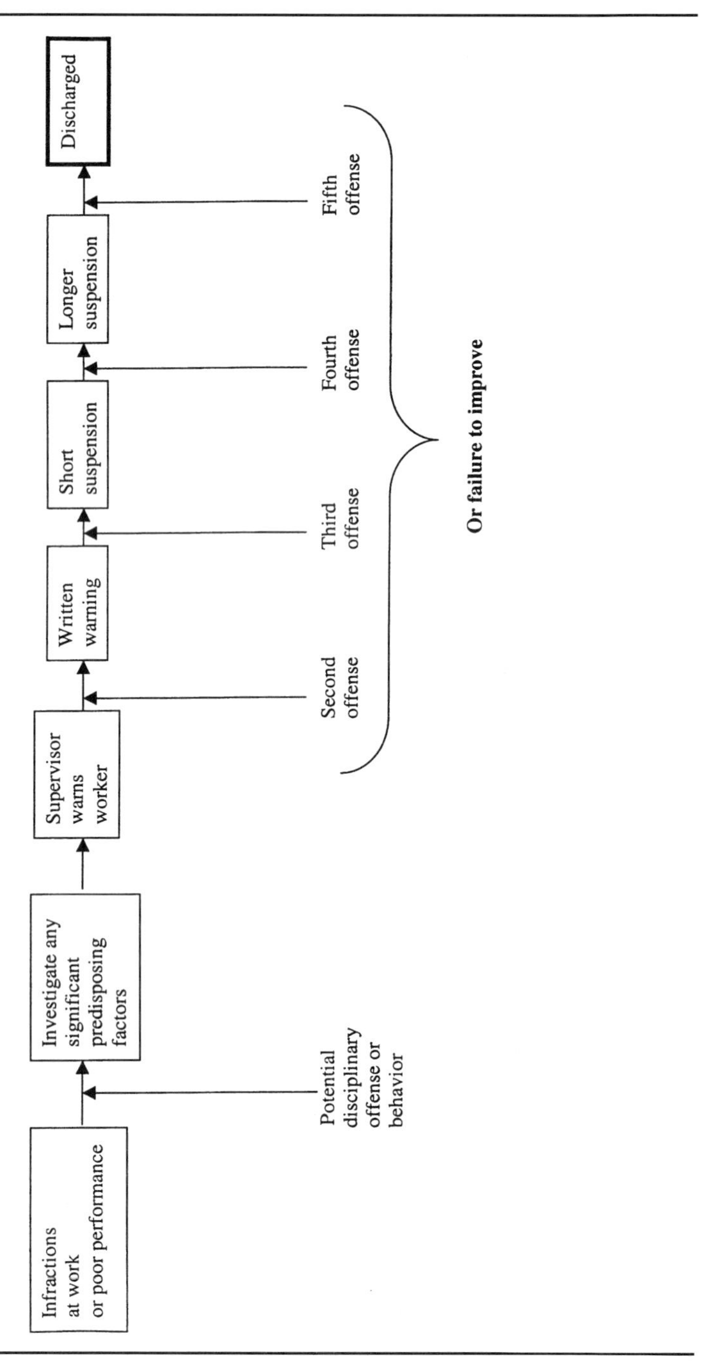

Figure 7.1. The Stepwise Disciplinary Procedure

strategy such as the one for correcting behavior discussed in Chapter 5; that is, a gentle, one minute reprimand. Subsequent to the encounter, many supervisors keep a workplace journal and note in it any significant worker disciplinary measures taken.

4. **The supervisor issues a written warning to the worker.** After an additional occurrence, the supervisor may "write up" the offense. The undesirable behavior and the response by the supervisor may be recorded in any one of several ways. The supervisor may send a brief, informal memorandum to the worker describing the offense. The supervisor also may write a longer, more detailed memorandum of the event, keeping one copy, forwarding a second to the worker, and providing the third for the employee's personnel file. The supervisor may also use a Disciplinary Action Report in which the complaint, the method of reducing the undesirable behavior, and the corrective action are described, as shown in Figure 7.2. If the workplace is unionized and disciplinary actions can be the source of a grievance, the form may be more detailed and composed of multiple parts: for the employee, the personnel department, the department head, and the union, as shown in Figure 7.3.

    - **The worker is given a short suspension.** Written notice of substandard behavior usually results in improved performance. However, this is not always the case. In such instances, a more substantive measure is called for: suspension. The suspension may vary according to the workplace and the nature of the unacceptable behavior. One large security program sends the employee home for the day the offense occurred without docking pay. The goal is to dramatically inform the worker that the employer is dissatisfied with the worker's behavior, but has faith that the employee's behavior can and will improve.
    - **The worker is given a longer suspension.** Should an additional offense occur, or should a pattern of documented poor work behavior continue, the supervisor may double the penalty of time suspended. The U.S. Postal Service, for example, usually suspends workers for seven days with loss of pay for the initial suspension. The duration of the penalty differs somewhat according to postal service zones and the severity of the offense.[3] Seven- and 14-day suspensions usually are without pay.
    - **The worker is discharged.** With an additional offense, the next step in the disciplinary process is to dismiss the employee.

Stepwise disciplinary procedures are appropriate for unacceptable but correctable behavior. In the event the employee committed a more serious offense, such as theft or violating serious work rules, the supervisor may consider preemptory discharge.

Disciplinary Action Report

Security Officer: ______________________ Date: ______________

Post: ______________________ Time: ______________

Violation:

____

____

____

____

Proper Procedure:

____

____

____

____

Corrective Action Taken:

____

____

____

____

Violation Reported By: ______________________ Date: ______________

Mgmt. Review By: ______________________ Date: ______________

Source: E.T. Guy, J.J. Merrigan, Jr., and J.A. Wanat (1981). *Forms for Safety and Security Management*. Boston, MA: Butterworth-Heinemann.

Figure 7.2. Disciplinary Action Report

WHY EMPLOYEES ARE DISCIPLINED

Infractions at work and poor performance represent the two leading causes for discipline that can lead to discharge. These two categories include many specific

Employee Misconduct Notice

To Personnel Department: **Date** ______________

Name of Employee ______________ Time No. ________ Dept. ________

The above-named employee has displayed the following misconduct, and has been warned that this misconduct will be entered on his Personnel Record.

MISCONDUCT (Check where applicable and specify details in section indicated below)

Smoking in Restricted Areas	☐	General Inefficiency	☐
Leaving Work Without Permission	☐	a) Quality	☐
Violation of Safety Rules or Dept. Rules	☐	b) Quantity	☐
Refusal To Carry Out Supervisor's Instructions	☐	c) Accuracy	☐
Irregular Attendance (Specify No. of absences to date)	☐	Discourtesy Toward Guest	☐
Frequent Tardiness	☐	Discourtesy Toward Fellow Employee (Mention other Employee)	☐
Violation of Eating Regulations	☐	Attitude	☐
Breakage	☐	Carelessness	☐
Poor Service	☐	Other	☐

Specify Misconduct in Detail ______________

Employee Comments ______________

Disciplinary Action Taken ______________

Signature of Supervisor

(Reprimand) (Layoff) (Other)

I acknowledge receipt of this notice

____

Signature of Employee

Original (White) to Employee
Duplicate (Blue) to Personnel Dept.
Triplicate (Pink) to Department Head
Quadruplicate (Yellow) to Union

Source: E.T. Guy, J.J. Merrigan, Jr., and J.A. Wanat (1981). *Forms for Safety and Security Management*. Boston, MA: Butterworth-Heinemann.

Figure 7.3. Employee Misconduct Notice

types of offenses. No published data are available on how security programs discipline and discharge workers. However, the U.S. Postal Service has codified unacceptable behavior into 26 categories of infractions, as shown in Table 7.1.

Table 7.1. U.S. Postal Service Disciplinary Infractions

Infractions	Number of Infractions	Percent of Total
Absenteeism	22,829	33.01
AWOL	10,488	15.17
Failure to follow instructions	8,601	12.44
Unsafe act or work habits	6,217	8.99
Poor work performance	5,010	7.25
Unauthorized absence from assignment	2,443	3.53
Tardiness	1,765	2.55
Delay or failure to deliver mail	1,510	2.18
Failure to protect funds/mail/property	1,170	1.69
Other	1,086	1.57
Disrespect to supervisor/customer	1,073	1.55
Scheme failure*	973	1.41
Insubordination	919	1.33
Altercation/assault/threat	857	1.24
Expansion of office time or street time	791	1.14
Failure to account for funds or accountables	757	1.09
Use/possession of intoxicants/drugs	566	0.82
Machine qualifications/proficiency failure	516	0.75
Falsification of record	363	0.52
Deviation from route	339	0.49
Pilfering/theft of mail or funds/property	294	0.43
Destruction/damage of mail or property	234	0.34
Crime (non-job-related off duty)	168	0.24
Crime (non-job-related on duty)	94	0.14
Falsification of application	56	0.08
Work slowdown/stoppage/strike	29	0.04
Totals	69,148	100.00

*Failure of the employee to demonstrate knowledge expected of the position.
Source: U.S. Postal Service Discipline Tracking System, 1987, Summary Report.

Formal disciplinary actions include letters of warning, suspensions for varying lengths, and removals. Of 69,000 disciplinary actions (representing about 9 percent of all employees), close to 60 percent of the disciplinary actions resulted in letters of warning, 30 percent were suspensions, and 10 percent resulted in discharge.

It is impossible to say whether the disciplinary pattern of the U.S. Postal Service parallels other programs. However, such infractions as absenteeism, absence from the post without leave, failure to follow instructions, the use of unsafe acts or work habits, and poor work performance are the six main reasons for disciplinary procedures in security programs.

LEGAL ISSUES FOR WRONGFUL DISCHARGE

The vast majority of employees who are discharged leave the workplace without taking further action against their employers. A few, however, will not leave quietly. They will make vigorous attempts to retain their jobs despite extensively documented records that led to the discharge. It is important for the supervisor to understand what the rights and obligations are of the employer at such a time.

At-Will Employment

For over a century, an employee who did not have an employment contract could not assume that his or her position was secure. In 1877, Horace G. Wood wrote *A Treatise on the Law of Master and Servant*, which observed that such an employee could be fired at any time for good cause, bad cause, or no cause at all. This influenced decisions in American courts and the "at-will" concept of employment was broadly adopted. Less than 20 percent of the U.S. workforce is covered by collective bargaining agreements. Under these conditions, an employer is often permitted to discharge a worker only for "good cause." By contrast, under the at-will-doctrine, employees may be discharged for whatever reason. (This policy can be superceded by state or federal statutory restrictions.)

Collective bargaining agreements began with the recognition of unions in the National Labor Relations Act of 1935. This law signaled that employees could bargain collectively with employers over a variety of issues, including discharge from work. The terms of such agreements were also binding for disciplinary procedures if specifically included in the agreement. Similarly, within government employment, Civil Service procedures came to govern how such workers could be subjected to disciplinary measures by supervisors.

Most for-profit and institutional employers have few, if any, written employment contracts. The exceptions are senior managers, research and development engineers and technicians with the care and custody of important proprietary information, and managers and others in creative and entertainment fields. The presence of such contracts may or may not specify the basis for dismissal by the employer. Written contracts without a minimum employment term specified are regarded as terminable at will by either party.[4]

In the security industry, contracts for personnel are uncommon, but do exist. One national security services firm requires all new security officers to sign an employment agreement.[5] Others have followed suit, usually for security workers who have had specialized recruiting and training at the contractor's expense. The agreements protect the relationship between the contractors and their assigned employees. Nonetheless, they recognize the duration of employment is a variant of an at-will relationship subject to termination by the employer for defined reasons.

Currently, three major other "exceptions" to the at-will doctrine exist: (1) breach of an express or implied promise, including representations made orally and in employee handbooks; (2) breach of the implied covenant of good faith and fair dealing; and (3) wrongful discharge in violation of public policy.[6] Supervisors concerned with possible legal problems related to discharge may wish to consider each of these issues in light of possible challenge to discharge. The following pages look at each of these three exceptions in depth.

1. **Breach of an express or implied promise, including representations made orally and in employee handbooks.** Express or implied oral representations made during pre-hire interviews and at the time job offers are being made and accepted can be considered oral contracts. Such contracts may be recognized by the court. Particular protection of the employment status occurs during times when the employer is contemplating furloughing or dismissing workers. However, discharges for violating the rules and regulations of the employer are not likely to be protected by such an agreement. The courts tend to recognize the language in employees' manuals to reflect a unilateral offer that is accepted by the employee. Once the employment has begun, "the policies embodied in those pronouncements become legally binding."[7] The basis on which employees may be subject to disciplinary actions, including dismissal, may be contained in a Personal Conduct Policy, as shown in Table 7.2. Such a policy is likely to change with the times, as employers do not wish to be burdened with a fixed policy that cannot be flexible to unforeseen circumstances. Therefore, when new measures are added to the policy, they should be communicated to all employees and added to the handbook.
2. **Breach of the implied covenant of good faith and fair dealing.** Under certain circumstances, dismissal of an employee is rendered more difficult. As noted earlier, violation of terms in the employee's

Table 7.2. Sample Personal Conduct Policy

The employee's manual acts like a unilateral contract with the worker. Conditions for personal conduct can change over time. The following are the rules and regulations for employees, including protective staff, issued by a hotel chain:

A. *Violations involving any of the following provide grounds for discipline up to and including termination*:

1. Supplying false or misleading information when applying for employment, or any time during employment.
2. Altering or falsifying hotel records, swiping the time clock for another employee, or having another employee swipe your time card, or other manipulation of attendance records.
3. Possessing, using, or being under the influence of illegal drugs or alcoholic beverages while on duty or in uniform.
4. Possessing weapons on hotel premises or while off hotel premises in the performance of hotel duties.
5. Abusing, defacing, or destroying hotel property or the property of guests or other employees.
6. Engaging in any act of violence or disorderly conduct, threatening or using abusive language or rudeness to a guest, supervisor, or co-worker.
7. Failure or refusal to follow safety or health rules and regulations or failure to report an accident which results in injury to any person.
8. Gross negligence, carelessness, or misconduct.
9. The conducting of non-company business, such as canvassing, collection of funds, pledges, circulation of petitions, solicitation of memberships, or any other similar type of activity during the working time of the employee doing the soliciting or being solicited.
10. Theft or unlawful possession of hotel property or the property of a guest, supervisor, or another employee, including lost and found items.
11. Immoral or indecent conduct or soliciting persons for immoral reasons.
12. Refusal or failure to perform assigned work, sub-standard guest relations, refusal or failure to follow a supervisor's instructions, or any act of insubordination.
13. Excessive absenteeism and/or tardiness. Failure to report to work on three consecutive workdays without proper notification will be interpreted as voluntary resignation and will result in immediate termination.
14. Unsatisfactory work performance.
15. Gambling on hotel premises.
16. Sleeping while on duty.

(continues)

Table 7.2. *(Continued)*

17. Working overtime without prior approval from a supervisor.
18. Unauthorized use of hotel facilities, including telephones and computers.

B. *In addition, violations involving any of the following acts will be considered just cause for remedial action, which may involve oral or written reprimand, suspension from work without pay, or dismissal (especially in the event of repeated violations).*

1. Failure to punch or sign in and out as instructed by your supervisor.
2. Consuming food or beverages in areas other than those designated for use during breaks. Gum chewing while on duty. Smoking by employees is prohibited in all areas of the hotel.
3. Failure to maintain a high degree of personal cleanliness at all times. Failure to wear prescribed clothing in good repair as well as appropriate company identification.
4. Failure to perform work assignments satisfactorily.
5. Failure to notify a supervisor prior to the start of the shift of tardiness or absence.
6. Being present on hotel property more than 15 minutes prior to the start of a shift or remaining more than 15 minutes after the end of the shift without express permission from a department head.

Employers with progressive disciplinary programs may wish to retain the right to terminate employees for particular conduct without the need for prior discipline. This policy should be stated expressly in the handbook.[8] Due to concerns for workplace security, an increasing number of employers state that workers who act violently, threaten others, or who without permission carry a firearm onto the employer's property are subject to immediate termination.

---

manual can result in termination for just cause. This is a circumstance in which the employer states that the employee will not be dismissed except on defined grounds. Just cause can be incorporated into the language of a written contract, oral agreement, and employee handbook or manual. In such a context, termination of employment for misconduct is only possible for *substantial* breaches.[9] A minor neglect of duty, an excusable absence, and a minor misrepresentation or rudeness have been ruled by the courts as not meeting the standard for just cause.

3. **Wrongful discharge in violation of public policy.** A violation of public policy exists when the employee is threatened with dismissal while legally protected for such behavior. A dismissal at such a time could be construed by the courts to be a "mixed-motive termination"

if the termination of the employee was motivated by lawful reasons (just causes) and also by unlawful reasons (violations of public policy). In such cases, the termination often is considered a violation of employee rights depending on the seriousness of these just causes. Regardless of such concerns, "an employer may discharge an employee if he has a separate, plausible, and legitimate reason for doing so."[10] At such times, the employer may have to demonstrate that it was not acting discriminatorily in derogation of a statute. This is discussed in greater detail later in the chapter.

SPECIAL DEFENSES AGAINST DISCHARGE

This chapter has focused upon just causes for which employment may be terminated. Apart for these reasons, it is important to identify areas in which federal statutes protect workers' rights in the event of discharge. These issues often are significant in security programs, which investigate such charges and may be involved at the moment of separation from employment.

Generally, employees who are being dismissed from their jobs for refusing to perform unlawful acts, exercising their rights under state law (such as pursuing workers' compensation benefits), and performing a civic duty (such as jury duty) are protected.[11] Another public policy exception to separation is when an employee is terminated for reporting illegal activity on the part of the employer. So-called whistleblower laws exist in about 40 states, but are not uniform. For example, in about half the states, only public employees are protected by these wrongful discharge acts. Public *and* private employees are protected in most of the remaining states. These acts protect employees from retaliation for reporting violations by the employer of laws or regulations, neglect of duty, and endangerment of public health and safety. Generally, such acts do *not* protect from discharge an employee who reports alleged violations of company policy, waste, or mismanagement.[12]

At the time of learning of an employee's allegedly justified complaint against the employer, security practitioners and Human Resource managers generally must respond to such assertions sympathetically and non-judgmentally. Employees are less likely to call authorities to complain of suspected violations of public policy at their workplaces if they feel that management has the internal means for hearing their complaints and achieving positive change, assuming the complaint is valid.[13] Employees who believe that the workplace is violating public policy may not only have the moral argument, but they could also pursue financial incentives if the organization subsequently were fined or paid damages for breaking the law (see Box 7.2). At such times, representatives of management should express their organization's commitment to uphold existing laws and then investi-

Box 7.2. Whistleblower Suits Can Pay Off

Numerous incentives exist for employees to expose dishonest practices committed by businesses against the U.S. government. In some cases, the alleged offender can be a business that derives revenues from activities reimbursed by government. In one example, an accountant for a small hospital turned a wrongful termination suit into a major federal action against giant healthcare management providers. James F. Alderson was the accountant for North Valley Hospital in Whitefish, Montana. He had managed the hospital's affairs for six years when the board decided to retain Quorum Health Group to manage the hospital operations, including accounting and financing. In selecting Quorum to manage the hospital, the board was impressed with the firm's cost-reporting system. Two months after Quorum took over, Alderson was asked by a Quorum manager if the hospital kept two financial reports. What Quorum meant, it turned out, was that it prepared one report of actual costs for internal use and a second aggressive cost report that would be submitted to the government for Medicaid or Medicare charges. In the event the aggressive claim survived a two-year audit period, the hospital could book the reserves as revenue. Alderson believed this process to be fraudulent. Soon after he was fired. He proceeded to institute a wrongful termination suit. In the discovery process, he collected evidence showing that the healthcare company systematically misrepresented expenses to the government, illegally increasing its reimbursement.

Alderson filed a lawsuit known as a qui tam, or false claims case, on behalf of the federal government. He notified the U.S. attorney general of this intention, filed the complaint under seal, and then sought to obtain support from the Department of Justice in his suit against Quorum and its previous parent, Columbia/HCA Healthcare Corporation. The process required over five years of effort by Alderson before the government lifted the seal and officially joined the case. In prosecuting the action, Alderson exhausted his personal funds. He became involved, however, with a law firm having previous successful experience with qui tam cases. When cases of this sort are settled in the plaintiff's favor, the share could amount to between 10 and 25 percent of any recovery, an amount which could total many millions.

Source: K. Eichenwald (October 18, 1998). "He Blew the Whistle, and Health Giants Quaked." *New York Times*, Sec. 3, p. 1.

gate the complaint. Whistleblowing actions are generally sustainable by employees and others who can demonstrate *all* of the following elements in their action[14]:

- The employee has particular expertise regarding the alleged violation of the law.
- The employee's charges relate to federal or state law or regulations clearly applicable to the employer, or to the code of professional ethics recognized as equivalent to state law and which code clearly is applicable to the complainant.
- A nexus exists between the complainant and the violation.
- The employee is not a high-ranking executive or manager who otherwise owes a special degree of loyalty to the company.
- The employee has a valid complaint.

LEGAL CASES OF PROPER AND IMPROPER DISCHARGES

Litigation helps mold the ways in which employers evaluate circumstances before discharging employees. Appellate level cases are instructive because they emerge from lower court issues and have been heard by the court of appeals or the supreme court within a state. They represent carefully considered arguments, though often on narrowly selected issues. Any decision by an appeals court becomes the law for similar situations within that state only. However, appellate decisions in one state often are broadly cited and affect policies in others. The following are a number of significant appellate decisions relating to the discharge process:

- **Discharge upheld after drug test dismissal was challenged.** A national hotel chain instituted a drug and alcohol testing program. All employees signed a consent and release form regarding their participation. Some time later, an at-will employee was randomly selected to be tested. According to the analysis, she was positive for an illegal substance. The employee denied drug use and asked to be re-tested at a different laboratory of her choosing. The employer refused, but offered the employee an opportunity to be re-tested at the original laboratory used by the chain. The employee refused this offer and was discharged. She subsequently sued the hotel and her supervisor for wrongful discharge in violation of public policy and other charges. The state supreme court rejected the appeal, stating that the plaintiff could not show that her employer violated a clear mandate of public policy in the discharge by insisting upon its own drug testing laboratory. (*Stein* v. *Davidson Hotel Co.*, 945 S.W.2d714 [Tenn. 1997][15])
- **Employee mistakenly fired over theft can sue employer.** A housekeeper in a hotel was arrested for theft for allegedly stealing items from a guest room. He spent six weeks in jail until his employer notified police that the missing items had been found. A Louisiana

appellate court ruled that the exclusive remedy provisions of the state's workers' compensation law did not prevent the housekeeper's suit for damages against his former employer. (*McGowan* v. *Warwick Corp.*, 691 So.2d 265 [LA. Ct. App. 1997][16])

- **Employee may be fired for violation of employee handbook.** A manager of May Department Stores Company, who had been employed there for 13 years, coordinated jewelry promotions. While visiting a store, she removed a large box of gold merchandise and took it across the store to lock it in a secured location. In the process, she left two other boxes unguarded. The store's jewelry manager was nearby at the time, but the promotion coordinator did not ask him to keep an eye on the exposed jewelry. When she returned after a delay, the two remaining boxes were missing. The next day she was called to the personnel office and was fired. Later, she sued for breach of an implied promise of "good faith and fair dealing." At trial, May provided a section from their employee handbook which stated:

  > Merchandise valued at $60.00 or more must be housed in showcases with locked doors. All other merchandise, including all 14K jewelry, must be kept in locked showcases or locked drawers. All jewelry showcases must remain locked at all times, unless a sales associate is attending a customer.

  The employee handbook also stated that failure to follow operating procedures can result in "corrective action and possible termination depending on the seriousness of the violation." The theft resulted in a loss of over $50,000. At trial, the plaintiff provided no evidence to suggest that she was being fired for "capricious," "unrelated to business needs or goals," or for "pre-textual" reasons. The court sided with May. (*Moore* v. *May Department Stores Company*, No. B043481, Ct. of Appeals of CA, 2nd dist., Div. 2, decided July 31, 1990, 271 Cal. Rptr. 841[17])
- **Fired guard has no claim when employee handbook gives employer discharge power.** At Peninsula Regional Medical Center, Maryland, three employees were attempting to secure a patient with leather straps. In the process, the patient bit a security guard on the wrist. The guard then struck the patient on top of his head. He was asked to leave the room and another security officer took over. In an investigation conducted by the security director and personnel director, written statements from eyewitnesses stated that the guard had struck the patient 15 to 30 seconds after he was bitten. The guard responded that his blow had been reflexive, in an attempt to prevent the patient from biting him again.

The guard was discharged for "committing an unnecessary act of putative retaliation." The guard sued for breach of contract, wrongful discharge, intentional interference with prospective relations, and other claims. The trial court granted summary judgment to the defendant. The plaintiff appealed, and the Maryland Court of Special Appeals sustained the discharge stating that the termination did not violate public policy. The special appeals court said that the plaintiff also failed to show that he acted in self-defense.

The plaintiff also charged that two different police departments declined to employ him after learning why he had been fired from the medical center. The court of special appeals rejected this claim noting that the plaintiff had signed a consent form provided by those departments that also released from liability those employers, such as the hospital, that provided information to them. (*Bagwell* v. *Peninsula Regional Medical* 655 A.2d 297 [Md. Ct. Spec. App. 1995][18])

- **Supervisor's comments about fired employee not considered defamatory.** An employee was dismissed from a convenience store and gas station after her supervisor was not satisfied with the employee's explanation why scratch lottery tickets repeatedly were missing during her work shift. The employee denied any wrongdoing and sued her ex-employer for defamation when other employees, including her husband, who was a part-time worker there, learned of the reason for her dismissal. The state Department of Labor Unemployment Division later concluded that she did not steal the tickets. The issue about whether she may have mismanaged ticket security remained open.

  The South Dakota Supreme Court found no evidence of malice on the supervisor's part in revealing to other employees that she thought that the former employee had stolen the tickets. In a split decision, the court's majority held that the employer, through the supervisor, did not make defamatory comments about the worker in responding to other employees' queries about the departure of their former co-worker. (*Petersen* v. *Dacy,* 550 N.W. 2d 91 ([S.D. 1996][19])

INSURANCE AGAINST WRONGFUL TERMINATION

In the event an employee institutes an action against his or her former employer for wrongful termination, the employer may expect its general liability insurance policy to help fund the defense. Such broadly written insurance coverage is likely to cover legal defense costs, related pre-trial expenses, and judgment against the defendant, if any. However, insurers may seek to deny coverage of wrongful termination claims in a standard form by claiming that the event was a non-occurrence.[20]

Standard form general liability policies usually pay for occurrence-type losses. In commercial general liability coverage, an occurrence is called an "accident," and includes continuous or repeated exposure to the same general harmful conditions. For the wrongful termination claim to be denied because it was not such an accident, the insurer must prove that the policyholder had a subjective intent to harm or injure the fired employee. Another related basis of non-coverage is that the policyholder "willfully" terminated the employee with a pre-conceived design to inflict injury. If the standard policy plainly and clearly excludes wrongful termination from coverage, the insured would have to look elsewhere for expenses related to the litigation. In conclusion, commercial general liability coverage is likely to support the insured's defense in wrongful termination claims. This assumes the insurer does not have a basis for proving non-occurrence.

PROCEDURES AT THE TIME OF DISMISSAL

Organizations differ widely on how discharges should occur. The methods of terminating unsatisfactory employees also vary widely among nations and cultures (see Box 7.3). Even in North American workplaces, strategies for dealing with employees who are being dismissed from their positions are diverse. The conditions can change according to the cause and seriousness of the dismissal, the rank of the employees involved, and the particular industry.

Box 7.3. How to Dismiss an Employee Japanese-Style

Terminating a worker's employment differs considerably depending on culture and laws. In Japan, for example, the concept of lifetime employment remains a goal of major employers, yet is presently crumbling as a workplace tradition. Generally, Japanese employers offer a severance package to individuals whom the company does not wish to retain. But what if the employee refuses? That's what Toshiyuki Sakai decided to do when he rejected a severance package of 2.6 million yen ($23,900) from his employer, the video-game maker Sega Enterprises Ltd. Sakai was told that his work was under par. He disagreed with that judgment. When he refused to resign and accept the severance package, he was transferred to the "Pasona Room." This room, named for the English word "personnel," was empty except for a desk, three chairs, a bare locker, and a telephone that received only incoming calls. Mr. Sakai was given no duties and had no personal possessions in the room. He was instructed in writing to report to the room from precisely 8:30 a.m. to 5:15 p.m. He was allotted 55 minutes for lunch.

Box 7.3. *(Continued)*

After two months, the personnel department formally recommended that he resign, and offered him a severance package 9 percent lower than the original one. About three weeks later, he heard through the union that Sega was firing him and would offer him a severance package that was 28 percent more than the original package he was offered. However, he continued to report each day to the Pasona Room. A month later, he filed suit against Sega, seeking to have his old job and salary restored. Seven days later, upon showing up for work, he was stopped by a security guard, who refused to allow him to enter. Sakai thus pursued his suit against his former employer from home. Meanwhile, Sega announced a plan to trim its workforce by one-quarter. Within a few weeks, most accepted Sega's severance package. "Everyone's afraid that they might be the next to be thrown into solitary confinement," commented Sakai. The worker elected to remain in the Pasona Room to protect his job opportunities at Sega or elsewhere.

Source: P. Landers (September 14, 1999). "Refusing to Move On." *Wall Street J.*, p. A1.

In most situations, dismissals are defined as permanent separations from the current employer. In many other occurrences, however, workers are placed on furlough—that is, temporary status without duty and pay. Furloughs occur when there is insufficient work or when other non-disciplinary reasons arise. Regardless of the basis for such action, security and Human Resources managers should anticipate the ways in which redundancy—temporary or permanent—might produce unwanted problems for management. The grounds for dismissal or furlough can be categorized as follows:

- **Economic downturn; retrenchment.** Organizations grow with actual success or the anticipation of it. Also, they can contract due to declining sales, decreased funding or support, recurrent financial losses, mergers that consolidate operations, uninsured or under-insured disasters, and other reasons. At such times, management often elects one of the most expedient means of reducing costs: cutting personnel. When economic factors cause staff reductions, management might respond to the situation in a variety of ways. Some employers allow workers to remain on the premises using their former offices and resources for a reasonable period of time as the basis of obtaining future employment. This lenient policy for the worker provides a humane way of helping the employee segue to the next opportunity.

At the time of redundancy, however, most employers would prefer that such workers leave the premises. This could mean after one week, two weeks, a longer time, or immediately. In such cases, employees are expected to leave soon after being informed of their termination by Human Resources or their supervisors. (Workers who resign from significant positions in organizations with high-value intellectual assets may be expected to leave their place of work shortly after giving notice.)

Security-minded employers worry about the potential of sabotage from discharged workers who remain on the premises prior to or after official termination. Common sense dictates that such individuals be treated with respect and dignity as they move their job search activities off premises. Fearful that such employees may harm company property, remove valuable assets, or create a confrontational situation, some employers expect security personnel to play coordinated and visible roles at the time of such dismissal. The use of security personnel to escort workers from their exit interview to their office to pick up personal effects and then, possibly, to the exit itself needs to be considered carefully before implementation. The practice can affect remaining employees negatively if no basis exists for treating the worker like a suspicious person. At such times, a worker being dismissed is likely to be angry and capable of irrational action. Security should be alert and responsive when dismissed workers are informed of the employer's decision, as attitude on the part of security matters.

For managers and executives it is increasingly common that outplacement services are provided. These services may include office space and resources, résumé services, and job counseling assistance.

- **Poor work and misconduct.** In cases of just cause for termination, it is normal for a security officer to escort the employee discretely to the exit. The security officer also obtains the employee's keys, identification, and any other materials belonging to the employer if they have not previously been collected. In the event that the materials are not obtained, the employee's final paycheck is usually held until the company's property is returned. The possibility of violent action from a discharged and disgruntled worker has emerged as a concern in recent years and is discussed later in the chapter.

THE EXIT INTERVIEW

The main purpose of an exit interview is to provide terminated employees with information on accrued wages and benefits, such as vacation pay.

Related matters like insurance coverage and pension options also need to be discussed. Terms of the employer's healthcare insurance coverage may be reviewed with options laid out for the employee. Such valuable information and assistance helps mitigate the shock and loss many newly unemployed persons feel.

The exit interviewer uses this opportunity to obtain opinions of company operations and management. This can also be a time to defuse possible hostility and correct employee misconceptions about the termination process.[21] The possibility of a lawsuit and potential claims may be ascertained and noted by the interviewer. Any angry sentiments or threats expressed by the employee should be received by the interviewer calmly. The contents of the unhappy worker's remarks should be recorded immediately following the interview. Such threats should be discussed with a security manager immediately after the exit interview.

Some issues covered in an exit interview are relevant to security. The employee is asked to return all property of the employer at the time of the interview. Thus, the employer's representative in Human Relations or security should be aware of the assets the worker possesses or otherwise has available. If some assets cannot be collected by management at the exit meeting, arrangements should be made for their prompt return. Experience shows that if the employer does not obtain all property under the control of the departing worker before the last paycheck is delivered, recovery of such property becomes problematic. This generalization applies to all employees who possess assets of the employer. For example, if a technical developer was permitted to work at home with a computer owned by the organization, a security manager might send a tactful officer—with the employee's knowledge and permission—to pick it up.

Unionized employees may have the right to a union representative at the exit interview. However, if the interview concerns previously discussed disciplinary or discharge matters, union representation is not required.[22]

Following the exit interview, most organizations promptly delete the employee's access code for physical entrance to the workplace. The former worker also is blocked from the Local Area Network (LAN) of the workplace, and telephone privileges cease. Security personnel at entrances and exits are informed promptly of workers who are discharged or quit.

DISMISSAL AND THE DISGRUNTLED EMPLOYEE

An employee of Pacific Southwest Airlines, a unit of USAir, was fired for stealing $69. His resentment mounted. In December 1987, he purchased a one-way ticket on a commuter trip from San Francisco to Los Angeles. He evaded pre-flight security controls in entering the plane. At 22,000 feet, he entered the cockpit and shot the pilot dead. The four-engine plane crashed, killing 43 people aboard, including the gunman and his former employer, who was a passenger.[23] Although this type of incident is exceptional, incidents of violence by terminated and disgruntled employees deserve attention in protection management programs.

The issue of workplace violence cannot be ignored by high-performance security practitioners or Human Resources managers. Incidents involving disgruntled employees who act violently are far less common, but should be considered seriously by security and Human Resources managers. Data on the frequency of such incidents is unavailable. However, verified cases appear in the media with frequency, keeping the issue alive. Table 7.3 discusses such incidents in greater detail.

Michael D. Kelleher, author of *New Arenas for Violence: Homicide in the American Workplace*, states: "The act of terminating an employee can be a dangerous undertaking, even after the actual termination itself has taken place . . ."[24] Predictions of future behavior can never be certain. The employee who seems calm at the time of dismissal may harbor resentments that build to a quiet fury over time and that may eventually trigger violent behavior. The following are guidelines for terminating an employee:[25]

1. The employee must be treated with respect, sensitivity, and dignity throughout the termination process.
2. If the termination involves a performance issue, the organization must ensure that performance standards are applied to all employees, without exception.
3. The timing of the termination process is critical. Most employers endeavor to avoid terminating the employee when he or she is undergoing stressful life situations, such as a divorce, illness, or the recent death of a close friend or family member.
4. Two members of management should always be present at the termination meeting, one of whom should be a security or Human Resources professional. This is particularly important if the departing employee is known to have a history of aggressive or violent behavior.
5. Expect the terminated employee to react emotionally. Try to understand the shock and pain of the process from the employee's point of view. Regardless of the emotional nature of the meeting, remain objective and calm. Try to keep the meeting focused on the issue at hand, always using a dignified, sensitive approach.
6. Act professionally in the termination meeting. Confine conversation about the termination to the business reasons motivating the organization's decision. Ensure that the employee understands what is happening and why it is happening. Do not assess blame or react in a judgmental manner.

Table 7.3. The Violent Disgruntled Worker

About two million individuals are victims of violent crime each year in the workplace. About 75 percent of these incidents are simple assaults, while another 20 percent are aggravated assaults. Incidents of violence towards supervisors, managers, co-workers, and others from disgruntled employees are few and are therefore not included in some workplace victimization studies.[1] Nevertheless, highly publicized examples of disgruntled and revengeful employees killing or injuring former associates and innocent bystanders at the workplace have received national attention. Theses incidents remind security practitioners and human resources officials that complacency about the risks can be dangerous. The following are some examples of such violent workplace incidents:

- An accountant for the Connecticut lottery at the state headquarters in Newington, Connecticut, failed at his attempts to be promoted and subsequently filed a grievance. Returning eight days early from a leave of absence for stress-related problems, he walked into the executive offices and stabbed one official, shot dead two others, and then chased the president into the parking lot and fatally injured him. The gunman then killed himself.[2]
- In Riverside, California, a former parks and recreation department worker was fired after working five years as a part-time chess coach. He instituted a wrongful termination suit for age discrimination and other causes. Four years later, before his claim had been heard, the ex-employee, now working for the postal service, invaded city hall and shot two city council members and two police officers there.[3]
- In Tampa, Florida, a worker for Fireman's Fund Insurance Company's local office returned eight months after being fired. He roamed the office building shouting: "This is what you get for firing me!" He killed three managers with his former firm and injured two more before killing himself later in the day.[4]
- In Walpole, New Hampshire, the former police chief shot to death the selectman who had forced him to resign and then killed himself.[5]
- A kitchen worker in the Denver, Colorado, suburb of Aurora returned a week after being dismissed from the Chuck E Cheese restaurant and killed four workers, including the night manager. A police investigator remarked that it appeared the gunman had "held a grudge over his firing."[6]

Sources:

[1] For Example: Violent Crime Strikes 2 Million People in the American Workplace Each Year" (1998). *Workplace Violence Report*, p. 1.

[2] J. Rabinovitz (March 7, 1998). "Connecticut Lottery Worker Kills 4 Bosses, Then Himself." *New York Times*, p. A1.

[3] D. Terry (October 15, 1998). "6 at City Hall are Shot; Ex-Worker is Accused." *New York Times*, p. A18.

[4] "Fired Worker Kills 3, Self in Fla. Bloodbath" (January 28, 1993). *New York Post.*

[5] Murder-Suicide Cited in Town Hall Shooting" (February 14, 1994). *New York Times*, p. A13.

[6] "Gunman Kills 4 Workers at Colorado Restaurant" (December 16, 1993). *New York Times*, p. A18.

7. Be honest with the employee. Ensure him or her that the matter will be handled in a confidential manner. Provide straightforward answers to questions important to the employee.
8. If a reason exists to suspect a violent reaction from the employee, be sure to have security personnel present at the termination meeting and in the presence of the employee when he or she leaves the premises.
9. Be prepared for the meeting. Have all documents ready for presentation to the employee. Have all benefit information ready for review and immediate delivery to the employee. Ensure that arrangements have been made for the employee to gather personal belongings and return company property after the meeting. Prepare and rehearse the meeting in advance so that all points important to the employee are covered.
10. Take any follow-up action necessary to ensure the continued security of the workplace (involving keys, password, and so on) after the departure of the employee.
11. Ensure that an effective outplacement program is available to the employee. A strong outplacement program often makes a significant difference in the transition process.
12. Ensure that the physical departure of the employee from the workplace is handled with dignity and in a confidential manner. No possibility of embarrassment or undue stress should exist in the departure process.

Such situations are difficult for management and employees alike. Preparation prior to the meeting can be the critical factor in short- and long-term success in managing the interaction. Further aspects of the process that may be emphasized include:

- **Timing.** According to Sandra L. Heskett, termination should be conducted late in the business day.[26] Many employers choose to plan such meetings so that they end after most of the employees have left for the day. Fridays are often, but not invariably, the day of choice to break the bad news, since they give the worker the weekend to recover.
- **Surveillance and investigation.** In the event of threats from an employee, the use of covert and overt surveillance may be desirable. Richard B. Cole writes:

  > This process requires the striking of a fine balance in recognizing that this individual has previously exhibited irrational and endangering behavior directed against the corporation or its employee(s), the absence of formal authority to remove the individual from the

> opportunity to further endanger, and the obligation to protect the employee and the workplace.[27]

Most protective functions do not have the capability of conducting such surveillance; therefore, an outside competent service may be retained. An overt surveillance team approaches the offending individual, advising him or her that they are present, what they intend to do, and how they intend to do it. They make it clear that such surveillance is believed to be allowed within the spirit of the law.

T.I.M.E. IS NOT ON YOUR SIDE

Consultant Gavin de Becker has described what he calls the T.I.M.E. syndrome, which occurs when management allows a growing situation to include Threats, Intimidation, Manipulations, and Escalation. De Becker comments: "When dealing with a difficult and violently inclined employee, T.I.M.E. is on his side, unless management acts quickly."[28]

This section began with some facts about a Pacific Southwest Airlines employee who was fired. Specifically, the reason was for taking $69 of the airline's bar cash. The employee's history was more convoluted. The worker had been with the airlines for 12 years. During this time, he was a thief, a drug user, and a drug dealer. He had been warned by his supervisor previously to shape up or face the consequences. Could this person have been "screened-out" at the time he was considered for hiring? The question is impossible to answer for certain. But couldn't an employee of this sort be disciplined and terminated much earlier?

Dismissing workers early is easier, with less emotional investment on the part of all involved, and with less perceived "unfairness" on the part of a supervisor or a manager. De Becker explains that workers often feel shocked and sense that they have been treated unfairly when facing dismissal. Simply put, managers who are reluctant to discipline or to terminate abusive employees are not astute. Joseph A. Kenney writes: "Employees who get away with rules violations often will push their luck in the future."[29] The courage to fire some employees early may prevent the supervisor and organization from remorse later.

SUMMARY

Fortunately, not all employees require formal discipline. However, some do, and the supervisor's task is to lead the worker into better behavior. To discipline effectively requires planning and awareness of the facts and options involved. Logical but inexcusable reasons for unacceptable behavior should not allow an oral admonition to get off track.

Unfortunately, some supervisors do not discipline, or do so ineffectively. They hurt themselves, the worker involved, and the entire organization by such recalcitrance or ineptitude. Supervisors can, however, learn progressive disciplinary measures to increase their effectiveness. Dismissals rarely lead to violence, yet the possibility cannot be ruled out. Certain precautionary measures can decrease the possibility of such violence occurring.

DISCUSSION AND REVIEW

1. In your opinion, why do some workers not achieve the minimally acceptable standards most of the time? What role does management have in dealing with this? What are the limitations?
2. Why should a supervisor collect all available relevant facts before approaching a worker to reprimand him or her?
3. What are the main reasons why supervisors fail to discipline? To what extent is senior management responsible for supervisors' failure to discipline? To what extent are supervisors responsible for their own lack of action in appropriate disciplining?
4. Describe the steps in the progressive discipline procedures.
5. What test must a plaintiff meet in order to have standing in a whistleblower case?
6. What role does insurance play against potential wrongful termination actions? How might an insurer seek to defend itself against such a suit?
7. When is an exit interview indicated? What are the gains for management? The risks?
8. What measures may mitigate the unlikely possibility of violent behavior from a disgruntled terminated employee?

ENDNOTES

[1] H. Fayol (1987). *General and Industrial Management.* Rev. by I. Gray. Belmont, CA: David S. Lake Publishers.

[2] The psychiatric diagnosis for such a person may be adjustment disorder. The predominant manifestation is the inhibition in work or academic functioning, occurring in an individual whose previous work or academic performance has been adequate. See: H.I. Kaplan and B.J. Saddock (1981). *Modern Synopsis of Comprehensive Textbook of Psychiatry/III.* Baltimore, MD: Williams and Wilkins, Ch. 23.

[3] U.S. Postal Service: Discipline Practices Vary (1989). Washington, DC: U.S. General Accounting Office.

[4] M.A. Rothstein, C.B. Craver, E.P. Schroeder, and E.W. Shoben (1994). *Human Resources and the Law.* Washington, DC: Bureau of National Affairs, Ch. 8.

[5] "Guardsmark Continues Fight to Have Employees' Restrictive Work Covenant Respected" (January 16, 1995). *Security Letter,* Part III, p. 1.

[6] M.A. Rothstein, C.B. Craver, E.P. Schroeder, and E.W. Shoben (1994). *Human Resources and the Law.* Washington, DC: Bureau of National Affairs, p. 422.

[7] Ibid., p. 425.

[8] Ibid., p. 427.

[9] Ibid., p. 437.

[10] J.B. Kauff, A.P. Rosenberg, and H.H. Weintraub (1981). "Terminating the Employment Relationship—Under Increasing Restraints," in *Employment Law: New Problems in the Workplace.* New York, NY: Practicing Law Institute, p. 162.

[11] M.A. Rothstein, C.B. Craver, E.P. Schroeder, and E.W. Shoben (1994). *Human Resources and the Law.* Washington, DC: Bureau of National Affairs, pp. 438–47.

[12] Ibid., p. 450.

[13] D.P. Westman (1991). *Whistleblowing: The Law of Retaliatory Discharge.* Washington, DC: Bureau of National Affairs.

[14] J. Barbash and J.D. Feerick (1981). *Employment Law: New Problems in the Workplace.* Litigation and Administrative Practice Series. New York, NY: Practicing Law Institute, p. 163.

[15] P. Leavitt (Ed.) (1998). *Avoiding Liability in Hotel/Motel Security.* 2nd Edition. Atlanta, GA: Strafford Publications, p. 294.

[16] Ibid.

[17] Ibid, p. 231.

[18] *Private Security Case Law Reporter.* (March 1996), p. 8.

[19] P. Leavitt (Ed.) (1998). *Avoiding Liability in Hotel/Motel Security.* 2nd Edition. Atlanta, GA: Strafford Publications, p. 290.

[20] C.E. Miller (March 20, 1989). "Wrongful Termination." *Business Insurance,* p. 27.

[21] J. Barbash and J.D. Feerick (1981). *Employment Law: New Problems in the Workplace.* New York, NY: Practicing Law Institute, Litigation and Administrative Practice Series, pp. 164–65.

[22] Ibid.

[23] *Security Letter* (December 15, 1987). vol-xvii, p. 1.

[24] M.D. Kelleher (1996). *New Arenas for Violence: Homicide in the American Workplace.* Westport, CT: Praeger, p. 131.

[25] S.A. Baron (1993). *Violence in the Workplace,* Ventura, CA: Pathfinder, pp. 103–104.

[26] S.L. Heskett (1996). *Workplace Violence: Before, During, and After,* Boston, MA: Butterworth-Heinemann, p. 85.

[27] R.B. Cole (1997). *Corporate Personnel Protection.* Springfield, IL: Charles C. Thomas, p. 343.

[28] G. de Becker (1995). In *The Complete Workplace Violence Protection Manual.* J.W. Mattman and S. Kaufer (Eds.). Costa Mesa, CA: James Publishing, Vol. 2, pp. 8–25.

[29] J.A. Kenney (1995). *Violence at Work.* Englewood Cliffs, NJ: Prentice Hall, p. 183.

8

ACCOUNTING CONTROLS AND BUDGETING

> Money for which no receipt has been taken is not to be included in the accounts.
>
> —Hammurabi

Managers monitor and regulate their programs in several ways. This chapter is concerned with one of the most important controls: the use of financial policies. Controlling purse strings involves numbers. This discussion is meant for readers who are not particularly numeric and who may never have taken a course in accounting. The principles involved are simple, but also fundamental to the success of any organization. Managers of security programs need to be comfortable with basic accounting processes in order to speak the language and understand the concepts raised by financial managers. The discussion thus takes into consideration simple notions a manager will need to understand in dealing with financially oriented irregularities. To create a context in which controls needed for security operations can best be understood, some basic principles of corporate finance will be discussed first.

FINANCIAL CONTROLS IN THE ORGANIZATION

All organizations have financial aspects associated with their activities. Guidance is provided by people who are dedicated to the management of money. The principal senior manager in for-profit and not-for-profit organizations is the chief financial officer (CFO), who may also have the title of treasurer and, possibly in small organizations, controller. The CFO's responsibilities include

budget analysis and forecasting, monitoring of accounts payable and receivable, salary and compensation projections and recommendations, internal auditing, investment of excess funds, and compliance with tax and regulatory issues. The CFO is also in charge of financial management, which includes capital raising, determining the mixture of debt and ordinary capital, helping to decide on investment opportunities, valuing businesses that might be acquired or that might be sold, and recommending dividends or capital payouts to shareholders. The office of the CFO maintains "books" or "journals" and financial records that are reviewed by an outside accounting firm, usually composed of certified public accountants (CPAs). Books or journals refer to major types of accounts—usually maintained electronically—which the organization uses for financial activities reporting.

The finance department is concerned with past, present, and future monetary issues. In public corporations, certain financial reports are widely available documents. For example, the finance department prepares an annual financial picture of operations of the previous 12 months. Quarterly (or in some cases semi-annual) reports are issued for publicly traded companies. Such reports are available on a more frequent basis as required by operations. These financial documents have similarities regardless of the type of organization involved, be they for-profit corporations, not-for-profit organizations, or government units. Two fundamental documents central to organizational control functions are the consolidated balance sheets and the consolidated operating (income) statements. In addition to the two historical reports, the modern corporation may also produce a consolidated statement of cash flow and other records as required for operating purposes.

The Evolution of Financial Controls

Accounting techniques have been important management tools for centuries, long before the modern corporation appeared. The Code of Hammurabi recognizes the significance of accounts. Organizations grew in complexity during the Renaissance, and with that growth came an increase in the number of persons who handled the organization's money. As a result, financial controls evolved to make sure that mistakes were not made or that assets were not misappropriated. By the 14th century, double-entry bookkeeping was used by the merchants of Tuscany in Italy. The first treatise on this topic was written by a friar, Luca Pacioli, and published in Venice in 1494.[1] Pacioli wrote in the vernacular of the region rather than in the Latin of the church. Consequently, the treatise became broadly useful in local commerce. This "Venetian" or "Italian method" of reporting assets with liabilities soon was translated into English, Dutch, German, Bohemian, and Russian. Today, this method continues to be used throughout the industrialized world as the fundamental procedure for stating the financial position of an organization.

The notion of double-entry bookkeeping is that any asset or resource has two aspects: its monetary value assets and the corresponding monetary claims on those assets (liabilities) according to who has a legal claim arising from it either as an owner or as a creditor. In bookkeeping, the duality of the accounting method is expressed by recording assets in one column and liabilities in another. A monetary value is assigned to all assets and liabilities in the organization. The balance sheet may be created at any time by totaling the two columns and subtracting the smaller total from the larger to indicate either a surplus or deficit; that is, showing either a positive or negative net worth.

The following are a number of formulas that describe the relationship between assets and liabilities in an organization:

Assets = Liabilities + Owners' Equity

therefore

Liabilities = Assets – Owners' Equity

or

Owners' Equity = Assets – Liabilities

Accounts can be increased or decreased either by credits or debits. Every credit entry triggers a debit entry; therefore, the terms are neutral in terms of significance. After a series of credits and debits are entered into the financial records, credits will equal debits. These are termed the permanent accounts. However, separate individual accounts with customers or suppliers may have their own debit or credit balances. These are temporary accounts and are consolidated with permanent accounts during periodic accounting reporting periods.

Consolidated (master) accounts represent the combination of constituent financial records, which are maintained separately. Normal balances of all accounts are positive, not negative. Total debits (increases) in an asset account will ordinarily be greater than the total credits (decreases). Therefore, asset accounts usually have debit balances. When an account that normally has a debit balance shows a credit balance, or vice versa, it indicates a probable accounting error or an exceptional circumstance. For example, a debit balance in an accounts payable account may suggest an overpayment. The rules of debit and credit balance help to identify irregularities in accounts, giving management a greater sense of control over operations. Table 8.1 shows a variety of account types with their normal balances.

Consolidated Balance Sheets

All organizations use balance sheets to determine their periodic financial condition. (Table 8.2 is an example of a consolidated balance sheet.) Assets

Table 8.1. Normal Balance of Accounts

Account	Increase	Decrease	Normal
Asset account	Debit	Credit	Debit
Liability account	Credit	Debit	Credit
Owners' equity account	Credit	Debit	Credit
Dividends	Debit	Credit	Debit
Income account	Credit	Debit	Credit
Expense account	Debit	Credit	Debit

Table 8.2. Example of a Consolidated Balance Sheet

	Current Year	Previous Year
Current Assets		
Cash and Cash Equivalents	——	——
Marketable Securities	——	——
Accounts Receivables	——	——
Inventories at Cost	——	——
Other Current Assets	——	——
Total Current Assets	——	——
Fixed Assets		
Property, Plant and Equipment	——	——
Land and Buildings	——	——
Machinery and Equipment	——	——
Less Accumulated Depreciation	(——)	(——)
Net Property, Plant and Equipment	——	——
Other Assets	——	——
Payments and Deferred Charges	——	——
Intangibles	——	——
Total Fixed Assets	——	——
Total Assets	——	——

and liabilities are constantly in flux, thus periodic statements allow comparisons to be made with earlier times. The term "consolidated" here and elsewhere implies that financial reports from separate operating units have been merged into it, creating a master financial statement for the whole organiza-

Table 8.2. *(Continued)*

	Current Year	Previous Year
Liabilities and Shareholders' Equity		
Current Liabilities		
Accounts Payable and Accrued Expenses	____	____
Notes Payable	____	____
Total Current Liabilities	____	____
Long-term Liabilities		
Long-term Debt	____	____
Other Long-term Liabilities	____	____
Total Liabilities	____	____
Shareholders' Equity	____	____
Common Stock	____	____
Non-voting Common Stock	____	____
Reserves for Dividends	____	____
Retained Earnings	____	____
Total Shareholders' Equity	____	____
Total Liabilities and Shareholders' Equity	____	____

The consolidated balance sheet is a means of determining whether the organization has a positive or negative net worth, reflected in shareholders' equity. Comparison with the previous reporting period permits the observer to determine the direction of the net worth.

tion. If the corporation is small and is operating as a single unit, the word consolidated is superfluous.

Balance sheets do not establish the value of the organization, as some assets may be omitted or given an unrealistic value. While balance sheets may vary in the items they include, generally the information will include the following components:

- **Current assets.** These are the assets on which the organization could obtain cash immediately or within a few days' time, if necessary. This is the strongest asset category because of its liquidity. Cash and highly liquid cash equivalents are the first current asset listed and are generally an important indication of strength of the organization. Next, marketable securities at cost, if any, are listed. (In most cases, the current market value will be different from the acquisition cost; this may be cited in the balance sheet or through a note.) Accounts receivable, representing invoices owned by customers, are also current assets. The allowance for any bad debt will be indicated. Finally, inventories at cost will be stated, assuming that they have a market value, as will any other current assets.

- **Fixed assets.** Most organizations will have invested in capital expenditures to meet their needs. Once acquired, these assets will appear as fixed assets. Capital expenditures are used for the purchase of assets that increase the utility of operations when the benefit is likely to extend over a number of years. Such expenditures will be subject to depreciation, which is the assessment of the asset's change in value due to use. An estimated expected usefulness of an asset is determined and then deductions from the original purchase price are made each year to adjust for this decline in value of the asset.[2] Depreciable assets include land, buildings, machinery, office equipment, furniture, and fixtures. Fixed assets specific to security programs include security systems, guardhouses, automobiles, golf carts, and sometimes security officer uniforms. Fixed assets are reported less accumulated depreciation. The schedule for depreciation relates to presumed lifetime use of the asset.
- **Payments and deferred charges, if any.** An example would be advanced payment of taxes and other credits earned.
- **Intangibles.** Assets like patents, trademarks, salesmarks, and goodwill are examples of these assets. (Goodwill is the value of a business that has been acquired that is greater than its asset value. Goodwill is written off by amortization to the profit and loss account over its useful life.)
- **Total assets.** This figure represents the summation of current, fixed, and other assets. Other adjustments to assets are possible. For example, if an operating unit has been discontinued, net assets remaining may be listed separately under "Other Assets."
- **Liabilities and shareholders' equity.** The other side of the balance sheet equation contains two parts: liabilities, which are economic claims on the organization, and shareholders' equity, which reflects ownership interests. Each is composed of several components.
  - **Current liabilities.** These are liabilities owed by the organization either immediately or usually within one year from the balance sheet date. They include accounts payable to trade vendors; notes payable to lenders; accrued expenses payable to employees, vendors, or others; and income taxes payable.
  - **Long-term liabilities.** These include mortgages, bonds, and other sums owed that do not need to be repaid within the next accounting period.
  - **Total liabilities.** This category summarizes current and long-term liabilities.

- **Shareholders' equity.** This section represents a claim by ownership to the net worth of the organization. This includes the value of claims from the capital contributed by shareholders or stakeholders. Also in this category are reserves from profits or surplus set aside and available for dividends or other use for shareholders. Retained reserves and profits are also included in this category. They represent the remainder after dividends and taxes have been allocated.
- **Consolidated statement of operations.** This important accounting report provides a picture of income and expenses over a previous stated period of time. An example of such a report is shown in Table 8.3. The statement of operations identifies the main types of expenses and charges against earnings that the enterprise experiences.

Table 8.3. Consolidated Statement of Operations

	Current Year	Previous Year
Net Revenues	——	——
Costs of Products and Services	——	——
Gross Profit	——	——
Selling, General, and Administrative Expenses	——	——
Depreciation of Intangible Assets	——	——
Other Expenses, Net	——	——
Interest Expense and Finance Charge	——	——
Earnings Before Income Taxes	——	——
Provision for Income Taxes	——	——
Earnings from Continuing Operations	——	——
Gain (Loss) from Discontinued Operations	——	——
Earnings (Loss) from Continuing Operations	——	——
Extraordinary Item	——	——
Net Earnings (Loss)	——	——
Earnings (Loss) per Common Share	——	——
Continuing Operations	——	——
Discontinued Operations	——	——
Extraordinary item	——	——
Net Earnings (Loss) per Share	——	——

- **Net revenues.** This identifies the sales recorded over the reporting period for all the goods and services provided by the organization, both for cash and on credit, minus any discounts taken. A series of deductions from this gross revenues are made, leaving a final profit on a per-share basis. The following categories are deducted from the net service and sales revenues:
- **Cost of products and services.** This category includes the total cost of manufacturing products and providing services. For a product that is manufactured, the cost of raw materials, employees, equipment and systems (before depreciation is excluded), and related expenses like transportation and storage are considered. For services that are provided, the amount includes salaries, bonuses, direct benefits of employees providing those services and their immediate managers, as well as relevant overhead and costs.
- **Gross profit.** This profit reflects the remainder after deduction of the fundamental costs in the operation. It is called "gross"—or "gross margin"—because numerous other expenses for operations are yet to be deducted from it, such as finance, administration, and taxes. The gross profit, however, can be a useful comparative tool for one year's performance with another's. The gross margin ratio is the gross margin as a percentage of sales. The only way a gross margin ratio can be increased is by increasing selling prices or reducing the cost of sales or both.

- **Selling, general, and administrative expenses.** Costs relating to sales and marketing personnel, advertising and public relations, plus all the general and administrative expenses of operating the enterprise, are recorded here. They include senior operating staff costs, legal and accounting costs, subscriptions, fees, donations, and a host of expenses not previously recorded. Subcategories may also be reported.
- **Depreciation of intangible assets.** This is the dollar value of the amortization of capital assets recognized in the balance sheet and deductible for the operating period. Accounting rules allow for the writing off of the cost of such purchases based on their estimated average lifetime use. Depreciation reduces the book value of the asset and is charged against income in the income statement.

- **Other expenses, net.** This category recognizes unconventional costs that the organization experienced for the reporting period. For example, if the organization had a minority ownership of a business that was not consolidated with other expenses, the amount could be indicated here.
- **Interest expense and finance charge.** Organizations normally have a line of credit; that is, funds available on a short-term loan basis from a bank. This category reflects the interest expense to

continuing operations. Changes from one year to another could relate to different debt levels and the variable costs of debt. Sometimes, long-term and short-term debt are indicated separately. Short-term indicates debt that matures within one year of the date of the financial statement.

- **Earnings before income taxes.** This amount represents the earnings before deductions for federal, state, and local taxes levied against the net earnings of a business. This may be significant if the organization has unusual tax consequences that could change from one year to the next.
- **Provision for income taxes.** For-profit corporations pay taxes. However, these can differ from year to year if the tax rate changes or if the organization has a loss carried forward from a past year that could reduce the tax basis.
- **Earnings from continuing operations.** Large organizations are dynamic, often selling or closing significant business units within a single financial reporting period. To aid comparison with ongoing operations, accountants distinguish continuing from discontinued operations for which there may be a gain or loss reported.
- **Extraordinary item.** This assumes that an exceptional event is being recorded, and presumably not likely to recur. For example, it could be due to the loss or gain from the early redemption of long-term liabilities, the settling of a significant lawsuit, the write-off of certain assets or costs, or other factors.
- **Net earnings (loss).** This figure indicates the post-tax earnings of operations for the year. However, the consequences for shareholders are not apparent from this amount and must be determined from the earnings (loss) per share.
- **Earnings (loss) per share (EPS).** As an example, a company that earned $10 million after tax and had 10 million shares outstanding would realize earnings of $1 per share. A note will indicate whether the number of shares has changed from one reporting period to another so that the consequences for the owner of a single share will be apparent. The company could have increased or decreased revenues substantially in a particular year with little difference in EPS. This can be because shareholder equity was reduced proportionately to the value of the new acquisitions, or the reverse in the case of divestitures.

Notes to the Consolidated Balance Sheet and Statement of Operations

All consolidated statements include a number of explanations about the organization that are relevant in the opinion of the independent accountant.

To become aware of fundamental and structural financial changes in the organization, perusal of these notes is vital (see Box 8.1). Such notes include a summary of significant accounting policies. This section describes important policies, including any changes made from the previous year. Other notes provide specific information on investment in affiliates, discontinued operations, valuation of types of financing held, leasing commitments, contingent liabilities (for example, significant possible losses or gains), retirement benefits commitments, and stock options. The notes to the consolidated financial statements also provide business segment information (assuming the corporation has different lines of business), income tax information (both domestic and foreign), a review of quarterly financial information, notes on major acquisitions, and information concerning capital stock and earnings per share. Revenues also may be reported separately for international and domestic activities.

Box 8.1. In Financial Statements, Notes Tell a Story

The main financial information in a corporation's annual report might seem encouraging, but in some cases, notes in the report reveal a serious problem that will affect the ability of the organization to operate. Only by reading and understanding the consequences of these notes may an investor or employee be alerted to impending disaster.

An example of a company that was able to "hide" certain accounting practices in its annual report is Crime Control, Inc. This Indianapolis-based business was incorporated in 1977, and was composed of small alarm companies owned by its two founders. In 1978, operating revenues were $685,000, with a pro forma (estimated) profit of 14%. Two years later, operating revenues had grown to $5,800,000, with a pro forma profit of 16%. Business grew steadily as the firm purchased accounts from other alarm businesses. This growth, up to that point, was made possible mostly by bank financing based on the positive sales and profit trend. In 1982, Crime Control issued an initial public offering, selling about 27% of equity, while the founding shareholders retained the remainder. The public market for Crime Control's common stock grew, enabling the alarm business to continue aggressively on its acquisition path.

This alarm monitoring business soared in sales and profits compared to the major businesses of the industry. Why? The answer relates to Crime Control's accounting policies. Alarm businesses generate income in two ways: from leases or sales of systems and from ongoing revenues derived from monitoring alarm signals. Both forms of income usually are reported in the year in which they occur. However, if security system leases are accounted for as sales-type leases under the provisions of the Statement of Financial Accounting Standards (SFAS) No. 13, a more aggressive account-

Box 8.1. *(Continued)*

ing method may be used. This is what Crime Control elected to do. It accounted for future anticipated years of revenues the first year the alarm contract was signed with a customer.

The company offered incentives to renew the rental rate at the expiration of the original lease with a savings of 10%. The company assumed that the bargain lease term would keep the customer for at least eight years. These sales-type leases were then accounted under SFAS since the eight-year lease term exceeded 75% of the estimated economic life of the equipment (10 years). Crime Control, with the approval of its accountants, Coopers & Lybrand, was able to report exceptional revenues and profits relative to other peer companies that used conservative reporting methods.

Due to the accounting assumptions and policies, the company grew quickly. The rapidly increasing "revenues" were booked. But they were not received, since they would not actually be paid by customers for years into the future. To keep the accounting game going as long as possible, Crime Control vigorously sought acquisitions, for which it paid generous amounts in stock or cash. Jealous competitors and astute investors could not understand how this alarm business was able to flourish compared with others in the industry possessing far greater experience. The answer was explained in the notes of Crime Control's annual reports. Few people bothered to read them and understand the consequences until it was too late.

Despite strong apparent revenues and profits, the company kept running out of money because actual revenues were weak. When investors finally realized the scheme, it was too late for most. Crime Control was liquidated and the remnants were purchased at a diminished value.

Source: *Security Letter* (February 1, April 1, August 1, September 4, November 15, 1984; July 1, 1985; July 1, 1986; April 15, 1987).

Statement from the Independent Auditor

All corporations are audited by independent accountants for tax purposes and to assure stakeholders and the public that the accounts and the process by which they have been created have been verified. Along with the annual financial statement, the accountants provide a statement usually stating: "In our opinion, such consolidated financial statements present fairly, in all material respects, the financial position" of the corporation being audited. In earlier years the auditor's report could provide a clue to any poor performance or potential trouble in the organization from the accountants' perspective. This brief statement generally does not serve that purpose any longer. However, independent accountants are likely to insist that any critical issue relative to the corporation's activity be mentioned, probably in the notes to the 10K form or annual report and in other publicly available documents. But such warnings may or not appear in the corporation's annual report.

The Significance of Change in Auditors

In publicly held corporations, auditors theoretically have a fiduciary responsibility, not to the corporate executives who hire them, but to the ownership of the organization. Failure of the independent auditors to identify to the shareholders and the public at large any substantial irregularity or fiduciary urgency, in the case of publicly held companies, can lead to civil action against the entire audit firm.[3] It is because of this fiduciary responsibility that the change in auditors for a publicly held corporation is normally a matter of public record. Shareholders vote annually on the appointment or reappointment of independent audit firms included as part of the annual proxy statement. The changing of one audit firm for another may be a normal and healthy development in which the corporation seeks fresh professionals to review their account. The change can also be one that reflects unwillingness to pay the fees requested by the existing audit firm; such a change thus represents significant cost savings.

Occasionally, the replacement of one audit firm for another signals the fact that the outgoing independent audit firm refused to report financial statements the way management wanted. The accountants may have wished to attach qualifications, or potential warnings, to their statements that would reflect unfavorably on the activities or prospects of their client. Should the audit firm refuse to back down, the client firm may opt to change its audit service.

The Securities and Exchange Commission (SEC)

In the 1920s and earlier, institutions and individuals who invested in stocks and bonds were frequently victimized by fraudsters who manipulated the market for their own benefit. Unfounded rumors—often fanned by scheming corporation officers themselves—might drive up the market price for stock long enough for insiders to liquidate their holdings before the market price crashed. To protect the public, the Securities Act of 1933 (15 USC 77a) required issuers of securities, and their controlling persons making public offerings of securities in interstate commerce, to file with an agency created to receive such information. The Securities and Exchange Commission (SEC) was established under authority of the Securities Exchange Act of 1934 (15 USC 78a-78jj). The SEC was designed to receive registration statements concerning financial and other pertinent data about the issuers and the securities being offered. In the U.S., it is unlawful to sell such securities unless a registration statement has been filed with the SEC and is in effect. Registration with the SEC does not suggest approval of the registration disclosure, nor is it taken to be as accurate. Further, investors are not insured against loss of their investments in common stock by any federal agency.

However, registration serves to provide information upon which investors may make informed and realistic evaluations of the worth of such securities. Today, about 10,000 public corporations are registered with the SEC, which provides a variety of timely information filed by the registered entities. Table 8.4 presents a guide to the filings of public corporations. To the general manager or security practitioner, this information represents readily available, accurate, and valuable information about corporations.

Table 8.4. Guide to Filings of Public Corporations

SEC Form #	Descriptions
8-K	Report of Unscheduled Material Events
10-K	A detailed annual accounting including comparison with previous years. It may be included optionally in the annual report to shareholders; otherwise, it is available to shareholders on request and from the SEC as mentioned below.
10-K405/A	Amended Annual Report
10-Q	Quarterly Report
10-SB12G/A	Amended Small Business Issuer Registration Statement
15-12B	Certification of No Change in Definitive Materials
DEF 14A	Proxy Statement
S-1	Initial Registration Statement. Includes Risk Factors.
S-3	Prospectus Filed for Secondary Offerings
S-4/A	Amended Business Combination Transaction Registration Statement
SC-8	Employee Benefit Plan Registration Statement
SC 13D	Ownership Statement
SC 13D/A	Amended Ownership Statement
SC 13G	Ownership Statement
SC 13G/A	Amended Ownership Statement
SC 14D1	Tender Offer Statement
SC 14D1/A	Amended Tender Offer Statement
SC 14D9	Tender Offer Statement

The Securities and Exchange Commission (SEC) requires issuers of securities and their controlling persons making public offerings of securities in interstate commerce to file registration and issue periodic activity statements. These provide the public with presumably accurate information on operations, including problems and opportunities. Filings may be examined at most SEC offices of at www.sec.gov.

Manipulation of Financial Statements

For-profit corporations have some leeway regarding the means by which some expenses may be reported. Corporate treasurers and independent outside auditors are bound by generally accepted accounting principles (GAAP), the accounting industry's body of widely recognized concepts and guidelines. Nonetheless, in the goal of managing earnings or hiding problems, many organizations use methods that do not violate the GAAP norms if reported to the public, but that suggest adverse circumstances. Sometimes, these reports go unnoticed. The following are some financial actions corporations may take to manipulate their official financial records:[4]

- **Writing off exceptional expenses.** The company decides to write off one or more failed activities, restructuring expenses, and other unusual costs. By eliminating excess expenses, future profits look better. Yet some companies have frequent restructuring write-off costs, suggesting an inability to produce a steady stream of earnings.
- **Smoothing quarterly profits.** Some companies experience a windfall, for example, from the sale of a major asset. But instead of reporting it in the quarter when the sale was achieved, the money is stored, typically in special reserves. Then when some bad news comes along, the company reports the special reserves as income to offset the loss.
- **Deferring costs.** Consider a company investing in a major new product. The expenses may last for several years before income is generated. Should management recognize the development expenses as they are incurred, or defer some of them until revenues start rolling in? The difference can have a substantial effect on profits.
- **Reporting revenues variably.** The most common way of accounting for long-term contracts is a method called *percentage of completion*. Management determines how much of the contract work has been completed, and recognizes the income and expenses related to that portion, even though a major part of the payment might not be received until the contract is completed.
- **Hiding inventory.** Businesses can make a quarter look good by shipping inventory to some customers even if they do not order it. (On a smaller scale, this same irregularity is committed by sales persons seeking to obtain a higher bonus during the sale period.) True, they may have to take some merchandise back later, but in the meantime, the report for the quarter will be better than it otherwise would have been. Such a manipulation of reality is most common in the month prior to the end of the financial year.

- **Dabbling with depreciation.** The useful life of assets can be depreciated over a different time period. For example, capital costs entailed for a new alarm monitoring account customer are expensed by some companies the year they occur. Other alarm companies, however, assume that the account has a lifetime of 2 to 13 years. The longer the write-off, the higher reported profits may be.
- **Combining one-time gains.** A company that buys and sells assets, separate from its major business, normally reports one-time gains or losses carefully segregated from normal operating income. But some companies argue that such regular gains should be included with normal income. This can distort the perception of whether the corporation is actually prospering.

Not-for-Profit (NFP) Organizations

The preceding discussion concerned corporations that are established with the intention of making a profit. However, an important portion of the economy is composed of NPF organizations, including charities, educational institutions, many healthcare and medical research facilities, religious organizations, and trade and research groups. Over one million such organizations are incorporated in the U.S. alone. These organizations have enjoyed federal tax exemptions since the passage of the first income tax law in 1894. Prior to that, such organizations were exempted from state property tax laws.

Despite such NFP status, most of the accounting and audit concerns of such organizations are identical to those of for-profit corporations. Similarly, the security risks to such organizations are largely equivalent.

BUDGETING FOR A SECURITY DEPARTMENT

Up to this point in the chapter, a macro view of financial activities in an organization has been presented. The next part of this discussion concerns programmatic details incorporated in the cost of doing business. The total cost of operations includes numerous departmental and programmatic activities that are intended to achieve the overall goals of the organization. One of these is security. Therefore, security activities require budgets in order to operate. Indeed, the importance of this topic is reflected in that many security directors find that they spend about one-quarter to one-half of their time on budget-related activities.

A budget is a statement of estimated revenues and expenses for a specified period of time. It is usually an annual plan of action, but it can also be set at quarterly or semi-annual periods. Budgets can extend for several years in the case of multi-year projects. A budget also refers to a sum of money allocated to a particular purpose or project for a specific period of time. Budgeting is inextricably involved with good planning, as it seeks to coordinate resources and expenditures.

The purposes of budgets are to:

1. Support planned operational activities with necessary financial resources.
2. Commit money to complete planned programs and projects.
3. Control allocated money.
4. Evaluate management effectiveness by noting how well resources are managed within previously set guidelines.

Annual budgeting is a process that extends over many months in large organizations. Throughout the process, planning and collaboration are vital between the security director, who is preparing the budget, and subordinates, who manage budget sub-sets. A security director with budget responsibility also will interact with senior managers, who will provide guidelines, raise questions about plans and proposals, and perhaps present obstacles in the annual budget approval process that must resolved. For organizations operating on a calendar-year basis, the following is a sequence common to the budget approval process:

Spring or summer: The finance department issues budget request instructions.

Two weeks later: Director consults with subordinates on next year's plans.

Two months later: Security director submits budget to finance department.

One month later: Budget is reviewed. Changes or explanations are requested.

One month later: Revised budget is approved. Consolidated budgets of all departments are presented to the board of directors for approval.

One month later: Budget approved, subject to minor revisions.

One month later: Final revised budget approved.

The budget process requires looking into the future to identify a variety of financial needs that conceivably could be growing while others are contacting and still others are being reorganized. Budgets must have details to show how money is to be allocated and spent. They also must be flexible enough to adapt to the dynamic contingencies that could arise in security programs. Some common types of budgets include:

1. **Revenue and expense budgets.** These indicate the revenues that a department might generate, if any, and the projected operating expenses for the budget period. If security operations generate income (see discussion later in the chapter), the estimate for such income appears. Each item on the expense portion of the budget reflects a particular monetary outlay calculated in advance. This type of budgeting, commonly called incremental budgeting, takes the expenses of the previous year as a baseline and uses them as the basis of proposed increments to represent programmatic changes, inflation, and merit pay increases.
2. **Capital expenditure budgets.** In this category, the department identifies capital costs for systems, equipment, vehicles, furniture, and fixtures. These capital costs are amortized (depreciated) over time. In contrast, supplies—and often security guard uniforms—are "expensed"; that is, written off the year they are purchased as expense items.
3. **Variable budgets.** These budgets identify fixed costs as well as expenses that may vary with needs. For example, if a program cannot be budgeted precisely because of factors that cannot be fully estimated in advance, a variable budget may be approved. This type of budget might be considered for a program not fully planned at the time an annual budget must be submitted.
4. **Zero-based budgets (ZBB).** This concept was developed by Texas Instruments in the 1970s, and was proposed as an alternative to the incremental budget process. ZBB questions all costs by setting the new year's budget at zero and forcing all operating managers to justify their expense requests.[5] ZBB forces managers to scrutinize all costs, rather then assuming that a little bit more each year—incremental budgeting—will satisfy the needs of the organization under constantly changing circumstances. ZBB requires that the security program "sell" its security services each year to the budget approval committee, as it assumes initially that no commitment exists to spend money on any activity unless adequate reasons justify it.

The Process of Budget Creation

Budget preparation and modification have changed drastically since the availability of software spreadsheets and specialized programs useful for the process. Gone are the days when budgets were created on accountant-type paper and items would be written across broad columns reflecting various payments or disbursements over a year's periods. These would then be totaled for each payment or disbursement category and then for the year as a whole. If the allocated budget equaled the total on the bottom right-hand corner, the process was considered a success. If not, the manager and staff had to review and revise projections to see where alterations could be made.

The same process still occurs, though procedures are greatly eased by software programs that produce running totals of the budget, instantly adjusting with each change. Each significant expenditure constitutes a line in the personnel and expense portion of the budget, as shown in Table 8.5. The manager reflects the budget changes that occur during the process. For example, if employees are granted pay increases at specific times, these are taken into account in the year's total plan by inserting the increase at the projected time.

Table 8.5. Security Program Line-Item Budget Example

Budget Line Item	Quarters			
	First	Second	Third	Fourth
Personnel Costs Including Benefits				
Director	——	——	——	——
Assistant Directors	——	——	——	——
Managers	——	——	——	——
Supervisors	——	——	——	——
Support Staff	——	——	——	——
Expenses				
Other Employees (Contractual)	——	——	——	——
Travel	——	——	——	——
Office Supplies	——	——	——	——
Uniforms and Laundry	——	——	——	——
Telephone	——	——	——	——
Training Expenses	——	——	——	——
Educational Costs	——	——	——	——
Insurance	——	——	——	——
Automobile Leasing	——	——	——	——
Automobile Repair and Maintenance	——	——	——	——
Consultant Services	——	——	——	——
Memberships	——	——	——	——
Miscellaneous	——	——	——	——
Total Expense Budget	——	——	——	——

Table 8.5. *(Continued)*

Budget Line Item	**Quarters**			
	First	**Second**	**Third**	**Fourth**
Capital Budget				
Security Systems	___	___	___	___
Automobile Purchases	___	___	___	___
Guard Structures	___	___	___	___
Two-Way Radios	___	___	___	___
Office Furniture	___	___	___	___
Other	___	___	___	___
Total Capital Budget	___	___	___	___
Corporate Overhead Charge	___	___	___	___
Total Budget Request	___	___	___	___

Budget planners allocate expenses for people, supplies, and major purchases over the length of the budget period. Numerous variations are possible to suit the recording and operating policies of particular organizations. The overhead charge, used in some organizations, is a charge management may impose on different departments reflecting a portion of the shared services provided by the organization to the department. A line-item budget of this sort is best loaded onto a computer program so that cost items may be changed at any time, with budgetary consequences being instantly reflected. Security services may also be included in the organizational overhead, which is charged to other departments.

Security program line-item budgets usually are completed on spreadsheets, which allows for specific indication of fund allocations. Personnel budgets frequently are expressed as line-item expenses in that each position is considered permanent, and funds are allocated for an entire period. The spreadsheet allows numerous adjustments to occur over time. When a user changes a figure, the program immediately updates the figures in all columns. The following are factors that effect change in a line-item budget:

- **Personnel costs.** Table 8.5 divides security program costs into four quarters. However, most programs allocate personnel expenditures according to pay periods; that is, weekly, biweekly, semimonthly, or monthly.
- **Expenses.** These predictable costs can be planned over an extensive period of time based on previous experience. They also represent variable expenditures. For example, plans for employees to attend conventions can be curtailed if projected costs get out of hand. In previous years, the "miscellaneous" category allowed

managers a safety valve for adjusting their budget according to contingencies. Today, the category is small or is eliminated entirely in many reports.

- **Capital budget.** This represents spending for purchases that have a lifetime of many years.
- **Budget emergencies and contingencies.** As one budget is being planned, a budget for the current year is operating. Further, the implementation of unexpected plans will cause budget changes that will have to be taken into consideration. Senior management expects operating managers to stay within their budgets without substantial deviation. Senior managers also may ask departmental managers to reduce their budgets on short notice. A reason for this could be to respond to an unexpected earnings shortfall or other reversal.

Managing the budget can be a challenge to security practitioners, who sometimes deal with emergencies that create budget overruns. At such times, the manager is nonetheless expected to "find the money" within the budget. This means that managers need to have the capacity to meet a contingency by cutting previously planned and allocated expenditures. Managers with budget responsibilities constantly analyze what cuts in programs or purchases could be made if they had to be made. Similarly, they consider how they would expand programs if additional resources were made available. At all times, it is vital for the protection manager to understand and communicate to senior management how the security function is contributing to the ultimate goals of the operations.

THE GOALS OF THE CORPORATION: PROFITS

For-profit organizations exist in order to make a profit. If a corporation does not achieve consistent profits, it will eventually face bankruptcy. If a corporation does not achieve sustained and adequate profitability, the providers of capital (shareholders, bond holders, lenders) will remove their money and use it where the return appears to be better and the prospects are safer. To achieve profit, the organization first must sustain fixed costs. These include overhead, such as space, utilities, and other startup costs. Such overhead costs exist whether the organization is just beginning, operating at a loss, or operating at a profitable level. In reality, overhead costs increase slightly as sales rise, but other factors—such as incremental production and service expenses—grow faster and in relation to increased production or service provided. Thus, considerable business must be generated before a breakeven point is achieved; only beyond that is a profit possible. Figure 8.1 illustrates the ways in which fixed costs and variable costs relate to profit.

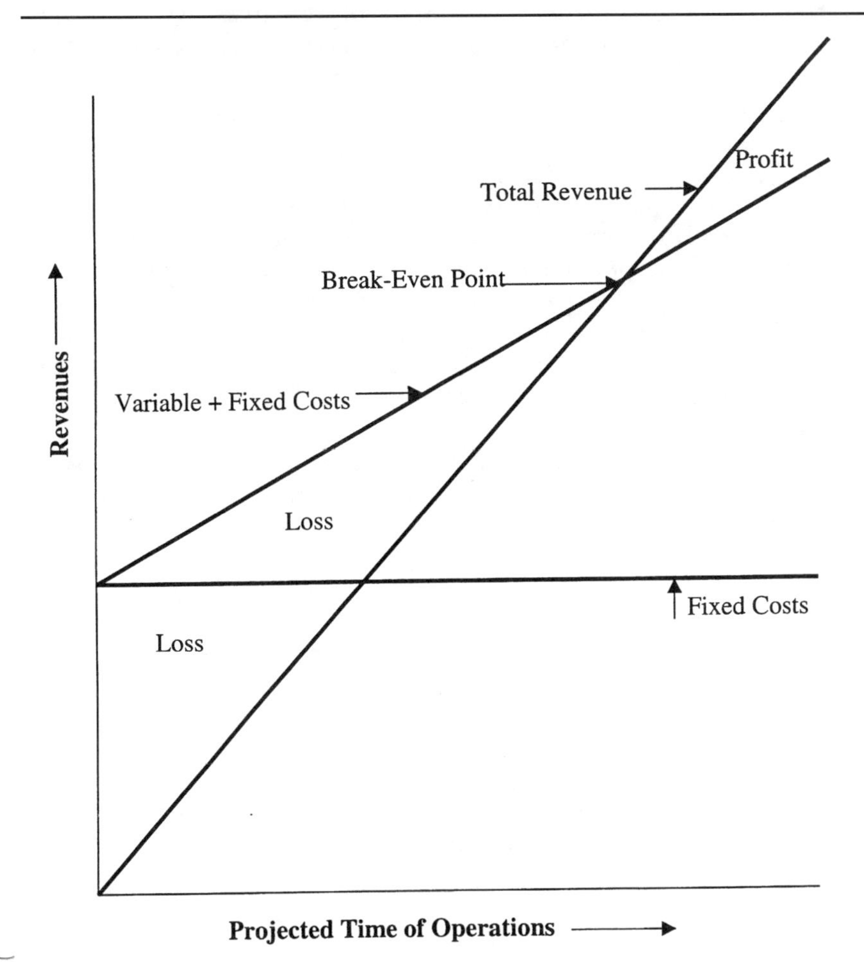

Profit is the objective of for-profit corporations, but it cannot be achieved until fixed and variable costs are met. These are not permanent, but vary constantly. As a corporation becomes more profitable, other organizations are attracted to compete to win a share of the profitable endeavor.

Figure 8.1. Fixed and Variable Costs Related to Profit

Achievement of a profit within an organization cannot be taken for granted. Considerable effort is required to earn a profit. Competition and changing market conditions invariably threaten a profitable operation. To sustain growth, corporations use money from profitable operations to fund new products or services. These might one day become profitable and offset earlier products or services that might decline in their financial return. However, some new products or ventures will not succeed, resulting in a loss of capital. The corporation must be managed so that such failures will not put the whole enterprise at risk.

New corporations and new ventures within a larger entity are not expected to make a profit initially. Developmental expenses are projected, followed by costs to produce the goods or offer the services. For example, in the pharmaceutical industry, a new patented proprietary drug may take 10 to 12 years to receive approval to enter the market. Then, only five to seven years remain in the life of a patent, during which the extensive investments in development can be recovered. After patent expiration, other firms can market and sell exactly the same product under their own name. Thus, the patience of investors is required for profit goals to be realized from any new product.

Security programs must operate with the same level of objectivity as any program in an organization. Every department considers itself to be critical to operations. It is more helpful to see each operating unit as a link in a chain. Any weak link threatens the entire structure. Directors of security programs need to understand the biases of financial directors toward the use of money for different segments of the operation. Programs are funded because they are critical, or at least desirable, to the goals of the operation. The security director needs to communicate how loss reduction programs contribute positively to the profit interests of an organization. This is possible when the manager understands the nature of the business and how an intelligently conceived program is justifiable and necessary to the organization's success. The following sections discuss concepts that are important in explaining or justifying the necessity for security spending.[6]

Return-on-Equity (ROE)

Return-on-Equity (ROE) measures the return on shareholders' equity and gives a measure of the company's return relative to equity; that is, shareholders' paid-in capital and retained earnings. This can be represented as follows:

$$\text{ROE} = \frac{\text{Net Income}}{\text{Equity}}$$

Return-on-Investment (ROI)

Financially oriented managers often put any use of capital to the test to determine what it represents to the organization. This is done using the

Return-on-Investment (ROI) measurement, which can be represented as follows:

$$ROI = \frac{\text{Net Income}}{\text{Equity + Debt}}$$

ROI is similar to ROE except that ROI includes debt. Both of these terms are used to estimate the performance of the organization as a whole, and both are used to indicate the payback of a capital expense, such as the purchase of a security system. Assume that a security director wishes to purchase a security system. Will the cost of the capital commitment pay back the investment required to obtain it, including debt costs, in a satisfactory period? Or is the investment unlikely to be a good use of capital?

Consider a facility with a $1 million per year budget for security, fire watch, and maintenance operations administration. Assume further that a new comprehensive security system costing $1 million can reduce personnel cost by 25%. Therefore, the facility could save $250,000 per year. Would the introduction of an electronic security-safety-fire communications system pay off sufficiently to pay for the cost of the system? An approximation of the value of such system could be calculated as follows:

$$ROI = \frac{\text{Savings in Personnel Costs}}{\text{Cost of System and Finance Charges}}$$

This would be calculated:

$$ROI = \frac{250{,}000}{1{,}000{,}000} = 0.25$$

Therefore, in this crude assessment, the investment would pay for itself in four years, with 25% being returned for each of the four years. Since the system would continue to be useful for several years longer, the loss prevention manager likely would have a fair to strong argument that the ROI is attractive to the company, and the purchase should be supported by senior management.[7] Other factors, such as the necessity of bringing the facility into code or linking the system installation with a wider capital improvement project, might tip the balance to a favorable vote to provide the financing. In many cases, a more detailed evaluation of the use of money for capital expenditure will be indicated.

A more detailed ROI scenario is provided by Walter E. Palmer, who proposes that the value of a security system—such as an electronic article surveillance (EAS) system for a retail store—be determined by first estimating the incremental cash flow.[8]

In this example, let's say that sales are at $10 million and are growing at an annual rate of 2 percent. The baseline shrinkage rate is set at 3 percent. The assumption is that the shrinkage reduction from the new system would be 25 percent of the baseline shrinkage rate, resulting in a savings of $75,000 the first year, as shown in Table 8.6. Because shrinkage is considered a cost, savings at retail must be calculated by the cost/retail ratio. In this example, the average cost of merchandise is considered to be 57% of the selling price, 0.57, and producing a savings at cost of $42,750. It is assumed that the management of the system will require expenses of $10,000, which are now added in.

Table 8.6. Cash Flow Statement to Assess a Capital Investment

Cost of Asset = $75,000	1st Year	2nd Year	3rd Year	4th Year	5th Year
Sales (+2% Each Year)	10,000,000	10,200,000	10,404,000	10,612,080	10,824,322
Baseline Shrinkage	3%	3%	3%	3%	3%
Baseline Shrink $	300,000	300,000	300,000	300,000	300,000
Shrink Reduction %	25%	25%	25%	25%	25%
New Shrink $	225,000	229,500	234,090	238,772	243,547
Savings (at Retail)	75,000	76,500	78,030	79,591	81,182
Cost/Retail Ratio	0.57	0.57	0.57	0.57	0.57
Savings (at Cost)	42,750	43,605	44,477	45,367	46,274
Less: Expenses	10,000	10,200	10,404	10,612	10,824
Less: Depreciation	15,000	15,000	15,000	15,000	15,000
Savings Before Taxes (SBT)	17,750	18,405	19,073	19,755	20,450
Tax (34%)	6,035	6,258	6,485	6,717	6,953
Net Savings	11,715	12,147	12,588	13,038	13,497
Plus: Depreciation	15,000	15,000	15,000	15,000	15,000
Cash Flow Investment	26,715	27,147	27,588	28,038	28,497
Cost of Investment	75,000				
Cumulative Cash Flow	(48, 285)	(21,138)	6,450	34,488	62,985

This illustration shows the payback on an electronic article surveillance (EAS) system that reduces shrinkage from 3% to 2.25%. Adjusting for sales growth of 2% per year and accounting for wholesale costs, expenses, and other factors, the payback cost of the system can be evaluated.

Source: W.E. Palmer (January 1998). "Return on Investment: Beyond the Conceptual." *Pinkerton Solutions.*

Depreciation of the value of the cost of the asset is determined as a straight-line write-off over five years, or $15,000. Expenses and depreciation are subtracted from the net savings.

The next step is to determine savings before taxes. Assuming that the tax rate is 34%, the net savings would be $11,715. But depreciation is added back in since it is an accounting device and does not generate an actual cash amount.

Finally, the addition of net savings and depreciation produces cash flow. This now allows the advisability of the project to be assessed from three types of analysis: payback period, net present value (NPV), and internal rate of return (IRR).

- **Payback period.** As noted earlier, the formula is:

$$\frac{\text{Cost of Project}}{\text{Annual Cash Flow}} = \text{Payback}$$

  If the system costs $75,000 and was expected to return $20,000 annually, the payback period would be $75,000/$20,000 = 3.75 years. Using the cash flow example, the project cost would be completely paid for in about two years and nine months, because $75,000 is equal to the first two years' revenues, plus $21,138. That remainder is equal to about .76 of the third year's revenues. The payback method of analysis may rank projects with shorter payback periods higher than those with longer paybacks. The disadvantage is that the straight payback method ignores the time value of money.
- **Net present value (NPV).** This method considers future cash flows of a project. Assume the company desires a 12% return on its investments. Using a financial calculator, the appropriate discount factor (DF) for the project was determined. Applying a 12% discount rate and multiplying cash flow by the discount factor, the present value of the future cash flow is identified, as shown in Table 8.7. The sum of the cash flows is $137,986. When discounted by 12%, they total $99,120. By substituting the original $75,000 investment, the NPV is $24,120. This assumes that all assumptions and projections are correct.
- **Internal rate of return (IRR).** The IRR is the cost of capital that would make the NPV for the project equal to zero. (The calculation of the IRR is complex and will not be discussed here.) If the IRR exceeds the cost of capital, the project is attractive. If the IRR is less than the cost of capital, the project is likely to be rejected.

The three methods each have their advantages. The payback method is easiest to compute. NPV and IRR require financial calculators, but these

Table 8.7. Discount Factor (DF) for Calculating Present Value (PV)

Year	Cash Flow	DF (12%)	PV
1	26,715	0.8929	23, 854
2	27,147	0.7972	21,642
3	27,588	0.7118	19,637
4	28,038	0.6355	17,818
5	28,497	0.5674	16,169
TOTAL	**137,985**		**99,120**

This is another way to determine whether the system is a good buy. This calculation takes into account the "present value of money." Cash flow is adjusted by a discount factor reflecting the cost of money. The present value of the money would decline each year. The EAS system is still an attractive project, but less attractive than it was under the payback period calculation.

Source: W.E. Palmer (January 1998). "Return on Investment: Beyond the Conceptual." *Pinkerton Solutions*.

methods are more accurate in identifying the time value of money. Typically, managers will consider different methods in making a decision. These issues are discussed in greater detail later in the chapter.

Capital Budgeting for Security Programs

Assume that a security program determines that it requires a computerized controls systems for a new facility. The security director would begin by preparing a written summary detailing the benefits and costs over the life of the system. The manager also will prepare a capital budget that will allow the parties involved to evaluate the proposal. They can then decide whether the capital proposal is attractive relative to competing uses of funds. Various ways of considering the value of the budget exist.

An important factor at the time such a decision is being made is the cost of capital. How it is determined can differ, and no single accounting method is used uniformly. In addition, the security director may face competition from other managers with their own capital requests. The attractiveness of capital expenditures for new systems, therefore, in part is that such systems may reduce ongoing costs over what is currently being paid. These benefits are calculated in order to identify factors related to savings through efficiencies. The following sections discuss widely used capital budgeting methods.

Payback Method. The payback method is the simplest and most widely used technique. It determines the earnings required by an investment in order to pay back the initial capital outlay. As seen in the previous example, this method is popular because it usually demonstrates a rapid repayment of capital, allowing management to reinvest the savings. However, this method does not consider the time value of money. In times of high or rising interest rates, this issue gains in importance.

Money invested earns interest, while money borrowed costs interest. Therefore, the money to be used for the capital budget has to be considered in terms of compounding and discounting. These terms are the opposite of each other. Compounding asks, "If money is invested at the current interest rate, what will it be worth in a certain number of years?" Discounting asks, "How much money should be invested at a given interest rate in order to achieve a particular amount at a defined period in the future?"

The payback method fails to consider the significance of compounding or discounting required for purchase of the capital asset. The longer time required for a capital investment to be paid back, the less interest senior management is likely to have in authorizing it. If the payback is less than the goal set by management, then the project has a good opportunity of being approved.

Initial-Investment Rate-of-Return (IIRR) Method. This method also overlooks the time value of money, creating a bias in favor of investments that yield a return quickly. The IIRR method considers the effects that taxes and depreciation have on the investment, a consideration overlooked in most payback method computations. However, the method does not identify operating cash flows that can be significant considerations. Senior management is more likely to approve a project if the IIRR is greater than the cost of capital to the organization.

Time-Adjusted Rate-of-Return (TARR) Method. This method considers discounted cash flow. TARR provides the interest yield predicted by the investment over its projected useful life. It is also called the internal rate of return (IRR) method. TARR is calculated using a personal computer with a spreadsheet program. If future cash flows are the same for each period, the calculations are performed easily. If the cash flows are uneven, then a trial-and-error process is necessary to arrive at the net present value. If the TARR is greater than the organization's cost of capital, senior management is more likely to approve the project.

Other Managerial Options. What if the savings projected from the capital expenditure are not sufficiently attractive to senior management to approve a capital purchase? In this situation, other measures may be available to the security planner. Assume that the security director advocated a new system but that a capital commitment is not available. One possibility would be to lease the system from the supplier or a leasing facility provided by the supplier. In this case, the system would not belong to the user. One advantage of this arrangement is that service of the system might remain the responsibility of the lessor. Another advantage is that the cost of the lease would be expensed each year, which could produce a tax advantage to the organization. The vendor may also offer flexible payment options to make the purchase attractive to the customer.

Another possibility, though one less frequently available, is that the vendor will provide the system free under certain conditions. The system must produce definable, certain savings. If the vendor can share in the savings, the vendor may be willing to "split the savings" with the customer; in effect, providing the system without cost but actually paying for it through cost reductions.

For example, an organization may desire a system to automatically turn off lighting, heat, and air conditioning if no individuals are in the area. The cost for such an energy management system including installation could be paid by the systems company. The customer then splits the energy savings with the systems company above the previous base level costs of power.

BUDGET DOWNFALLS

Budgets are often management's main way to gauge performance, but they can also block managers from shifting resources to take advantage of opportunities and can distort long-term planning.[9] Budgets also concentrate on spending, but may fail to identify what is really important to the customer.

Security services, like research and development, accounting, and Human Resources, are sometimes regarded as expense centers, not profit centers, by persons who fail to grasp the interrelatedness of organizational units. This mentality puts security at a disadvantage if measures are not found to identify value for the whole organization and communicate it effectively. Well-designed, relevant security measurements can be translated into value-added services that are critical to the entire operation. This can be achieved, for example, by demonstrating savings gained by the lack of costly problems that beset other organizations with inferior security programs. Lower turnover, fewer costly litigations, employee and customer satisfaction, and attractive insurance premiums are other signs of value provided by high-performance security programs.

SECURITY AS A PROFIT CENTER

If organizations seek to create profits, how can proprietary security services be crafted into a profit center? In many cases, security programs have no options to create profits in the same way the principle business of the organization does. It would be unwise for some organizations to attempt to develop new sources of income from security programs. However, in other situations, such possibilities exist. The following are a number of such services:

- **Alarm services.** A technology firm was required by federal contract to maintain an advanced proprietary security system. The firm was able to provide services to nearby non-competitive corporations that were attracted by the advanced standards of the system. These corporations preferred to contract with a well-regarded neighbor than with a distant security alarm service business or to provide the service for themselves.
- **Investigative services.** A clothing manufacturing concern developed a team of highly proficient investigators to solve a series of complex internal loss problems. When the initial issues they were hired to resolve were brought under control, they had less work to do. The security director then made their services available at professional fees to vendors and customers who had become aware of the industry-specific knowledge and expertise of the investigators. A profit center was created.
- **Parking revenues.** The security department of a healthcare facility took responsibility for the management of parking garages and lots. The program was able to keep a portion of revenues for discretionary purposes. This led to improving and expanding parking use, which produced further revenues.

FORENSIC SAFEGUARDS TO INTERNAL FRAUD

The cost of fraud and abuse in the workplace is undoubtedly high. In fact, the *Report to the Nation on Occupational Fraud and Abuse* estimates this loss level at 6% of the monetary value of all goods and services.[10] This totals over $400 billion annually if extrapolated into the U.S. gross national product.

Richard C. Hollinger and John P. Clark conducted an important study of workplace deviance in 1983,[11] which included surveys of 47 corporations in three communities; 9,431 employees; 247 top executives; 30 labor unions and employee organizations; and a variety of other organizations interested in business-related crime. One-third of all employees self-reported that they had taken property from the workplace. White-collar crime is a far greater financial burden on companies than robbery, larcenies, and auto theft combined. Crime that occurs by manipulating financial assets in publicly held corporations as well as not-for-profit institutions is a major issue throughout the world. Security management should thus establish controls that seek to mitigate it.

Generally Accepted Accounting Principles (GAAP)

All corporations have financial records audited by independent auditors. The purpose of this practice is to ascertain that the financial practices and records of the organization follow conventionally established accounting concepts and principles. The outside independent auditor (normally, a certified public accountant) works with the CFO and designated personnel, such as controllers and internal auditors, to review financial procedures. Independent accountants insist that their tasks do not go beyond verifying the accounting methods and providing related counsel to their clients. They maintain that the audit process is not intended to detect fraud and embezzlement, though auditors frequently do uncover serious defalcations from their efforts. Yet the failure of regularly conducted audits to identify a pattern of financial deviance or irregularity can lead to legal action against the accounting firm in the event an undetected or unreported embezzlement takes place despite regular audits (see Box 8.2).

Box 8.2. Limitations to Independent Auditors

Most examples of fraud and embezzlement in operations are not encountered by independent auditors. They are discovered by accident or are brought to the attention of management by whistleblowers. Forensic investigators and fact-finders may then confirm the assertions of improprieties.

Audit firms maintain that it is the responsibility of management to establish financial controls. Supreme Court decisions have supported this view and the number of lawsuits that auditors have been forced to defend decreased in the 1990s. Nonetheless, numerous celebrated cases of accounting improprieties attest to the fact that auditors can be negligent. Jack Bologna and Paul Shaw present a list of some of the companies cited in these cases:

American International
American Biomaterials Corp.
Cenco, Inc.
Coated Sales
Crazy Eddie
Datapoint
Equity Funding
ESM Government Securities
Four Seasons Nursing Homes
Gucci American, Inc.
H.J. Heinz
Leslie Fay
North American Acceptance
OPM Leasing
Penn Square Bank
Pepsico
PharMor
Regina Vacuum Cleaners
Rocky Mountain Undergarment Co.
Sahlen Associates
Saxon Industries
Stauffer Chemical
Stirling Homex
J. Walter Thompson

Box 8.2. *(Continued)*

Mattel, Inc.	U.S. Financial
McCormick Spices	U.S. Surgical
Miniscribe	ZZZZ Best

This list does not take into consideration "regulated" industries like banking, brokerage, and defense contracting. The causes for these accounting improprieties are numerous: pressure for performance; industry competition; quality and integrity of management; and the goal-setting process. In most cases, the quality of auditing and the ethics of those involved allowed the improprieties to continue longer than what otherwise could have been the case.

Source: J. Bologna and P. Shaw (1997). *Corporate Crime Investigation*. Boston, MA: Butterworth-Heinemann; C.J. Loomis (August 2, 1999). "Lies, Damned Lies, and Managed Earnings." *Fortune*, p. 75.

The financial information presented in the periodic statements is prepared according to widely respected principles of accounting to assure that individuals external to the enterprise—such as shareholders, creditors, government agencies, and the general public—have accurate, relevant information. The same information is useful to management in directing the organization. In planning any future activity, management evaluates past financial statements for relevant past activities.

Fraud, Embezzlement, and Security

Within an organization, internal auditors are charged with monitoring the controls and checking systems. They are created to assure that fraud (the conversion or obtaining of money or other assets by false pretenses) and embezzlement (misappropriation of entrusted assets with the intention to defraud the legal owner) are detected early. Accountants and financial investigators are involved in the specialized process of investigating potential or actual financial dishonesty. This type of activity is carried out by forensic accountants or investigators who often have specialized training in conducting such investigations, collecting relevant facts, and preparing an action for criminal or civil prosecution.

Security directors are frequently involved with the financial functions of the organization. This can include creating or consulting on the establishment of checks and controls within the operations, testing such controls, and investigating any financial crime. Should a crime occur, the security director with forensic expertise will interface with financial management on aspects of the investigation. In some cases, the security director will retain and supervise forensic investigators in their fact-finding. The internal financial officer most often concerned with possible fraud or embezzlement is the chief internal auditor. By contrast, the controller is the chief accounting officer of an organization involved with financial reporting, taxation, and possibly auditing.

Separating Tasks: A Powerful Tool Against Fraud

One of the most fundamental controls created by security practitioners and internal auditors is to separate functions within the organization, as shown in Table 8.8. By separating functions, a potential offender finds it difficult to commit fraud or embezzlement alone. Such crimes may still occur, but they require the compliance of abettors to succeed. This type of offense is called collusion and is much more difficult to control than a single rogue acting alone. However, even a single individual acting alone can cause huge losses to the organization over extended periods of time. It is for this reason that security-conscious managers seek to minimize the possibilities of such crimes occurring by separating routine functions. While this principle is particular relevant for financial controls, it has significance in other managerial venues as well.

Table 8.8. Separating Functions to Improve Security

Function A is separated from . . .	Function B
Financial examples	
Accounts payable	Accounts receivable
Payroll preparation	Payroll reconciliation
Records custody	Use of records
General	
Authorizing	Transaction processing
Receiving	Sending
Purchasing	Approval of purchases
Ordering	Verification of order
Custody	Accounting
Vendor proposal	Vendor approval
Computer programming	Computer operations
Library management	Use of library materials
Employee employment offer	Independent review before final offer

Separation of duties is a powerful security safeguard against internal theft and fraud and embezzlement. To prevent collusion further, management institutes a third activity: verification of the process by an independent third party.

A typical financial control separates check preparation from check authorization. But a third step—independent verification of the process—makes the likelihood of financial fraud less possible. Security practitioners are involved in the creation of controls that make financial deviance more difficult and that review and modify such measures regularly to reduce risks from changing vulnerabilities.

SUMMARY

Accounting controls and budgeting are among the most potent measures used to direct the operations of entire organizations as well as individual programs. Often, the accounting mentality focuses on "making the numbers" rather than achieving long-term goals of the organization. Understanding the nature of such controls, nonetheless, is vital to interpreting the value of security operations for the entire organization. Measures should be identified so that security can be judged according to important organizational goals and values. In some situations, security programs can create fresh revenues for an organization. However, such measures should not divert security management from its principal tasks for its own organization.

DISCUSSION AND REVIEW

1. How have accounting techniques changed in recent years? How do they resemble practices in Italy during the Renaissance?
2. Explain why the concept of debits and credits equaling each other is a useful bookkeeping control.
3. Discuss the importance of notes to consolidated balance sheets and statements of operations.
4. How has the Securities and Exchange Commission (SEC) improved reporting measures for publicly held companies? What are the weaknesses in SEC procedures?
5. What are the merits of zero-based budgeting compared with incremental methods?
6. Why is a series of break-even reports unsatisfactory for a corporation in the long run?
7. Compare and contrast three ways of determining the value of a capital investment that produces reduced losses or costs for operations.
8. Cite examples of separation of controls in addition to those discussed in the text.

ENDNOTES

[1] G.A. Lee (1984). "The Development of Italian Bookkeeping 1211–1300," in Christopher Nobes (Ed.), *The Development of Double Entry: Selected Essays*. New York & Lon-

don: Garland Publishing, p. 25; F.L. Pacioli (1996). *Double-Entry Book-Keeping*. P. Crivelli, Trans. London: Institute of Book-keepers, p. 8.

[2] Two types of calculating depreciation may be used. First, the straight-line method, by which the asset value (less estimated scrap value) is written off by equal installments over its estimated life. Second, the reducing-balance method, by which depreciation for any year is a certain fixed percentage of the balance at the beginning of that year. Accounting policy maximizes the earning power of assets as much as possible by selecting the most advantageous financial policies for use.

[3] Independent auditors are expected to identify accounting irregularities; most irregularities are discovered accidentally or are reported by whistleblowers. Still, auditors are always subject to litigation for negligence if they reasonably fail to detect and report abuses that a diligent audit might be expected to uncover.

[4] G. Hector (April 24, 1989). "Cute Tricks on the Bottom Line." *Fortune*, p. 193. H. Schilit, in *Financial Shenanigans*, provides another list of top 10 accounting tricks, which includes many of the abuses in Hector's article, but adds some new ones: recording revenues early; capitalizing costs; changing the way inventory is valued; and swapping debt for equity. H. Schilit (1994). *Financial Shenanigans*. New York: McGraw-Hill.

[5] A.H. Conrad (1997). *Zero-based Budgeting*. Monticello, IL: Council of Planning Librarians.

[6] L.E. Hargrave, Jr. (1999). *Plan for Profitability! How to Write a Strategic Business Plan*. Titusville, FL: Four Seasons Publishers.

[7] R.L. DiLonardo (1997). "Financial Analysis of Retail Crime Prevention," in M. Felson and R.V. Clarke, (Eds.), *Business and Crime Prevention*. Monsey, NY: Criminal Justice Press.

[8] W.E. Palmer (January 1998). "Return on Investment: Beyond the Conceptual." *Pinkerton Solutions*. p. 17.

[9] T.A. Stewart (June 4, 1990). "Why Budgets Are Bad for Business." *Fortune*, p. 179.

[10] J.T. Wells (1997). *Occupational Fraud and Abuse*. Austin, TX: Obsidian Publishing Co., p. 35.

[11] R.C. Hollinger and J.C. Clark (1983). *Theft by Employees*. Lexington, MA: Lexington Books.

Additional References

R.A. Brealey and S.C. Meyers (1991). *Principles of Corporate Finance*. New York, NY: McGraw-Hill.

E.C. Bursk, D.T. Clark, and R.W. Hidy (Eds.) (1962). *The World of Business*. Vol. I. New York: Simon & Schuster, p. 60.

P. Ghemawhat (1991). *Commitment: The Dynamic of Strategy*. New York, NY: Free Press.

M.E. Porter (1980). *Competitive Strategy: Techniques for Analyzing Industries and Competitors*. New York, NY: Free Press.

S. Tully (April 26, 1999). "The Earnings Illusion." *Fortune*, pp. 206–10.

9

OPERATING PERSONNEL-INTENSIVE PROGRAMS

Quis custodiet ipsos custodes (Who will guard the guards themselves?)
—Juvenal, *Satires*

Security programs concern themselves with people, technology, and procedures. Earlier chapters considered theories of workplace productivity and particular strategies of selecting, training, motivating, and generally managing employees in security functions. This chapter considers the macro level of managing people in functioning programs. It looks at two significant categories of security employment—security officers and investigators—and the programs that make them productive and efficient.

Like all aspects of protective management, personnel use has experienced considerable change in the past years, and continues to experience change today. Contemporary security officers are more likely to be contract workers than permanent proprietary employees relative to past decades. Similarly, investigative resources in industry and government have increased, and both contract and proprietary investigative services have expanded to fit these new workplace needs. Moreover, the scope of investigative tasks has broadened considerably.

THE PROPRIETARY/CONTRACT EMPLOYEE DEBATE

For most of the 20th century, security workers were permanent proprietary employees with the same expectations and corporate relationships as any other employee. Beginning early in the century, and most notably since the early 1950s, security guard positions have increasingly been provided by contract services. In fact, by the year 2000, about 60 percent of all security officers were provided by contract services, and this trend continues to grow. Although proprietary and contract personnel both have their own distinct advantages, it is important that all factors be weighed in making this decision, as shown in Table 9.1.

The main reasons cited for contracting-out security employees are as follows:

- **Less total cost.** Employers generally believe that it costs less to contract-out for services than it does to employ a full-time staff. In determining the level of savings, management must be certain to include in-house costs for acquiring, monitoring, and administering such services. Savings for liability insurance may also be factored in since the contract firm will provide its own coverage. (This does not mean, however, that *all* liability for negligent guard services passes to the contract service.) Nevertheless, taking such costly factors into consideration, management typically expects a net savings of 5 to 20 percent from the previous cost level.

  Rather than focus solely on economic benefits, however, employers considering converting from proprietary to contract services should perform a comprehensive cost/benefit analysis to determine whether the advantages are significant. In doing so, it is important to remember that fringe benefits are the responsibility of the service provider, as are benefits to attract and retain staff.

Table 9.1. Contract or Proprietary Security: Weighing Factors

Factors Favoring Contract Security	Factors Favoring Proprietary Security
Less Total Cost	Personnel Retention
Administrative Ease	Perception of Greater Quality of Employee
Criminal Records Screening	Greater Site Knowledge
Recruiting and Vetting Transferred	More Flexible Controls
Training Transferred	Greater Loyalty to the Employer
Supervision Transferred	Reliability of Service
Specialized Liability Insurance	Cost Savings
Specialized Protective Experience	
Personnel Scheduling Flexibility	
Less Likelihood of Collusion and Fraternization	
Emergency Staff Available	

In some cases, management will have other compelling reasons for deciding to contract-out or not to contract-out that are more significant than cost savings. Dennis Dalton, a consultant experienced in security services conversions from proprietary to contract, says: "All too often executives make the decision to convert based solely on the economic gain. What they fail to consider are the dynamics involved in displacing long-term, loyal employees."[1]

- **Administrative ease.** The contract firm is responsible for recruiting and vetting security employees. They should be highly proficient at this task. The security services firm handles routine details for contract security employees, similar to those for proprietary workers. Nonetheless, the security services firm's client generally maintains a residual duty to review personnel files and conduct interviews of workers who will be assigned positions of responsibility within the workplace. Taxes and benefits are the responsibility of the service provider.
- **Criminal records screening.** In some states and geographic areas, the law dictates that security personnel be screened through criminal justice databases for the presence of convictions that would bar them from working in the field. Such databases are not always available to private sector employers. Many services providers are thus able to assure employers that security workers have no evidence of significant criminal records in the jurisdictions where such information is checked.
- **Recruiting and vetting transferred.** The process of recruiting and vetting new security personnel is the responsibility of the services provider. This relieves the client from the primary responsibility for this process.
- **Training transferred.** The security vendor is expected to have the commitment and expertise to train security personnel to the level required for the contractor's needs. Additionally, the contractor may provide specific additional training to meet the needs of the assignment.
- **Supervision transferred.** Security officers usually are supervised by the services provider.
- **Specialized liability insurance.** Security services providers normally should possess comprehensive liability insurance as a safeguard against potential lawsuits. Even in the event that the contractor has some liability in a judgment, the amount is likely to be divided between two parties, lessening the burden on each.
- **Specialized protective experience.** The security services provider should serve as a general resource, as needed, in security matters.

Security services firms may share their practices and resources, acting as informal consultants on procedures and policies.

- **Personnel scheduling flexibility.** When the contractor requires additional personnel for special purposes, such as a meeting or an untoward event, the services contractor can add employees on short notice. Similarly, if an employee fails to meet the needs of the contractor, the employee often may be replaced rapidly.
- **Less likelihood of collusion and fraternization.** Contract security personnel are hired and managed by a separate organization from the clients they serve. This managerial separation makes collusion less likely than if security workers were proprietary staff members. Similarly, the likelihood of fraternization—extensive socialization between security and non-security personnel—is reduced. This principle is codified in Section 9(b)(3) of the National Labor Relations Act, which gives employers the right to terminate voluntary recognition of a non-guard union. That is, the employer of a unionized workforce may require that security officers be members of a different union than the primary union.
- **Emergency staff available.** If the client requires additional security personnel for brief assignments, security services may provide extra officers as needed. This could be less burdensome than hiring additional personnel on a temporary basis.

Now that we have discussed the main reasons why organizations contract-out for security services, let's look at the factors cited by organizations that have instead elected to retain a proprietary security service:

- **Personnel retention.** Security directors generally prefer to have low employee turnover. This is due to many reasons, including the time and cost of recruiting, training, and guiding workers. Proprietary employees are more likely to remain on the job longer. However, this is not guaranteed, and some security services providers also can point to extensive longevity with some of their employees.
- **Perception of greater quality of employee.** Many employers believe that proprietary workers, in general, are superior workers because they are attracted to the normally better compensation and career opportunities within a proprietary organization. However, some employers ascertain that the quality argument is relative to the circumstances and that, given attractive inducements by a security services provider, the quality of service is not likely to be substantially different over time.
- **Greater site knowledge.** An aspect of greater worker retention is that such employees are likely to know people, procedures, and principles better than those with a shorter tenure do. Logically, this produces more reliable service.

- **More flexible controls.** In proprietary programs, personnel usually may be transferred from one location to another as a condition of employment. Contract employees also may be shifted with ease from one site to another. However, some security directors believe that this process is easier for employees who are permanent workers.
- **Greater loyalty to the employer.** Many directors of proprietary programs believe that staff workers are more loyal than contract workers are. This view, which they support with anecdotal evidence, cannot be quantified, though the argument is appealing on its surface. It implies that proprietary employees make extra efforts to provide quality service. However, it can be argued that contract workers have reason to be loyal to the place where they are assigned as well.
- **Reliability of service.** A few contract security firms prove to be disappointing after an initial period of meeting standards. According to Randolph D. Brock, a security director and former principal of a security services firm, "Contractors tell clients what clients want to hear and believe, and say they do but they really don't, opting instead to maintain control of the relationship and dictate terms and conditions."[2]
- **Cost savings.** Normally, managers expect to achieve significant savings from contracting out for security guard and patrol services relative to their in-house equivalents. However, in some cases—particularly with smaller programs—savings will not be achieved, and the organization would save money by staffing its own program internally.

As indicated in this discussion, there tend to be more reasons for contracting-out security services than for retaining them as proprietary services. Indeed, the direction has favored growth of the contract sector for several decades. Yet many organizations have retained proprietary security because they conclude that it is the best policy for their organization. The decision to convert from proprietary to contract or vice versa should only be made after careful consideration of all the factors involved.

Combined Proprietary and Contract Staffs

Many security directors conclude that proprietary and contract services have complimentary benefits. Therefore, they include both types of services in their operations strategy. Similarly, some analytical directors will contract with more than one outside service. This permits qualities to be compared with different service vendors. It also serves as a means of an ongoing assessment of performance of each type of service unit.

The search for enduring, cost-effective, reliable security services challenges managers. Consultant Dennis Dalton notes that seven out of ten security directors for the largest companies in America identified "finding and retaining a really quality-driven contract security agency" as one of the three critical factors in their programs.[3] There are, therefore, many reasons why organizations seek to obtain the perceived quality of a proprietary staff with the flexibility and depth provided by the contract sector.

CORE EXPECTATIONS OF SECURITY OFFICERS

Security officers have numerous obligations to their employers, each of which is of critical importance. The following are the security officer's main obligations:

- **Deter.** The presence of adequately trained security personnel deters crime. The vast majority of potential offenders restrain themselves from disorderly or criminal conduct in the presence of security personnel. This is because the opportunities for identification, arrest, conviction, and civil suits are too high for the crime to be attractive to a potential offender. In deterring illegal and disruptive behavior, security personnel provide visible security that makes the public feel safer and more confident about being where they are. This function may be the most important quality provided by security personnel in most circumstances.
- **Delay.** In the event that offenders commit a crime, security personnel may delay their successful flight, leading to apprehension.
- **Detect.** When an incident occurs or when procedures are not followed properly, security officers can ascertain the violation quickly, mitigating or reducing the chances of loss.
- **Respond.** Security personnel are trained to respond to detected incidents or calls for service while on duty. They are expected to take action at such times to protect people and property and to make the public feel safer.
- **Report.** In the event of an incident, a report from an independent observer, like a security officer, provides important information for management. The report serves as a possible factor in changing internal procedures, a basis for an insurance claim, or as possible evidence in an arrest or lawsuit. Reporting normally is completed as soon as possible following a response. Supplementary reports may be prepared as additional pertinent details concerning the incident are discovered.

Other Important Expectations

The security officer is expected to maintain a suitable appearance, as this reflects positively on his or her employer. Also, personnel who make good impressions are likely to have more positive experiences with the public, thus improving their own self-esteem at work. Security workers often are the first individuals the public meets in the workplace. They often serve as providers of general information to the public and are in a position to enhance the public's impression and regard for the organization in the process.

Protective personnel who have contact with the public need to be sensitive to individuals who may be impatient, angry, confused, incoherent, and possibly mentally ill. While encounters with such people are exceptional, they do occur. Dealing with an unpredictable public must be anticipated in training. For the most part, security personnel best serve themselves and the public by being friendly, sincere, and having relevant knowledge about their workplace. All security employees—proprietary and contract alike—need to make an ongoing commitment to advance the interests of the employer while protecting people, property, information, and other assets.

Finally, the honesty and integrity of security personnel are of utmost importance. Juvenal's query, "Who will guard the guards themselves?" appears facetiously in his *Satires*. But a serious implication is clear: Protection personnel should be above ethical reproach.

Non-expectations of Security Officers

Regardless of proprietary or contract status, security officers should not be expected to perform tasks that are not part of their job descriptions. Unless required by an emergency or specifically requested by a supervisor, or both, the security person is not expected to undertake duties normally performed by other employees. This is in part because such diversionary activity prevents the security officer from performing his or her intended functions—that is, deterring, delaying, detecting, responding, and reporting security-related incidents.

Additionally, security personnel are not expected to take unreasonable risks in their tasks or to take on potentially hazardous tasks. Further, security personnel must not usurp the duties of sworn public law enforcement officers. (The exception is a security worker who is also an off-duty sworn law enforcement officer who must take such action in the context of official requirements.) However, in all case, if a simple, brief action by a security officer can correct, improve, or facilitate a situation, such action should be taken. This is preferred even if the task would normally be performed by another worker or at times even if the action appears to be against written policies and procedures (see Box 9.1).

Box 9.1. Fire and Smoke in an Emergency Response

The Emergency: A contract security officer was performing a nighttime patrol of a new high-rise luxury office building, soon to be ready for use. On one floor, the security officer smelled what he thought was smoke coming from under locked doors. The floor was not occupied at the time as the space was being prepared for tenants. The security officer quickly returned to the base and obtained the key for the floor, which had been left with the security officer by decorating contractors who were preparing the new space.

When the officer returned, he entered the floor and found the source of the smoke. Decorators had been staining the new wood panel walls of a conference room. Inexplicably, they had left a heaping pile of used, soaked rags on the floor. The rags were smoking and would have combusted. The security officer found a portable fire extinguisher and applied the contents of the extinguisher to the pile. He then called the fire department, which arrived eight minutes later. Firefighters further spread and doused the rags and ventilated the area.

The next day, the agent for the building's owner called the contract guard firm to commend the firm for the officer's action. However, the security company manager was less than enthusiastic. Written policy directs security personnel to call the fire department *first* when discovering a fire condition. Then the security officer may return to the location and see what prudent measures could be taken. (Of course, if an actual fire was encountered, the security officer is expected to send an alarm, retreat safely, and then stay available to direct firefighters to the source of the problem.) The contract security manager felt conflicted: Should the security officer be rewarded, or disciplined, or both?

The Resolution: The contract security manager discussed the situation with the building owner's agent and a fire inspector. All agreed that the stated written policy was correct: that is, security personnel should call the fire department first when a fire condition is encountered. However, they also agreed that exceptions do exist. In this case, a trained security officer with good common sense determined that personal risk was acceptable in attacking the potentially dangerous situation immediately. The few additional minutes saved by not contacting the fire department immediately were better used in attacking the fire risk immediately. However, praise for the guard was subdued by the conflict between a sensible written policy and the equally reasonable action taken by an alert security officer.

Expectations of Investigators

Also called detectives and fact-finders, investigators are persons who systematically and thoroughly examine and make inquiries into an event. According to J.J. Fay, an investigative survey is "an in-depth probe or test check of a specific operation or activity, usually conducted on a programmed basis, to detect the existence of crime or significant administrative irregularities."[4] Investigators or fact-finders undertake a wide variety of inquiries. Most organizational investigations deal with business or civil issues, but the private sector may conduct criminal investigations on behalf of itself or at the request of the public sector. A clear understanding between the two is imperative, and is explained in Table 9.2.

Most private sector investigations concern specific incidents. A loss of assets, contract dispute, or crime may have occurred and facts need to be collected to stop the loss, resolve the contract issue, and, conceivably, conduct a portion of the criminal investigation. Most investigations for security operations, however, involve civil and contractual issues and are not criminal in nature. Retailing is an exception. Such investigations cover a huge range of possible topics, which grow as the nature of commerce evolves. An example is investigation for diversion fraud. This type of loss is a major activity, but scarcely existed a generation ago (see Box 9.2).

Table 9.2. Criminal versus Civil Investigations

Factor	Criminal	Civil
Plaintiff	The State (public sector)	Private and Non-profit interests
Prosecutor	The People	The Victim
Main purpose	Punishment of the Guilty	Redress of Injury
Investigations	By or on Behalf of the State	By the Victim
Sanctions	Jail, Prison	Damages to Victims
	Fines	Corrective Action or Behavior
	Specific Corrective Activity	
Conviction	Beyond a Reasonable Doubt	Preponderance of Evidence
Appeals	Possible by Defendant	Possible by Either Party

Fundamental differences exist between criminal and civil litigation. A defendant may be sued criminally, civilly, or both, in which case different plaintiffs will bring charges. Private sector investigations normally serve the interests of plaintiffs and defendants. However, private investigators may be hired by the government, when indicated, to collect evidence on behalf of the public sector in criminal cases and administrative issues.

Box 9.2. Investigating for Diversionary Fraud: A Pharmaceutical Case History

---

Often, manufacturers have price agreements that differ for various members of the distribution chain. As part of the agreement to distribute a product, an organization agrees to specific terms with the manufacturer or distributor that usually limit the area into which certain products can be marketed and sold. When such an agreement is broken, the manufacturer is deprived of rightful profits. Similarly, distributors in the areas undersold by rogue distributors lose profits and goodwill. Investigations often identify the source of such illegal practices and stop them.

The pharmaceutical industry frequently is victimized by diversionary fraud. (Similar examples could be cited from numerous other types of manufacturers including software vendors.) Here is how it works: A pharmaceutical company may charge distributors in the U.S. a particular price for its products. The price of the identical drug may be considerably less for distribution, say, in Central or South America or Africa. This is due to competitive, governmental, and humanitarian reasons. A foreign distributor may order the product from the manufacturer in the U.S. at the favorable foreign price and then scheme to divert the product back into distribution channels within the U.S. This undercuts profits to both the manufacturer and American distributors of pharmaceutical products.

When the U.S. pharmaceutical manufacturer discovers price undercutting in its normal distribution channels, an investigation is in order. Fact-finding is needed to ascertain the losses and identify the likely source of the product diversion. Investigators must become familiar with manufacturing codes and packaging variations in order to identify diverted products. They must gain access to pharmaceutical buyers in hospitals, drug chains, and other distributors in order to develop leads. Often, they must seize the back-channeled merchandise. At other times, they purchase products that do not belong on the premises of the organization they are visiting and hold such materials as evidence against the offender.

By collecting all the facts possible in a case, the investigators, in cooperation with management, are able to quantify the extent of the diversionary fraud. This may be the basis for a civil action against the offenders. When the facts are indisputable, the chances for a resolution favorable to the victimized manufacturer are good to excellent. Yet no recovery would be possible without the creative and persistent efforts of investigators—proprietary or contract—evaluating the problem and collecting facts.

Source: *Security Letter* (July 16, 1984). Part I, pp. 2–3.

Investigations can extends into personal concerns as well as civil and contractual issues.[5] The following are types of investigations undertaken in the private sector:

- Accidents (aircraft, vehicular, industrial, construction, personal)
- Adjustments (in the case of claims for liability)
- Antitrust activity
- Asset location
- Breach of contract (to determine facts and possible damages)
- Competitive information
- Consulting
- Conversion (controlling another's property)
- Copyright and trademark violations
- Cyber-crime
- Electronic countermeasures
- Employee background (also called vetting)
- Espionage
- Fire incidents
- Fraud
- Injunctions
- Insurance claims or counterclaims
- Inventory shortages and shrinkage
- Locating lost individuals
- Marine investigations
- Mergers and acquisitions
- Negligence
- Personal injury
- Patent infringements
- Polygraph examinations
- Product liability
- Property and equity claims
- Public records searches (vital statistics, assets, credit, crime, debit, education)
- Security surveys
- Sexual harassment
- Surveillance
- Trial preparation
- Undercover operations
- Workers' compensation
- Workplace violence

Investigations can occur for *any* reason that seems to make economic or strategic sense for an organization. However, investigative work depends upon the experience, training, and abilities of persons available. In many cases, a business or institution anticipates its needs and is staffed with a large cadre of competent investigators. This is the case, for example, in the financial and insurance industries, which face many repetitive types of investigative activities. Most organizations will not possess such depth of available investigators. In addition, the varied nature of the types of business, contract, and institutional problems makes keeping a staff of suitable investigators more difficult. Therefore, organizations frequently turn to independent investigators who specialize in certain types of investigations.

Investigations by the private sector sometimes are an important part of criminal prosecution. While it is the duty of the state to prosecute alleged offenders, it is frequently in the interest of private or not-for-profit organizations to conduct at least part of the criminal investigation. Often, a public prosecutor will not accept a case on behalf of the state unless the evidence is sufficiently compelling. Specifically, the state wishes to ensure that an offense has in fact occurred, that the offender has been identified, and that evidence collected for use at court is strong. A private investigator normally does not act as an agent of the state and must not imply this. However, he or she may develop the case substantially and turn it over to the police and public prosecutors.

The following are criminal cases in which the private sector plays an active role in the investigation prior to arrest, arraignment, and trial:

- Arson
- Bomb threats
- Burglary
- Cargo thefts
- Computer crimes
- Conspiracy to commit crime
- Criminal defense
- Cyber-crime
- Embezzlement
- Employee dishonesty
- Extortion
- Fencing
- Forgery
- Fraud
- Insurance fraud
- Kidnap and ransom cases
- Motor vehicle thefts
- Narcotics/drugs violations
- Organized crime
- Shoplifting
- Substance abuse

- Terrorism
- Theft (personal, commercial, institutional)
- White-collar crimes
- Workplace violence

The Importance of Investigations in IT Crimes

Protection-related issues constantly evolve. In the 21st century, risks associated with information technology (IT), including communications, have taken on great significance. Chapman and Zwicky note: "The Internet is a marvelous technological advance that provides access to information, and the ability to publish information, in revolutionary ways. But it's also a major danger that provides the ability to pollute and destroy information in revolutionary ways."[6]

The IT revolution has produced a new category of deviance called "cyber-crime."[7] Such crimes include extortion, boiler-room investment and gambling fraud, credit card fraud, pyramid schemes, fraudulent transfer of funds, telephone fraud, and sex crimes. Additionally, denial of service, privacy invasions, attacks by high-energy radio frequency guns, commercial software piracy, and attacks by computer viruses all require systematic investigation to identify offenders and reduce chances of future recurrence. The alleged inadequacy of IT security can result in civil litigation. In one case, *Schalk* v. *Texas*, the defense argued successfully that the plaintiff had not taken sufficient due care to protect information that was stolen from a computer system.[8] As with other types of offenses, IT crimes and attacks require post-incident and pre-trial investigations. Generally, computer crime investigators and auditors specialize in this area.

Non-expectations of Investigators

While exceptions exist, the investigator generally is not expected to arrive at a final conclusion relative to the point of the investigation. Rather, the supervisor of the investigation, or another person for whom the investigation is being performed, should review the facts and all other pertinent information and then make a conclusion about the central question. For example, an investigator may collect extensive information, both positive and negative, relative to an applicant in a background investigation. But the final decision on extending or not extending an employment offer is best left to others. The opinion of a seasoned investigator generally is welcome, but the hiring decision is ultimately the responsibility of a person who considers more factors than those unearthed by the investigation alone.

An investigator also has ethical and legal obligations in providing services to the employer, which also extend to individuals who may be subject to the investigation. For example, federal rules of entrapment apply to private security agents.[9] In particular, it is unethical and illegal for an undercover security agent to originate the idea of committing a crime. Such felonious notions should originate solely with suspects. An undercover agent can appear to be a willing participant in a criminal or improper act, but he or she must not step over the line and actually initiate a crime.

Trends in Proprietary and Contract Security

The necessity for security in the workplace has grown for several reasons, as discussed earlier in this book. In the year 2000, the aggregate employment of security personnel in the U.S. reached an estimated 2.1 million persons. This compares to about 700,000 employees in law enforcement at the local, state, and federal levels. This trend represents a long period of growth in private security in absolute terms and also relative to law enforcement. Meanwhile, the structure by which those security services are being provided has also changed.

PROPRIETARY SECURITY STRATEGY

Private guards employed to protect individuals and private property have been used at least since the time of the Egyptians, when guards protected tombs and tomb sites.[10] In Psalm 127, Solomon observes the importance of security in urban life: "Unless the Lord builds the house, those who build it labor in vain. Unless the Lord watches over the city, the watchman stay awake in vain."

With the arrival of the Industrial Revolution, the scope of private protection grew enormously.[11] Security personnel were needed because industrial operations had increased beyond small, discrete businesses into those involving hundreds of workers, sometimes working round-the-clock shifts. Further, as industrial production increased in scope, work became more specialized. Whereas the protective function initially was generalized, it later became the task of individuals assigned to that particular process. Agents were hired to protect people and assets when risks were greatest.

Proprietary security began in the early 19th century in the U.S. as a means of general deterrence against crime, vandalism, and fire. In the second half of the 19th century, security agents protected industrial facilities against threats like external and internal theft, vandalism, and sabotage, especially during times of labor unrest. During World Wars I and II, private security was responsible for heightened property vital information control, both as an anti-espionage measure and as a means of protecting industrial know-how. In the last thirty years of the 20th century, physical protection remained an important facet of the security industry, while new concerns about IT-oriented risks and threats to intellectual assets shaped the way in which proprietary security is structured in the workplace.

Several aspects of monitoring a proprietary security program have been covered in previous chapters. The following section will consider further operational management aspects of security personnel activity.

Scheduling Requirements

In the process of analyzing security tasks to be performed, security managers must determine the number of personnel required. The complexity of this issue is based on the size of the program, its geography, and the level of training required. In earlier years, security directors faced the tedious task of preparing scheduling plans on paper forms or chalkboards. The scheduling information was constantly being changed; therefore, the information had to be erasable to accommodate the new changes. Security directors frequently delegated the scheduling details to clerks who kept master records and distributed them to supervisors and area managers on a periodic basis. Derivative reports dealing with shift rotations, work locations, and event scheduling were made from the master schedule.

In recent years, software has become available that is written specifically to help schedule security personnel.[12] Scheduling time has been cut, saving considerable clerical costs in the process; schedules can be displayed graphically and changed with ease; and relevant information on security personnel can be instantly available. Management may review assignment plans over the Internet, while information can be linked to other management and financial controls software packages, further reducing time and costs. Figure 9.1 shows an example of such a software package. These software packages can be learned in a short time. Managers and supervisors can create useful supportive reports such as those indicating guard availability and unavailability; scheduling conflicts; and assignment or customer work history data. They can also create "barred-from-customer" conflict lists and seniority lists. Automated officer check-in systems allow managers the opportunity to check that security personnel are at their posts.

Management's support for such software packages is enhanced by their additional features. Costs are saved because the schedules are prepared much faster than the previous manual process. However, further cost savings result from the reduction of overtime by better monitoring of hours and separate links to payroll, invoice creation, accounts payable, accounts receivable, and general ledger software programs.[13]

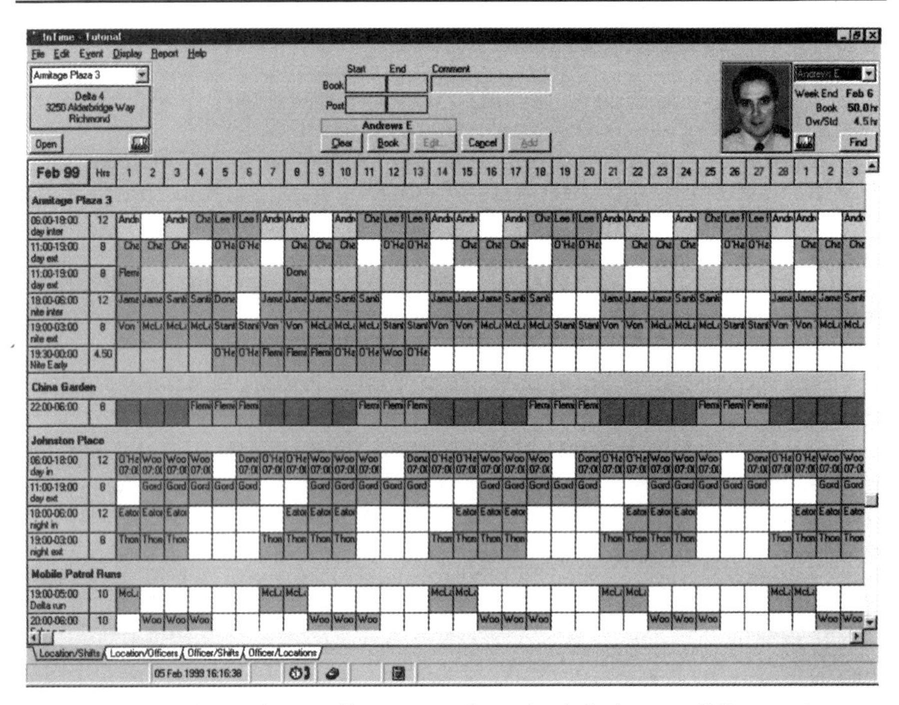

This program shows how officers may be scheduled over different times at various locations.

Source: InTime Solutions.

Figure 9.1. Software Programs Specifically Written for Security Operations Facilitate Operations and Control.

Determining Personnel Needs for Posts

As a rule of thumb, managers require the equivalent of 4.2 to 4.5 persons to be available to staff each post on a 24-hour 365-day basis, assuming a 40-hour workweek. Complications in planning occur because of vacations and other schedule changes, such as illness or personal time off. Consequently, an adequate pool of full- and part-time security personnel needs to be available to fill the requirements of large security programs with 24-hour posts.

Salary and Compensation

Money is not the only motivator, but it is the most important one. Managers have to keep within their allocated budget guidelines, constantly improve services, and control payroll costs. The average (median) hourly pay for security guards varies widely based on geography alone. In some programs, average pay for security personnel will be consistently below average. In others, it will be within the mid-range, and in still others, average compensation will be above average.

Managers can determine the average hourly pay for security officer services from data collected by the Bureau of Labor Statistics of the U.S. Department of Labor.[14] These reports are prepared by the Office of Compensation and Working Conditions in cooperation with the Office of Field Operations and the Office of Technology and Survey Processing. The data may be months old when a manager accesses the latest report for a particular region and therefore the rates need to be adjusted for any circumstances that may have changed. The information is national in scope and covers private and public positions. Scores of employers provide reasonably accurate base salary data to the survey, thus revealing the normal hourly compensation for 150,000 to 250,000 security officers. Managers can turn to this source to judge their own compensation programs relative to other employers in varying regions.

Compensation ranges for security supervisors, investigators, specialists, managers, and executives are equally important in determining the costs of operating a department. Such information can be obtained from compensation studies specializing in the security, protective services, and law enforcement categories.[15]

Managers can reduce operational costs for salary and benefits processing by using software dedicated to the tasks. Alternatively, the activity can be turned over to an outside service bureau.

Insurance

The actions of security officers, investigators, and other protective personnel are covered under the organization's comprehensive insurance. The risk of errors and omissions from security performance normally is defended by the insurer on behalf of the insured. In the event that security services are contracted-out, savings on liability coverage should be realized.

CONTRACT SECURITY SERVICES

As previously discussed, the trend in recent decades has been for operating entities to contract-out numerous types of services. The scope of such services is broad and includes diverse temporary help, technical services, entertainers, drivers, hospitality industry workers, janitorial workers, home healthcare workers, agricultural workers, and property managers. However, one of the largest contract services is security personnel. In fact, it may also be the earliest organized service industry. The Pinkerton Agency contracted-out armed security guards to packing plants and commercial houses beginning in 1858.[16] By the end of the 19th century, scores of security services businesses could be found in urban America.

The concept of "employee leasing" has many attractive features in that the leasing firm must undertake numerous accounting and administrative duties on behalf of the workers.[17] Requirements for the administration of records and benefits is regarded as particularly burdensome to small businesses or those with temporary operations. The employer might have to:

- Maintain tax deduction records
- Compute tax liabilities and make timely bank deposits
- Respond to garnishment from the court or other taxing bodies
- Conduct audits for workers' compensation benefits
- Conduct audits for unemployment claims
- Comply with COBRA legislation (1986) requiring employers with more than 20 employees to offer continuation of healthcare coverage in the event that an employee is terminated or experiences another qualifying life event
- Reconcile employee paychecks
- Post records for state and federal taxing agencies
- Prepare workers' year-end W-2 filings
- Maintain vacation and other authorized leave benefits
- Maintain payroll and tax changes
- Answer questions from workers on regulatory, pay, and benefits issues

The employer must provide such services to proprietary workers as a matter-of-course, and this includes security personnel. However, when such duties are transferred to an outside organization, the primary organization is relieved of these duties and related costs. Yet contract security programs are differentiated from proprietary ones on more substantive issues than who will be responsible for routine administrative issues. These issues were discussed earlier in this chapter.

SELECTING CONTRACT SECURITY SERVICES

The process of selecting a contract security service may be uncomplicated for small and simple organizations. However, it can be a formal and extensive process for large, complex organizations. The following sections discuss these two scenarios.

Small, Simple Programs

Employers requiring one or more security officer for an assignment that involves visible patrol begin the process by preparing a report or memorandum in which the specific tasks of the required security personnel are outlined. This need be only two or three pages long, but it should provide the basic vital information required by a contract security firm to make an informed proposal. This report should include the number of security personnel required; the type of vetting to be performed; the training to be completed prior to assignment; the duties to be performed; the nature of experience desired; the hours to be worked (including any exceptional circumstances); the type of uniform desired (military style or blazer and slacks); the insurance to be provided; and the field supervision to be offered.

Additionally, the prospective employer should consider the level of pay desired for the security personnel, including projected increases over the length of the contract. This information will be sufficient for the prospective bidders for the contract to prepare a proposal. Many other nuances to the selection process may come up later, but the basic material just described will start the process for the contract proposal for a protective program requiring contract security officers.

The next step is to interview three to five security guard firms. Typically, a client wanting security officer services will seek to meet with one to three national security services firms and one to three local, independent security services providers. In some areas, the number of choices will be limited. In addition, many established security firms may not choose to pursue the business of an operation that offers little opportunity for growth and that does not provide the contract firm with its gross profit target.

The proposed service providers may be selected by references from colleagues who manage existing locally respected programs and from local membership lists of professional organizations like the American Society for Industrial Security and the National Association of Security Companies. It is also advisable to peruse relevant business directories or source books for names of service providers. An initial letter to the identified companies can provide details on the security services required and invite an expression of interest by a particular date. In some cases, security services firms will learn of the search on their own and ask management for the opportunity to be considered.

The next step will be to meet with the prospective bidders and provide them with full details on the assignment. For a new security services requirement, it is probably best to meet individually with the prospective bidders. The security firms—individually or as a group—should be able to see the exact area where the security personnel will work as part of a site visit. The prospective security services firms should have time to complete their surveys and have equal access to all relevant details. A date for the completed proposals must be set.

When all the proposals have been received, the client compares and contrasts the various submissions. Sometimes, a modified comparative scale is used to judge the security firm by various objective and subjective criteria. The client may briefly visit the office and training facility of the prospective services provider, and may wish to check any references. Often, considerable negotiations occur on finer points of the agreement before both parties agree to the terms. When the winner is selected, a contract is signed. The contract is usually drafted by the client, but may be provided by the security services firm and must be carefully reviewed by the client, possibly with a lawyer or industry consultant.

Large, Complex Security Programs

Large organizations with formal structures require a much more detailed process to identify and award a contract for security services than the one just described. This is particularly true for institutions and the government, which often issue a request for proposal (RFP) when making selections for security services providers. (Corporations also use RFPs, or modifications of them, when selecting security venders.) Larger government or institutional contracts usually attract numerous hopeful bidders. Thus, RFPs are used both to ensure fairness and to defend against claims of unfair awards. The process is invariably structured with firm dates and specific demands for replying to the RFP. An outline for a formal RFP follows. It provides insight into the types of issues that have occurred in the past when organizations have contracted for protective officer services, experienced difficulties, and wrote specific demands into a formal agreement.

RFPs require effort and expense to prepare. The questions and concerns of bidders should be anticipated in advance, thereby facilitating the selection process. Often, the document begins with a Solicitation Summary, which provides proposed instructions, submission requirements, and conditions. These include such factors as:

1. **Statement of purpose.** This states what the task would be for the required security personnel.
2. **Client contact person.** Usually, a single individual is identified for all written or telephone inquiries or contacts. If a potential vendor requests interpretation of the RFP, this request should be placed in writing. The client reserves the right to respond to any and all such requests, sharing answers with all responders, usually via fax or e-mail.
3. **Submission.** The final proposal usually is sent to a different office and contract officer than the client contact person identified to handle inquiries.

4. **Proposer conference.** A proposer's conference is set for a particular date, usually in a conference room where all representatives can be accommodated. Proposers may be requested to notify the agency at least five days in advance if they plan to attend. Proposed vendors are under no obligation to attend such a conference, but normally they do. Attendance is usually taken at such times.
5. **Letters of intent.** Proposed vendors interested in submitting a proposal are requested to submit a letter of intent to the agency.
6. **Submission requirements.** This section summarizes what the prospective vendors must prepare and submit in order to be considered. The number of copies of the proposal and the date and hour on which the sealed proposal package must be received by the client is stated.
7. **Modification or withdrawal of proposals.** The costs in the proposal are deemed to be irrevocable until the contract award, unless the proposal is withdrawn or modified prior to the time or date set as the due date for the proposal and in accordance with the RFP.
8. **Post-opening withdrawals.** The client may allow a potential vendor to withdraw its bid only after the expiration of a stated number of days after the opening of the proposals. Such a withdrawal must be in writing and in advance of the actual award.
9. **Late proposals, withdrawals, or modifications.** To prevent charges of unfairness, clients generally do not make exceptions for late proposals, withdrawals, or modifications. An exception is when any modification of a successful proposal makes the terms more favorable to the client than those initially presented. Occasionally, a client may allow a vendor to join the process after it has begun. Such a vendor would be expected to meet all stated submission deadlines and requirements pertaining to other vendors.
10. **Proposers' right of appeal.** The process whereby proposers may protest and appeal decisions regarding the solicitation and award of a contract is identified.
11. **Payment policy.** The agency states when it expects to pay proper invoices, which set forth the description, price, and quantity of services rendered with the appropriate documentation appended. (A generation ago, some security service providers billed clients monthly, in effect giving the client improved cash flow. In recent years, weekly or bi-weekly billing has become the norm.)
12. **Amendments to the RFP.** If amendments to the RFP occur, all proposers will receive such notifications and must verify that they have received all addenda issued.
13. **Discussion with proposers.** The client may wish to conduct discussions with proposers who have provided the most responsive proposals. However, this discussion is not necessarily required prior to a contract award.

14. **Procurement policy rules.** The RFP may be subject to the rules of procurement by the government agency involved in the bid. In the case of a private organization, it may be stated that the decision to award is left solely to the discretion of the client.
15. **Fairness and ethics.** If any potential vendor feels that unfairness, favoritism, or ethical improprieties have occurred in the proposal process, the vendor may contact the client's attorney designated by name to receive such information.

Element of a Comprehensive RFP

Following the Solicitation Summary, the proposal then provides considerable depth on the nature of the work to be accomplished by the proposed contracted services. A table of contents may begin the section, listing detailed aspects of the proposal covered in the document. The following is the outline of an expanded RFP, indicating considerations that could be important in completing a mutually beneficial contracting-out process:

1.0 **Introduction**

1.1 **Purpose.** The name of the proposed client and the nature of the security services required is explained. For example, "the client is requesting proposals from fully licensed firms in the business of providing trained, uniformed, unarmed male and female security officers who have had at least five consecutive years of experience in furnishing such services to large institutions, corporations, or to government units." The specific nature of the experience expected to be demonstrated by the security services firm is also mentioned.

The length of the contract is indicated (for example, three to five years). The cost proposal must state a rate for each year of the contract. However, the client may wish to retain continuance of the contract for one to three 12-month extensions beyond the initial period at the client's discretion. The date at which the contract is to begin is indicated.

1.2 **Background.** This section describes the nature of the client's organization and states the importance of security in the view of management. Further details on the type of work generally provided by the organization may also appear in this section or be made part of the RFP packet provided to interested proposers.

1.3 **Specific facilities.** Next, the RFP will list all the facilities to be covered in the contract. It will include address, type of activities, number of employees working at the location, and any special features of concern from a security standpoint. If any physical changes in the nature of these facilities are anticipated within the life of the contract, they should be identified.

1.4 **Security issues.** A statement or section about the nature of the most significant security issues requiring management may be included here. For example, such issues as fire watch, employee theft, outsider theft, meeting and greeting the public, and emergency response capabilities are quite distinct activities and require that the security firm consider them thoughtfully in arriving at a bid proposal that would provide security officers who had the required skills.

1.5 **Services needed at the facilities.** Following from the previous section, this section identifies the serious problems that have occurred in the facilities in the past. There is potential risk if the RFP fails to identify serious crimes and incidents that have occurred, say, in the past three to five years that would affect the nature of the staffing required. Specific details do not need to be presented in the RFP; however, the document needs to identify the major issues that have been the focus of security in the past. If the facility is new, the RFP will identify the nature of the problems management expects the security services firm to be able to manage. How the client has dealt with these incidents elsewhere in the past may also be explained.

2.0 **Scope of services.** In RFPs for large and complex operations, it is possible for the organization to organize the work for one contract firm, several firms, or a combination of contract and proprietary services. The way management expects to divide such services may be identified here.

2.1 **Minimal specific tasks and requirements.** The writers of the RFP are not in the position to instruct the security services firm on how to do their job. Rather, the RFP writers can describe what the job involves and ask how the security supplier plans to meet the client's security objectives. For example, the RFP may identify specifically the salient issues of greatest concern to it:

- Protecting staff, customers/clients, vendors, and visitors against malicious injury.
- Protecting the premises against theft, pilferage, vandalism, damage, or destruction.
- Permitting only authorized persons to enter protected areas.
- Reporting to the client all violations of regulations that are the nature of written reports.
- Operating and monitoring a comprehensive closed-circuit television and communications monitoring center.
- Operating and using X-ray, walk-through, and hand-held metal detectors at entrances to the facilities.

- Patrolling areas of the facilities including perimeter walls and fences, building exteriors, parking lots, roofs, main floors, corridors, stairwells, restrooms, and basements.
- Observing and reporting at assigned locations activities that can lead to improved security performance.

2.2 **Guidelines for guards.** In this section, the client indicates the minimum expected guidelines of the services. The following are examples that may be cited:

- Contract security personnel must view the safety of all site employees, visitors, vendors, and others as their main duty.
- Security officers shall challenge all persons entering protected premises for proper authorization and identification prior to entry.
- X-ray, walk-through, and hand-held metal detectors and other physical security devices, where provided, may be used to process individuals who enter the facility.
- Doors, windows, and other portals must be secured when required.
- All incidents must be reported to the local director or site supervisor.
- A security log book must be maintained.
- Contract security personnel must adhere to any site policies as prescribed by the local manager or other officials in charge.
- Contract security personnel should safeguard from damage all equipment, systems, and property on the premises.
- Contract security personnel must provide only general information to the public, such as directions and locations of various offices. At no time will contractors' employees be permitted to discuss with the public operating activities at any of the client's sites. Such a discussion may serve as a breach of security of the site and of the organization.

2.3 **Requirement for incident reports.** A policy on incident reports may be expressed:

- Unauthorized intrusion, trespassing, or other illegal entry onto the site.
- Any criminal or unlawful act which has been committed on the site.
- Any assault, altercation, or confrontation that results in any injury.
- Any emergency responses to the site by fire, police, emergency medical, or government agencies.

- Any safety or health hazards observed by the security officer.
- Any exceptional incident that could require a claim against management or that would require further investigation on the part of the client's management.
- No information contained in an incident report or brought to the attention of the security officer at the site shall be disclosed to third parties without written consent of site management.

2.4 **Vendor responsibility.** The following vendor responsibilities may be mentioned in this section:

- **Continuity of services.** It is vital that security services be provided without interruption to the client. Accordingly, the security services provider must propose how to assure coverage in the event of strikes, work stoppage or slowdowns, or in other situations in which services and operations may be disrupted.
- **Provision for female security officers.** The security services provider may be expected to provide a certain minimum number of female officers depending on the locations and the nature of services to be provided.
- **Uniforms and equipment.** The following uniform and equipment guidelines may be mentioned in this section:

    a. The contractor must furnish and ensure that each security officer wears a uniform in compliance with any state and local regulations. Security personnel should present a neat and orderly appearance at all times during the performance of their work. Proposers should identify details of their uniform options in the proposal. Alternatively, the client may prescribe special clothing to be worn by security personnel and may indicate whether the client will pay for the clothing or if it is to be included in the proposer's fee.
    b. At no time shall firearms, knives, or other unauthorized instruments or tools that might be used as a dangerous instrument or weapon be carried by security personnel.
    c. Supervisory personnel shall wear a uniform that distinguishes them from subordinate security personnel.
    d. Security personnel shall exhibit identification credentials, as prescribed by the contract, in order to gain access to the facility for the performance of work.
    e. Each security officer shall be furnished with a two-way radio, beeper, or cellular phone. Inoperable devices must be replaced immediately. The equipment shall be in working order at all times. The provision to safeguard and recharge

the equipment remains the responsibility of the security services provider.

f. Security personnel responsible for the operation of security systems shall be held responsible for the systems' security and care. Any unreported or unexplained damages to such systems shall be deemed the service provider's responsibility.

3.0 **Guidelines for proposal preparation.**

3.1 **Proposal content.** The proposal specifically must explain how certain tasks will be handled. This content should include:

- The specific security tasks that will be performed by security officers.
- The monitoring and supervisory control of the workforce.
- A description of the training program to be provided security officers and supervisors.
- A security management assessment for the location—or part of the location—needed to adequately secure the facility.
- The internal controls that will be exercised by the security services firm over the officers and their supervisors.
- The security officer standard measures required to provide adequate security.
- The measures to be taken to protect employees and visitors against malicious injury and to protect the premises and property against theft, pilferage, vandalism, damage, and destruction.
- The means of securing and safeguarding documents and records.
- Procedures for reporting, preparing, and maintaining logbooks, forms, and records.
- Procedures and methods for excluding unauthorized persons.
- Access control and security procedures for storage and equipment.
- Details of uniforms to be furnished to and worn by security officers.
- Any other security measures appropriate to the client and its sites.

3.2 **General information.**

3.2.1 **Conflicts.** If the proposer feels that any part of the proposal appears to be in conflict with another part, or if an aspect of the work to be performed is unclear, the proposer should ask the potential client for clarification. If relevant, this elucidation will be shared with other proposers.

3.2.2 **Inspectors and tests.** The proposer should be aware that all personnel and equipment may be subject to inspection, examination, or test by the client at any time during the course of the contract. The client shall have the right to reject unqualified personnel who do not comply with the guidelines and requirements of the contract. Similarly, unsatisfactory equipment or materials may be rejected by the client.

3.2.3 **Supervision by contractor.** The proposal shall indicate in what manner a representative will supervise the work that his or her staff is performing to ensure the firm's complete and satisfactory performance in accordance with the terms of the contract. The representative shall be authorized to receive and put into effect promptly all orders, directions, and instructions from the client.

3.2.4 **Adequate and competent supervision.** The proposer shall provide, as part of its contract and at no additional cost to the client, a visiting site manager. The site manager should visit all facilities at least once during an eight-hour tour to assure that quality security services are being provided. The site supervisors shall have a minimum of two years experience as supervisors in security-related or law enforcement positions prior to being assigned to the client.

3.3 **Additional requirements.**

3.3.1 **No arrest policy.** Security services personnel must not make any arrests without the expressed consent of the director or assistant director of security. Security services employees shall not sign a complaint on behalf of the client or sign any request to do so by governmental authorities.

3.3.2 **Tour limitation rule.** Security services employees should not be on duty in excess of 16 hours per 24-hour period or in excess of 60 hours per week. Each security officer shall have a minimum of 24 consecutive hours off each week. Any hours in excess of the above-stipulated maximum shall not be billed to the client.

3.3.3 **Overtime pay policy.** In the event of an emergency or other contingency requiring guard service for a period in excess of the stipulated shift, payment for such services shall be at the same rate as the standard contract rate. This provision will not apply if the contractor has *not* been notified at least eight hours prior to the start of the shift, in which case the contractor shall receive the rate of one-and-one-half times the regular rate for that shift. If overtime is caused by the security services firm and is not attributable to the client, the service firm shall not be entitled to any overtime rate.

3.3.4 **Training agreement.** The proposer must provide a training program for its employees at no cost to the client. All employees must complete the training program prior to assignment. Such proposed training must consist of [a stated number] hours of training. The curriculum for the training and the resources used must be approved by the client prior to beginning such training. The client retains the right to provide part of the training prior to assignment.

3.3.5 **Right to audit.** The client reserves the right to audit relevant security services providers' financial records related to the contract to assure compliance.

3.4 **Standards for workforce.**

3.4.1 **Education and background requirements.** Proposers should offer security officers and supervisors who meet the following minimum requirements: at least 21 years of age; high school diploma or GED; at least three years, alone or in combination, of (1) satisfactory prior work experience in a security-related field or human care-related area; (2) military service; or (3) satisfactory completion of college-level study. Security officers with records of criminal convictions, other than minor traffic violations, will not be satisfactory to the client. In fulfilling the contractor's obligations under this requirement, the contractor shall comply fully with all laws of the state and disclose pertinent information to the client. Security officers shall be in good general health without abnormalities that would interfere with the performance of security duties, and must be capable of performing normal and emergency tasks requiring moderate to arduous physical exertion. Medical fitness is to be determined by a medical examination conducted and documented within 90 days prior to entry on assignment. Security officers must speak and write English intelligibly. Security officers must be citizens of the United States or possess acceptable alien registration documentation. Documented proof in the form of true certified copies (i.e., birth records, diplomas, military discharges) and previous employment verification should be maintained in a permanent personnel folder for officers. The folder will include training records.

3.4.2 **Hiring and replacing personnel.** The contractor shall be required to provide the client with the names of all newly hired officers. The contractor shall provide a certification for each individual officer stating that he or she meets all contract requirements. The contractor shall provide the certification with the officers' folders for review prior to the beginning of their employment at the client's site for the first time.

3.5 **Information on the security services provider.** The contractor shall provide the client with comprehensive information on the fitness of the contractor to perform the proposed contract. This information shall include historical information on the contractor; an organizational chart; résumés of the contractors' principals and managers in charge of the contract; a list of the board of directors; a list of current contracts including a contact name and address and the date on which service began; and letters of support/recommendation concerning the contractor.

4.0 **General guidelines for submitting proposals.**

4.1 **Letter of intent.** Prospective vendors shall submit a letter indicating their intent to submit a proposal by the deadline stated in the Solicitation Summary at the beginning of the RFP. Letters of intent are not binding, and proposals received by the stated deadline from proposers who did not file a letter of intent will still be considered. The letter of intent should identify the name of the proposer, plus the name, location, and telephone number of an authorized representative and any proposed subcontractor.

4.2 **Application deadline.** Proposals should be filed by the close of business on the date of the deadline stated in the Solicitation Summary. Proposers who mail their proposals should allow sufficient mail delivery time to ensure receipt of their proposals by the deadline. Delivery is the sole responsibility of the proposer. All appropriately filed proposals will be acknowledged in writing.

4.3 **Questions.** Written questions are to be submitted to the client contact listed in the Solicitation Summary. Answers to all questions, as well as copies of the questions, will be provided to all proposers unless, in the opinion of the client, a question is of such a nature that it is proprietary to the asking proposer.

4.4 **Proposals or declinations.** It is requested that the client receive responses, either proposals or declinations, from all parties receiving the RFP.

4.5 **Restriction on contact with the client's employees.** From the issue date of the RFP, all contacts with the client's personnel must be cleared through the client contact.

4.6 **News release.** The proposer shall make no news releases pertaining to this project without prior client approval.

4.7 **Proprietary information.** All proprietary information submitted in the proposal which the proposer desires to remain confidential shall be indicated clearly by stamping the word "Confidential" on the top and bottom of pages on which such information appears. For those proposals that are unsuccessful, all copies of such confidential information shall be returned to the proposer.

4.8 **Contract award.** The client reserves the sole rights to judgment and acceptance of the vendor's proposal. After the proposal(s) has/have being selected, the name(s) of the successful applicant(s) will be disclosed. Upon selection, the successful applicant(s) will be required to execute a contract with the client. In general, contracts will be awarded to the qualified proposer whose proposal is most advantageous to the client in terms of quality, cost, and other factors. The contract to be entered into between the client and the successful proposer shall contain negotiated provisions based on the specific requirements set forth in the RFP, and on the successful proposer's treatment thereof, as contained in the proposal.

4.9 **Reservation.** Notwithstanding anything to the contrary, the client reserves the right to reject any and all proposals received in response to this RFP; select for contract or for negotiations a proposal other than that with the lowest net cost; wave or modify any informalities, irregularities, or inconsistencies in proposals received; negotiate on any aspect of the proposal with any proposer; negotiate with more than one proposer at the same time; and terminate negotiations if a satisfactory agreement is not reached.

4.10 **Site visit.** Proposers must inspect the client's named location(s) prior to the submission of their proposals.

4.11 **Oral presentation.** Proposers may be required to give an oral presentation to accompany their written submission.

4.12 **Incurring costs.** The client is not responsible for any pre-contract activity or costs incurred by applicants in the preparation of their proposals.

5.0 **Selection of vendor.**

5.1 **Method of selection.** The client will evaluate all proposals and select the proposal(s) which it deems most beneficial. The quality and appropriateness of security operations will be evaluated first. Proposals meeting minimal acceptable standards will then be considered for overall costs. The final selection will be based on the combined merits of both the quality and pertinence of security services and the fee proposal. The evaluation will include the following criteria:

- **Organizational capability and quality (25 points).** Prior experience of the proposer in similar undertakings and the quality of such work must be demonstrated. Proposers should submit evidence of managerial effectiveness in this field for the previous five years or more. Proposers also should provide documentation regarding their proposed team/organizational structure to oversee the scope of work required in the RFP.
- **Understanding the need of the client (10 points).** This criterion is to be scored based on the proposer's ability to appropriately use required staffing, resources, and planning to address the nature of the client's unique needs.

- **Recruitment strategy and planning (10 points).** Proposers should submit a complete plan for the recruitment of qualified security officers and supervisors for the duration of the contract.
- **Qualifications and experience of the proposer's personnel (10 points).**
- **Training plans, curriculum, and training capability (10 points).**
- **Supervisory control (10 points).** Proposers should submit detailed job descriptions of the supervisory position, including span of control, and the method of disciplinary action for security personnel assigned to them.

5.2 **Rating system.** Proposals best meeting the minimal acceptable requirements on a 75-point scale will then be considered for their overall cost proposal.

6.0 **Payment.** The client agrees to pay the vendor on a timely basis from weekly invoices, which shall be submitted accompanied by attached original time sheets signed by the site director or designee.

7.0 **Liquidated damages/adjustments of compensation.**

7.1 **General provisions.** Upon the occurrence of any of the acts or omissions listed below, liquidated damages may be assessed daily against the contractor in the amounts indicated for each occurrence and for each day, starting from the day the occurrence commenced until the day the irregularity is corrected. The amount of assessment will be paid by the contractor or deducted from the contractor's invoices. (The dollar amount is not a standard and is included for reference purposes only.)

7.2 **Liquidated damages: $350 per day.** Liquidated damages may be assessed against the contractor in the sum of $350 per day per occurrence for each of the following acts or omissions:

- Failure to provide a security officer who meets the criteria specified in the contract.
- Failure to provide a site supervisor who meets the criteria specified in the contract.
- Failure to maintain complete personnel records folders for employees specified.
- The contractor's employees engage in a strike, work stoppage, or slowdown at the client's premises (a fee is assessed for each employee).

7.3 **Liquidated damages: $200 per day.** Liquidated damages may be assessed against the contractor in the sum of $200 per day per occurrence for each of the following acts or omissions:

- Failure to provide a visiting site manager who does not visit at least once during every tour as specified in the contract.

- Failure to provide a site supervisor at each client location for each shift, covering seven days per week.
- Failure to provide security services in an emergency or other contingencies for a period in excess of the stipulated shift hours as specified in the contract.
- Failure to notify the client of the names of security officers newly hired and assigned and to provide their personnel folders for review as specified in the contract.
- Failure of the contractor's employees to obtain approval from the client prior to signing a complaint on behalf of the client.

7.4 **Liquidated damages: $100 per day.** Liquidated damages may be assessed against the contractor in the sum of $100 per day per occurrence for each of the following acts or omissions:

- Failure to train guards required for a post or shift as required by the contract.
- Failure to maintain complete records of all hours each security officer assigned to the client's premises is engaged in, for which work is computed on the basis of actual hours worked.
- Failure to assign a correctly dressed security officer.
- Failure to replace any security officer within eight hours upon request by the client.
- Failure to submit oral or written reports of incidents occurring on the client's premises to the client.
- Failure to provide each security officer with required and working equipment.
- Failure to report missing fire extinguishers, smoke detectors, or hazardous conditions as specified in the contract.
- Failure to properly maintain the security officer location logbook as specified.

7.5 **Incomplete shift penalty.** Failure to provide a security officer at a specified client location on time or the early departure of a security officer from a specified client location will result in an agreed-to hourly assessment.

7.6 **Improper assignment penalty.** If the contractor assigns a security officer to the client's premises who, it is later determined, has a criminal record, the total paid to the contractor for the security officer's services shall be deducted from the contractor's invoices.

7.7 **Return of a previously terminated worker.** If a security officer is dismissed from one client location and shows up for work at another without the client's permission, the total amount paid for the officer's services from the date of dismissal will be refunded to the client plus $100 per day for each infraction.

7.8 **On the job negligence.** If the client's security system or property is damaged or stolen as a result of misuse or negligence by the contractor's employees, the contractor will be held liable for replacement or repair costs of the items.

8.0 **Notification requirements.** Vendors must notify the client and obtain advanced approval for any change, addition, or termination of major contract components and any change in the staffing by the vendor needed to serve the client effectively.

9.0 **General contract provisions.** The contract entered into between the client and the successful proposer(s) shall contain negotiated provisions based on the specific requirements set forth in this RFP and the successful proposer's treatment thereof, as contained in its proposal.

10.0 **Submission of Federal Employer Identification Number (EIN).**

11.0 **Insurance.** The contractor must have a minimum of $10 million of general liability insurance at the time the contract is awarded. The contractor must provide a certificate of insurance listing the client as being insured under the vendor's policy. The contractor's insurance policy must be written with an insurance carrier with a rating of at least "A" from A.M. Best. In the event the insurer is reduced to a lower rating, the vendor must take timely action to replace the insurance coverage with an insurer rated as at least "A" as judged by A.M. Best. Additionally, the vendor shall maintain specific adequate coverage for any vehicles to be used in the function of the contract.

12.0 **Reports.** The vendor shall maintain records and make reports as may be required by the client, including information needed for computerized data systems.

13.0 **Prime contractor responsibilities.** The selected vendor(s) will be required to assume sole responsibility for the fulfillment of the resultant contract(s).

14.0 **Sub-contracting.** No part of the work covered by this RFP shall be sub-contracted by the successful applicant(s) without prior approval from the client.

15.0 **Reservations.** As the need requires, the client reserves the right to *increase or decrease* the number of personnel to be authorized at the protected locations subject to the contract without limit. In addition to furnishing security services at the client's locations listed in this RFP, the contractor may be required to provide services at other client facilities not stated in the RFP.

16.0 **Length of contract.** The contract is for a 36-month period to be extended for a second 12-month period at the same rate for the initial period (or for whatever rate and policy the proposer wishes to achieve).

17.0 **Termination.** Either party to the agreement reserves the right to cancel any or all sites with 30 days' written notice.

18.0 **Equal Employment Opportunity.** The contract is awarded subject to applicable provisions of federal, state, and local laws and executive orders requiring affirmative action and equal employment opportunity.

Determining Final Costs

In addition to analyzing quality and competence of service (75 points), the final contract in this sample RFP is to be awarded based on cost considerations as well (25 points). The client determines the projected number of hours of coverage for security officer services. In the example of the above RFP, the client asks that the contractor anticipate a variety of expenses and incorporate them into the price to be charged for each hour scheduled by a security officer. In this instance, the vendor must consider a large number of ancillary costs—officer overhead, profit, selling and proposal preparation costs, uniform and equipment expenses, supervisory costs, insurance and other amounts—into the hourly proposed rate.

In other cases, proposed vendors and clients may identify other factors that could be expensed separately. For example, if vehicles are required to patrol the facilities, they may be provided by one party or the other, accounting for differences in payment.

Other Considerations

The RFP discussed above leaves much to be negotiated between the prospective vendor and the client. To the vendor, successful agreement on numerous fine points may be the difference between profit and loss on the actual contract performance. To the client, the difference can be between projected costs and costs that exceed budgeted targets. Clearly, a spirit of harmonious goodwill between both parties is needed throughout the life of the contract to resolve issues that normally occur. The RFP above provides for "liquidated damages," or cash penalties, against the vendor for specified breaches of the contract. Wise clients do not desire a punitive environment in which such measures occur with any frequency. If the client has screened the vendors carefully, service will meet expectations without the necessity of frequent cash penalties.

The design of the security officer's job relates to quality of performance. A guard who sleeps on a post faces dismissal, yet some jobs are designed with such little stimulation that it is difficult for someone not to sleep, especially if the security officer is sitting during an evening shift.

Meaningful tasks, built-in variation, and regular visits by supervisors can maintain the discharge of duties at an optimal operating level with high alertness.

Continuous Supervision

Whether the security contract is small or large, it is incumbent on the client to maintain an ongoing evaluation of security services. Observations on the quality of services—good and bad—should be constant. Feedback may be provided orally, in written form, or both to the contractor. It should be understood that a satisfactory level of service can lead to a continuing relationship between the parties.

RETAINING SERVICES OF PRIVATE INVESTIGATORS AND CONSULTANTS

Over 8,000 private individual investigators and investigative firms are licensed in the U.S. Countless others work under the supervision of experienced, licensed investigators. Many of these are solo practitioners work only occasionally as assignments come along. A few are organized, deeply staffed, and have offices throughout North America and beyond. Some practitioners are generalists; others specialize in a particular field. The *Security Letter Source Book* identifies over 35 different categories of investigators and security consultants; doubtlessly others will emerge as needs occur.

An employer requiring an investigator for a specific assignment must identify candidates who appear to have the experience, training, and ability to undertake the assignment. This is usually achieved by interviewing candidates for the assignment and describing in general terms the work needed to be undertaken. Competent investigators will outline how they would undertake the process, what resources they would need, the amount of time required, and the approximate cost for the service.

Compensation for investigators, consultants, or their agencies can be structured in different ways. The methods include a per-project basis in which the fee includes all personnel and out-of-pocket costs; a per-project basis with ancillary costs billed extra; as well hourly rates plus extra expenses. The cost of an investigator or a consultant will be marked up two-and-a-half to four times the actual hourly rate paid to the investigator. This divergence pays for the overhead and benefits for the investigator or consultant and provides sufficient extra to subsidize the investigator during times when assignments are few. Generally, the fee is related to time, skill, and difficulty for personnel involved in the investigation or consultation.

SUMMARY

Personnel represents the largest cost in most security programs. Consequently, the optimally performing manager will seek to assure that security functions are achieved with the minimum number of people required. Proprietary and contract security services both have advantages; therefore, managers sometimes plan to employ both in large operations. Software programs specifically written for security applications have improved accountability and decreased costs since their introduction. Investigations are currently playing an ever-widening role in both civil and criminal processes. A request for proposal (RFP) is a bureaucratic means of selecting a security services vendor. However, the process sets out a fair basis for identifying the best security company for the client and also to determine the most favorable arrangements for the client.

DISCUSSION AND REVIEW

1. What factor is "all too often" not considered by management when the possibility of converting from proprietary to contract guard services is considered?
2. What is the significance of possible collusion between employees and security personnel? What do many security directors feel is a measure that combats collusion?
3. How important is making people feel safe in security programs?
4. Outline the critical differences between criminal law and civil or contract law as discussed in the chapter.
5. What new types of investigation have emerged in recent years?
6. Security officers who are on "barred-from-customer" lists can be monitored most easily by what type of management tool?
7. What are convenient means by which security directors can determine compensation ranges for security officers in various geographic areas?

ENDNOTES

[1] D.R. Dalton (1991). *Managing Contract Security Services: A Business Approach,* Fremont, CA: Mill Creek Publishing, p. 5.

[2] D. Dalton (September 1994). "Looking for the Quality-Oriented Contractor." *Security Technology & Design,* Vol. 4, p. 6.

[3] Ibid.

[4] J.J. Fay (1987). *Butterworth's Security Dictionary,* Boston, MA: Butterworth-Heinemann, p. 102.

[5] A. Buckwalter (1984). *Investigative Methods,* Boston, MA: Butterworth-Heinemann.

[6] B. Chapman and E. Zwicky (1995). *Building Internet Firewalls.* Cambridge, MA: O'Reilly & Associates, p. 1.

[7] D. Parker (1998). *Fighting Computer Crime.* New York, NY: Wiley Computer Publishing.

[8] Ibid, p. 428.

[9] J.K. Barefoot (1995). *Undercover Investigations.* 3rd edition. Boston, MA: Butterworth-Heinemann, pp. 92–94.

[10] A.R. David (1986). *The Pyramid Builders of Ancient Egypt.* Boston, MA: Routledge & Kegan Paul, pp. 68–69.

[11] M. Lipson (1975). *On Guard: The Business of Private Security.* New York, NY: Quadrangle/The New York Times Co.

[12] Examples include CCS Security Guard Management System from Complete Computer Service, Ltd., Farmington Hills, MI; InTime Officer Scheduling from InTime Solutions, Inc., Burnaby, BC; and Security Management Systems (SMS) from Valiant Communications, Inc., Woodbury, NY.

[13] W.L. Winston and S.C. Albright (1997). *Practical Management Science: Spreadsheet Modeling and Applications.* Belmont, CA: Duxbury Press; L.J. Krajewski and L.P. Ritzman (1993). *Operations Management: Strategy and Analysis.* Reading, MA: Addison-Wesley. For example, see *Valiant SMS.*

[14] *Security Letter* (October 14, 2000). Part II. This information is also available at www.securityletter.com.

[15] S. Langer (2000). *Available Pay Survey Reports: An Annotated Bibliography,* 5th edition. Crete, IL: Abbott, Langer & Associates.

[16] F. Morn (1982). *The Eye That Never Sleeps.* Bloomington, IN: Indiana University Press, p. 98; also: J.D. Horan (1967). *The Pinkertons: The Detective Dynasty That Made History.* New York, NY: Bonanza Books, p. 50.

[17] J. Willey (1988). *The Business of Employee Leasing.* San Bernardino, CA: Employee Leasing Consulting Group.

Additional References

ASIS (1998). "ASIS International Presents Introduction to Security for Business Students." Alexandria, VA: ASIS International.

E.P. Kehoe (1994). *The Security Officer's Handbook: Standard Operating Procedure.* Boston, MA: Butterworth-Heinemann.

R.J. Meadows (1995). *Fundamentals of Protection and Safety for the Private protection Officer.* Englewood Cliffs, NJ: Prentice Hall.

Office of Federal Protection and Safety (April 1984). *Contract Guard Information Manual.* Washington, DC: U.S. General Services Administration.

10

OPERATING PHYSICAL- AND TECHNOLOGY-CENTERED PROGRAMS

> The ultimate purpose of any security system is to counter threats against assets and strengthen associated vulnerabilities.
>
> —Joseph Barry and Patrick Finnegan

Most of the costs of security operations are personnel costs created by security services. But it is the nature of management to drive down costs whenever possible. An important means by which security operating dollars can be made more effective is through the judicious use of physical- and technology-centered programs. These are concerned with physical security measures and electronic technology—often computer based—used to safeguard people, to reduce chances of theft, to evaluate ongoing operations, and to safeguard assets against damage or loss. Well designed and executed, such operations may decrease the number of personnel required to implement and maintain a high-performance protection program. If conceived and implemented poorly, however, physical- and technology-centered programs can produce unsatisfactory results. Further, if badly conceived and implemented, such initiatives can produce a sense among workers that management is "putting systems above people." Clearly, physical- and technology-based measures should enhance the use of protection personnel and other resources, not detract from them.

We begin this chapter by briefly citing the theory of situational crime prevention concepts first encountered in Chapter 1, and then consider the use of non-personnel based resources to mitigate the risk.

SITUATIONAL CRIME PREVENTION: A PHILOSOPHY OF CRIME REDUCTION

Law enforcement and criminal justice practitioners who create crime prevention programs usually are concerned with a set of defined priorities. These include community programs, juvenile deviance, coordination with prosecutors and the courts, as well as police staffing and environmental and technological strategies. By contrast, persons concerned with security operations management for a corporation or institution have little short-term control over such factors as where the facility is located, policies and programs related to juvenile offenders, the responsiveness and leadership of local criminal justice programs, and how the police do their job. These differences have influenced the ways in which law enforcement and private security firms tend to view the causes of crime and disorder.

Over the years, legions of criminologists and social critics have written on supposed "root causes" of crime, and the social disorganization and individualism that perpetuates deviance.[1] These same writers largely fail to consider why a few individuals in a particular social, ethnic, political, economic, and even familial situation commit crimes while most do not. In the end, the manager concerned with reducing losses does not likely speculate on what is neither quantified nor provable, but rather concentrates on what can be accomplished in security programs based on the results of accepted research activities. In part, this means deterring or suppressing crime rather than focusing on apprehending and prosecuting violators.

For example, a facility can be designed to make it less amenable to loss. In many circumstances, however, the manager faces situations in which changing the facility design—using architecture and engineering methods to create spaces that are less amenable to crime, loss, or injury—is not an option. Instead, personnel, procedures, physical measures, and technology must be altered to prevent or mitigate losses. The good news is that security managers today are generally involved early in the design considerations of a new facility so that their insights into loss control can be implemented. (This process is discussed near the end of this chapter.) Several architectural and engineering firms have loss-prevention specialists who stay abreast of protective and life safety measures so that they may be designed into new facilities. Other specialized consultants offer services that lower property risks to existing sites through better security planning and design.

In the 1970s, the architect Oscar Newman studied public housing in New York City and elsewhere and determined that crime rates vary according to territoriality, surveillance, image, and environment.[2] "Territoriality" refers to the sense of possession by residents or workers of an environment and the tendency of people to defend this territory against those who would commit criminal within or near the area. "Surveillance" relates to the ability of people within buildings to view people outside their immediate environment. "Image" refers to the general reputation of a place. And finally, "environment" refers to the nearby area that renders the zone safe or unsafe.

Defensible space, therefore, defines an area where surveillance is extensive, the image is positive, and the nearby environment is safe and protective of residents and visitors. Newman's theory produced a design strategy called Crime Prevention Through Environmental Design (CPTED), discussed in Table 10.1. Defensible-space theory concludes that crime may be reduced by improving surveillance of public areas, demarcating private versus public space, and improving the image and environment of the area.

Research on defensible space concepts sometimes concludes that the concepts are not always successful at reducing crime because they fail to take into consideration the cognitive processes individuals use to adapt to physical environments.[3] Patricia and Paul Brantingham analyzed crime

Table 10.1. Crime Prevention Through Environmental Design (CPTED)

CPTED argues that changing the environment through design can make certain types of crimes less likely to occur. Antisocial and criminal behavior will not disappear, but the frequency will decline because the environment is less hospitable to potential offenders This is because the area seems better protected by its owners and thus the would-be criminal is more likely to be detected and arrested. The principle is based on research involving residential buildings, although the same principles relate to commercial and institutional property. CPTED is based on four principles:

Principle	Methods of Applications
Territoriality	Property looks cared for; broken windows are repaired; graffiti removed.
	Residents are seen making improvements or enhancements to their areas.
	Access control discourages unauthorized visitors and deters their entry.
	Controlled space is differentiated clearly from nearby transitional zones.
Surveillance	Residents can observe outer areas from within their buildings with clear lines-of-sites to call for assistance in the event it is needed.
	Hallways and public areas are designed to be open and non-constraining.
	Cul-de-sacs and hiding places in public areas are designed-out.
	Closed-circuit television (CCTV) and modern alarm systems are likely to be in use.

(continues)

Table 10.1. *(Continued)*

Principle	Methods of Applications
Image	The property has a favorable image in the area and is looked at as being well maintained and cared for.
	Events and activities are programmed to increase use of public spaces.
Environment	The area immediately beyond the property—nearby buildings, streets, retail space and parks—are equally well cared for.
	Communications systems permit persons in both congested and isolated areas to call for assistance when needed.
	Conflicting activities—such as a playground for toddlers and a basketball court—are separated.
	Street furniture, sitting areas, and fountains are designed to serve locals while not attracting vagrants.

rates by occupation and economic specialization and have determined that crimes likes murder and assault occur in areas of economic decline and neglect, whereas white-collar crimes occur in areas in which a high number of potential victims exist.[4] They also found that certain environmental changes increase public use and decrease fear.

More recently, CPTED has evolved to include the concept of situational crime prevention, which argues that crime may be reduced in a particular area when aspects of the environment are changed, often involving little cost or effort. For example, making it harder to commit a crime by modifying the environment—by installing better lighting, broader surveillance, CCTV, and alarms that will call police to the scene quickly—can decrease crime in an area.[5]

THE RISK VERSUS COST RATIO

The level of security in a given area can depend upon many factors. Therefore, a range of options should be evaluated for their pertinence to a given situation. A broad spectrum involving widely varying degrees of risks and controls exists. At one end, controls are absent and risks for loss are high. At the other end, the reverse is true. The thesis of this book, indeed the view of many security practitioners, is that weakness is eventually exploited. Therefore, lack of adequate security increases the likelihood of losses. Further, as assets increase in value, the potential for loss also grows, as shown in Figure 10.1. The following sections look at the continuum of security conditions and their management relevance.

Security Conditions	Protectionless	Low level	Medium	High level	Exceptional	Failsafe
Factors	No overt measures; psychological, environmental crime deterrents only	Impedes and detects some external activity	Impedes, detects, assesses some external activity and some unauthorized internal activity	Impedes, detects, assesses most unauthorized external and internal activity	Impedes, detects, assesses, and repels most unauthorized external and internal activity	Prevents access except under constricted circumstances; usually not economic
Risk Factors	High ←					Low
Cost Factors	Low					→ High

Risk of losses and the cost of security measures have a reciprocal relationship. Low protection has low cost, but invites higher risk of loss. In response, the cost of security can increase.

Figure 10.1. The Risk versus Cost Continuum

- **Protectionless security.** The author's grandparents lived in a small, safe community. Their front door was rarely locked. When they left their house for a trip, they locked it, but left the key in the front door keyhole "in case someone needed to get in." This seemed to make sense to them. Security depended upon the fact that residential burglary was rare at that time and place. Those who might be inclined to commit such a crime did not systematically survey the neighborhood to see what residences were unlocked and were easy targets. In time, the younger generation convinced their parents and grandparents that wisdom dictated that the key should be elsewhere and it was henceforth placed under a nearby flowerpot.

  Such protectionless behavior still exists in some residential areas. However, contemporary organizations realize that reasonable and adequate measures must be taken to protect their operations. The orderly and lawful behavior of others cannot be assumed. That means implementing appropriate security measures to protect the value of the assets located there.
- **Minimum security.** With little effort, this type of system impedes some unauthorized external activity, which is achieved by physical barriers and locks.[6] This level of protection is adequate for some houses, but not for commercial or institutional activities.
- **Low-level security.** This system impedes and detects some unauthorized external activity. Doors and windows may be reinforced and a local alarm system may be installed.
- **Medium security.** Here the system impedes, detects, and assesses most unauthorized external activity and some unauthorized internal activity. This is achieved by the use of advanced alarm systems and possibly animals and unarmed security officers.
- **High-level security.** Relying on greater capital investment, adequate personnel, and well-considered procedures relative to the previous category, this system impedes, detects, and assesses most unauthorized external and internal activity. In addition to features found in lower levels, this level of security can include CCTV, advanced perimeter and interior (volumetric) security systems, highly trained and supported security officers, and management dedicated to constantly seeking programmatic improvements.
- **Exceptional security.** This advanced protective status impedes, detects, assesses, and neutralizes most unauthorized external and internal activity. This is achieved by tamper-resistant, complex systems, and highly trained and vetted personnel capable of responding promptly to any alarm condition and most threats.
- **Failsafe security.** This is a conceptual level of security in which serious losses over an extended period of time are highly unlikely

because of the exceptional defenses that prevent such occurrences. Such a level of security is costly and impractical in most situations because the controls are deliberately restrictive and time-consuming to overcome even for those with some authorization over control and custody of the assets to be protected. Failsafe security is unlikely to permit losses, but its nature also deters normal economic activity.

Sales and marketing–oriented managers often battle against measures that inhibit the development of protective controls. Meanwhile, security directors face the task of justifying and obtaining the maximum level of protection reasonable for the situation required. Lack of any security is not an option; neither is failsafe security. The strategy for security operations is to find the right level of security measures to satisfy constantly changing requirements.

Determining the reasonable protection level for an organization involves evaluating four types of issues: the types of risks faced by the organization; the likelihood of their occurrence; the impact they would have on operations; and the resources reasonably available to identify the risks. This is a topic of considerable importance in justifying security programs and their expenditures. Some managers prefer to create programs based on their and their associates' experienced judgment. Others use a matrix of risk types, likelihood, and impact, which can lead to an estimate of resources needed. Other managers use software programs that guide the user into identifying risks and considering various possibilities of occurrences.[7] These are converted into strategy during the planning period to meet management's objectives.

WHY PHYSICAL SECURITY IS IMPORTANT

Physical barriers have been used as a means of protection for centuries. Along with animals and human sentries, walls, fences, weapons, and locks have always been important means of protecting people. Early humans chose their housing with defensive features in mind, and communities were formed to take advantage of mutual protection. Safes and vaults became important ways of protecting assets in early market centers when their owners could not be present and in locations where hiding places were limited. The ancient Egyptians, meanwhile, developed the pin tumbler lock.[8] Early and medieval communities protected themselves from foreign armies as well as organized gangs with walls and fortifications. Such physical security measures had many advantages: they represented one-time-only costs; they were usually reliable and worked well for long periods of time; and they achieved their objectives by deterring or reducing opportunity for unauthorized entry.

J.J. Fay defines physical security as "that part of security concerned with physical measures designed to safeguard people, to prevent unauthorized access to equipment, facilities, material and documents, and to safeguard them against damage and loss."[9] The term encompasses measures relating to the effective and economic use of a facility's full resources to meet anticipated and actual security threats. Concerns of physical security planners include design, selection, purchase, installation, and use of physical barriers, locks, safes and vaults, lighting, alarms, CCTV, electronic surveillance, access control, and integrated electronic systems. The term physical security includes physical barriers, mechanical devices, and electronic measures. Typically, systems involve a combination of two or more distinct measures to protect people, physical assets, and intellectual property.

Security operations planners sometimes think first of physical security in their protection strategies. Several reasons support this tendency: physical security substantially requires a one-time cost only; physical measures are usually clearly visible and deter unlawful or unwanted acts; care and upkeep are limited; specific standards have been set in many cases to guide the security planner on decisions; and physical security measures are uncomplicated to purchase, install, and care for. However, the primary goal of security measures is to protect people, not physical or intellectual assets.

Technology can be used as a powerful tool in well-conceived security programs. It can perform complex monitoring operations and control features that individual security personnel cannot control. As a result, contemporary high-tech implements permit a higher level of confidence in protective programs than in the past. Just as changes in communications, sensing, and computing have affected society at large, these developments also have re-shaped the means and quality by which security services are performed. Indeed, a security planner learning of a new technological development is likely to wonder how it can be applied to enhance operating security programs tomorrow, if not today.

The following section on security countermeasures to loss is neither comprehensive in its scope nor detailed in its discussion. The discussion does, however, serve to provide an overview of physical security measures to consider when creating or assessing protective programs. A final section in this chapter offers guidelines for designing and constructing a complex integrated system with the aid of security engineers or consultants. In addition, the notes at the end of the chapter provide resources in which to further pursue individual topics.

SELECTING SECURITY COUNTERMEASURES TO REDUCE LOSS

The security planner should consider a variety of countermeasures if an event is likely to produce a significant loss to the organization. No single measure will be adequate, more than one option should be looked at. The term "concentric circles of protection" reflects the concept that numerous protective measures separate the outer environment from the innermost protected locations. In planning the measures to be taken, thought is always given to the appropriateness, utility, and cost of the procedures, equipment, and personnel required to meet the expected objectives.

Effective countermeasures may serve one or more of the following objectives: deterrence (that is, preventing or discouraging unwanted action); delay or denial (that is, impeding or stopping an unwanted action); and detecting (that is, discovering or ascertaining the significance of a possible security breach) (see Box 10.1).

Box 10.1. Physical Security Countermeasures to Loss—Strengths and Relative Cost

	Strengths			
Countermeasures	**Deter**	**Delay**	**Detect**	**Relative Cost**
Facility Design	x	x	x	Low
Animals	x	x	x	Low
Barriers	x	x		Low
Security Glazing	x	x		???
Signs	x			Low
Locks, Keys, Containers	x	x		Low
Lighting Systems	x			Low
Closed-Circuit Television (CCTV)	x		x	High
Intrusion Detection Systems	x		x	Moderate
Access Control Systems	x	x	x	High
Alarm Systems	x	x	x	High
Robotic Systems	x			Very High
Communications	x			Moderate
Information Security Systems	x		x	Low to High
Contraband Detection		x	x	Very High
Fire Detection and Life Safety			x	High

Source: *Introduction to Security for Business Students* (1998). Alexandria, VA: ASIS.

Facility Design: Location, Area, and Architecture

From earliest known times, humans have improved their safety and security by considering their location and using its protective environmental advantages. Natural barriers that provide an intrinsic protective value include bodies of water, marshy areas, deserts, mountains, and hidden areas such as caves. These naturally occurring physical characteristics were enhanced with walls, fences, bars, moats, ditches, cleared spaces, and other adaptations. Medieval cities, for example, were often built on hilltops in order to take advantage of the superior lookout provided there. (Examples of such hilltop villages include Carcassonne in France and Urbino in Italy.) The lake dwellers in Switzerland built their towns on stilts for security. Fortified dwellings expanded over time, eventually becoming castles. These were developed partially to protect residents from endemic warfare, as well as organized bands of thieves in Europe threatened the safety and security of small communities.[10]

Locations that had natural defense characteristics, often strengthened by structural barriers, served to define property boundaries; to control access to restricted or privileged areas; to delay and impede unauthorized entry; to channel and restrict the flow of traffic; to facilitate the identification of possible intruders or threats to the area; and to provide for efficient use of security personnel or other guardians.

In the contemporary organization, location matters just as much as in earlier centuries. In crime prevention research, the study of location holds considerable importance. Areas with high personal and property crime and urban problems will lose residents, commerce, industry, and institutions. Further, new organizations will be reluctant to locate to such crime areas without considerable inducements. Nonetheless, it is possible for an organization to locate in such areas and thrive with appropriate physical security measures. Before committing capital investment in a new location, a security-conscious manager must carefully consider location and area. This includes collecting crime data, studying law enforcement resources and culture, and determining how these factors relate to the security measures that must be put in place.

Animals

Some of the earliest protective sentries were animal, not human. Livy describes how geese on one of the hills in ancient Rome sounded an alarm when Gaullic invaders sought to surprise an army encampment at night from a steep and unprotected side of the hill.[11] In current applications, geese have been used sometimes to protect NATO facilities.[12]

For centuries, dogs have played an important role as guards. Guard dogs usually patrol inside fenced areas and buildings without a handler, and are often used in facilities with no evening security personnel or workers, such as at retailers, car dealers, construction yards, and distribution facilities. In other circumstances, guard dogs work with handlers. Because of their superior sense of smell and acute hearing, dogs play a large role in searching for lost or hidden persons, contraband, and explosives.[13]

Breeds of dogs vary in their ability to be useful for security purposes. Animals can be categorized as high or low in such diverse characteristics as reactivity, aggression, trainability, and capacity for investigation. Dogs used in security and police work for bomb or drug detection are trained by Pavlovian methods. The handler trains the dog every day to find hidden explosives or drugs and then feeds and praises the dog with each success. Security planners interested in using dogs for security tasks must select personnel who will be trained with the animals and will be committed to their welfare.

Barriers

Barriers may be constructed to further the protected area. For example, a body of water or difficult-to-penetrate shrubs may provide psychological and distance deterrents. Manufactured fences also provide an important barrier for physical security.[14] If a security planner determines that a fence is desirable, related issues may be raised: Will one fence be enough, or will two fences with a patrol space between them be better? Will the fence have clear areas around it so that a good line-of-sight is maintained? What impediments to climbing will be used? To digging under the fence?

While fences may be made of many materials, chain-link fence is commonly used due to its availability, flexibility, cost, established use, and ease of installation. In security applications, fences are typically no. 11 American wire gauge or heavier, with 2-inch mesh openings. They are usually 7 feet tall and are topped by three strands of barbed wire or razor ribbon evenly spaced 6 inches apart and angled outward 30 to 45 degrees from the vertical. Since attackers may pass under the fence, the bottom may be designed so that penetration is difficult.

To many security planners, a fence may seem like an attractive security option: chain-link fencing is a widely used visible deterrent requiring little maintenance. However, fences do have shortcomings and should be regarded as being able to provide only temporary deterrence. As Gigliotti and Jason note:[15]

> Regardless of how elaborate fences may be, they still offer only a modicum of security. Fences are necessary, but investments in this area should be kept to a

minimum as the money can be better used on other components of the total system.

Consider that someone wanting to pass from one side of a fence to the other has three options: they can go under the fence, through it, or over it. It is possible to go under a fence by digging a hole beneath it, though the time and effort required to successfully accomplish this make this approach onerous. Going through the fence is possible, as wire cutters can cut out an area for someone to pass through in a few minutes' time. Finally, someone can go over the fence. Research at Sandia National Laboratories, Albuquerque, for the Department of Energy, determined that trained individuals with penetration aids like ladders, sheets, carpet fragments, and wood planks can scale over a fence in 5 to 10 seconds depending on the penetration aid used and whether someone was assisting in the passage or not.[16] The average untrained individual would not be able to cross a fence so quickly, and the presence of the fence would discourage many attempts. However, the inherent weaknesses of fences indicate that the security planner must think beyond this structure to make a facility safe from penetration. In addition, most fences present an unwelcoming, rigid impression, which may be unacceptable to an organization. Some facilities with medium-to-high security vulnerabilities have decided not to use fences, but rather to rely on other means of protecting facilities.

An alternative to metallic fences is the use of plants such as hedges to serve as a natural fence. Such measures are limited to facilities in which the hedge will not be needed for a few years as it grows to the proper height and depth so that it can serve its intended protective function. Perhaps the most widely used protective hedge is *Trifolium orange*, which grows in many types of soils, putting down deep roots in time so that even a jeep would have trouble driving through it.

Security Glazing

Glass is a transparent and brittle substance composed chiefly of silicates and an alkali fused at high temperatures. It is manufactured in many types. For security purposes, glass is often fused with layers of plastic, usually polyvinyl butyral (PVB), although laminates of polycarbonate have a robust protective value and are also transparent. Thousands of glazing configurations have been created, but only a few meet security standards. Some products can withstand bomb blasts, others can withstand burglary attempts and bullet discharges. Well-selected security glazing can provide performance, control, and cost savings. In some cases, a transparent film placed over ordinary glass can increase bomb blast resistiveness.

Signs

Warning signs have an important deterrent effect. Placed at the perimeter of a protected facility, they discourage would-be offenders from a variety of unwanted behaviors. The excavated ruins of Pompeii revealed that two millennia ago, homeowners sought to deter possible predators with prominent signs, some made of mosaics, warning *cave canem* (beware of dog). That same message—but in English—still appears on the fences of construction lots or the door of garages and distribution facilities where guard dogs are found.

Much more common are warning signs that indicate that trespassers will be arrested and prosecuted or that electronic security systems are in use. Such signs are an indication of guardianship; that is, the owners and operators of the facility are aware of risks and have taken measures to protect assets. Such signs should be placed around the perimeter so that potential perpetrators from any direction will be warned of the protective measures being taken.

Signs usually represent a one-time cost. Although they will not stop determined thieves, they will signal such individuals that they must move quickly due to the threatened risks of being detected and apprehended.

Locks, Keys, and Containers

As mentioned previously, locks were one of the earliest manifestations of physical security. The art of the locksmith has been respected over the centuries for its beauty, practicality, and necessity. Locks remain an integral part of contemporary physical security planning. Locks, along with their keys and the containers of which they may be a part, have many benefits for security programs. Simple to use, they are complicated to make. Involving a one-time cost, they may be used repeatedly with reliability over years of service. Locks and keys may meet different levels of security according to requirements of the location. They are easy to employ and can be designed into containers, furniture, doors, and machines with ease.

A disadvantage of mechanical locks is that they provide no evidence of who accessed the lock over its previous uses. This drawback is eliminated by electronic locking systems now available. These systems may be opened with cards or tokens that are inserted or brought near (proximity lock) a sensor, which in turn opens the door.

The security planner should concentrate on a series of easy-to-understand principles in deciding what locks to use and why.

Key-Operated Locks. Most locks use "tumbler mechanisms" to operate. That is, the key enters the keyhole and moves the variable tumblers into a straight line so that they then can turn the lock cylinder. If the correct key is not inserted into the keyhole, all of the tumblers will not be in a straight line (shear line) and the lock cylinder will not be able to rotate around the shell of the lock.

The most widely used key-operated mechanism is the pin tumbler, for which a wide variety of security levels is available. As every observer of action programs on television or the movies is aware, pin tumbler locks—despite their ubiquity—may be picked; that is, they may be entered without a normal key by manipulating the tumblers to the shear line so that rotation can occur. Picking is one of several ways by which mechanical pin tumbler locks can be defeated (see Box 10.2).

In addition to pin tumbler locks, other mechanisms are available, including magnetic or optical locks. The security planner may wish to evaluate and compare the strengths and weaknesses of these other mechanisms with pin tumbler locks.

Box 10.2. Pin Tumbler Lock Security

Pin tumbler locks are the most widely used mechanical lock type. Despite their popularity, this type of mechanism may have—depending on the lock type—inherent weaknesses of which the security planner should be aware.

Factor	Consequence
Picking Tools	Widely available for sale.
Picking Skills	Taught in legitimate locksmith schools, but also in mail-order instruction courses.
Age of Lock	With use over time, pins and keys get worn down, making them easier to pick.
Number of Pins	A three-pin tumbler has about 130 combinations; a six-pin tumblers has about 65,000. However, for technical reasons, the number of possibilities is actually lower. Yet the principle remains: more pins, higher security.
Angle of Pins	Pins that are vertically aligned to the cuts of the key are easiest to pick. Those that are aligned on different planes are extremely difficult or impossible to pick.
Control of Master Key	Pin tumbler locks may be subdivided into master, sub-master, and other divisions permitting key control. However, if a master key is lost, stolen, or inappropriately copied, all the locks in the protective systems may have to be changed at great cost and inconvenience.

Lock Hardware and Mountings. For moderate- to high-security applications, locks are mortised; that is, they are installed within the core of the door or locking device rather than on its surface. Mortised locks can be changed by a locksmith when needed, thus maintaining the level of security after the previous lock setting has been damaged or compromised and must be replaced.

A high-security lock on a low-security door, inserted into a weak doorframe attached to a plasterboard or glass wall, is poor means of providing security. The intruder will bypass the lock and instead attack the door, frame, or adjacent wall. Thus, all of these parts must have comparable resistance to attack; otherwise, the security objective will not be achieved. Bolts and locks must be inserted so that they do not represent a temptation to the would-be intruder.

Vaults and Safes. Vaults are windowless enclosures with the walls, floor, roof, and one or more doors designed and constructed to delay penetration. Safes are containers, usually with one or more locks, and are smaller than vaults. Both vaults and safes are constructed with tool-resistant steel as well as brick, concrete, stone, tile, or similar masonry. The lock may be either electric or mechanical, with other locks placed on inside containers.

Underwriters Laboratories (UL) provides standards for the burglary resistiveness of vault doors. For example, UL808 signifies protection against expert burglary attacks by cutting torches, fluxing rods, portable electric-powered and hydraulic tools, and common hand tools. UL also promulgates standards for safes. A typical standard (UL687) is for Class TL-30X6. This signifies a combination locked chest or safe designed to offer protection against entry by common mechanical tools for 30 minutes on all six sides. The safe weighs at least 750 pounds or is equipped with suitable anchors to the floor or other surface.

It is important to note that burglar resistance and fire resistance are not the same thing. Vaults and safes that are fire resistive demand a minimum thickness for walls on floors where they may be located. The National Fire Protection Association establishes minimum standards for the type of materials required to meet two-, four-, or six-hour protective classifications.

Lighting Systems

Violent and property crime, disorder, and accidents occur disproportionately at nighttime or in poorly lighted areas. Good lightning therefore represents one of the greatest deterrents to crime, disorder, or unauthorized access after dark. Dark commercial areas that undergo improved lighting become accessible to more people and stimulate use. The technical quality, energy costs, and longevity of different lighting systems vary widely, although standards exist for minimum-security lighting (see Box 10.3).

Protective lighting should permit the public—including security officers on patrol—to easily see physical features in their immediate environment. Light should be evenly intense along the patrol route. Illumination may be directed toward the outer area where unauthorized people may seek to approach a facility. When buildings are to be protected, lighter colors and unobstructed areas for clear vision are valuable.

Security planners also are conscious of the need for standby and movable lighting to supplement normal lighting conditions. Emergency lighting may supplement standby and movable illumination and is used during times of power failure or other occasions when normal systems are inadequate. Normally, local public utilities are the primary sources for power, but all comprehensive security plans anticipate periodic, unpredictable outages. In such cases, alternative power may be provided by standby batteries or gasoline-driven generators.

Box 10.3. Minimum Security Lighting Standards

Formal standards specify the minimum lighting required for different security applications. The topic can be complicated because of the irregular ways in which surfaces are illuminated at night. That is, a particular spot may meet minimum standards in one place, but a few feet away, the light may be inadequate. Security practitioners take readings with light meters over several spots to determine whether illumination is satisfactory. This device measures illumination in footcandles as a standard, where a footcandle is defined as the amount of light shining on a square foot of surface from a single candle one foot away. Generally, measurements are taken three feet off the surface or on the surface itself.

Surface	Minimum Security Illumination
Perimeter Fences	0.5 footcandle on either side of the fence
Building Exterior	0.5 to 2 footcandles on the surface
Potential Hazardous Area	1 to 3 footcandles minimum
Parking Lots (Covered)	5 footcandles at about 6 feet above the surface
Entrances	10 footcandles on ground level

Source: Illuminating Engineering Society of North America (1993). *Lighting Handbook*, 8th edition, New York, NY: IES.

Closed-Circuit Television (CCTV)

Television transmission that does not broadcast TV signals but rather transmits signals over a closed-circuit via an electric wire or fiberoptic cable is called a closed-circuit Television (CCTV) system.[17] These systems are invariably part of integrated security systems, which combine CCTV surveillance with other countermeasures.

The first generation of CCTV cameras used for security applications relied upon cathode ray tubes (CRTs). These are vacuum tubes in which electrons emitted by a heated cathode are transmitted via a beam toward a phosphor-coated surface, which then becomes luminescent. CRTs have different performance qualities, requiring the systems designer to select different types of tubes according to the circumstances encountered.

Since about 1990, security planners have shown a marked preference for a new generation of camera: the charge coupled device (CCD). This is a camera that uses a chip—a solid-state semiconductor imaging device—that transfers information by digital shift register techniques. Chip cameras have numerous advantages over CRT technology. As a result of their light weight, CCDs present less demand on their environmental housing and on motors that pan-tilt zoom the mechanism. They are smaller in size, lending themselves to aesthetic demands of the environment as well as use in covert surveillance. Of greater significance is that picture quality is superior with no loss of definition at the edges of a visual field. CCDs are also rugged. The smearing and blooming that plagued tube cameras and created burned spots do not occur with CCDs. Additionally, chip cameras offer good value with a lifetime use several times that of a CRT model.

Monitors. CCTV systems involve more than cameras. Monitors are devices for viewing a television picture from the output of a camera. The monitor may display the video signal directly—live from the camera, from videotape or other stored media, or from special effects generators. CCTV monitors are designed to work with closed-circuit systems, as shown in Figure 10.2. Often, but not invariably, they possess better performance characteristics than those made for consumers, and their price generally reflects this. Digital monitors are available in standard sizes of 5 inches, 9 inches, 12 inches, 15 inches, and 19 inches, with the 9-inch screen being used most widely. Larger-size screens are used when the application divides the screen into multiple images or when a security officer wishes to move an image from a small screen to a larger one for better visibility.

The images may be transferred from the camera to the monitor via coaxial cable (commonly RG59U or RG11U), fiberoptics, and, increasingly, wireless means such as radio frequency (RF) or microwave transmissions. Signals can also be transmitted via telephone lines, making it possible to monitor signals over the Internet.

Source: Methodist Hospital, Indianapolis. Photo courtesy of Gyyr.

Figure 10.2. Hospital Monitoring Room (Surveying Over 200 Camera Locations)

Recording Devices. CCTV images collected for security purposes often are recorded and archived temporarily. Real-time videocassette recorders (VCRs) convert the signal from a video camera onto magnetic tape. During playback, it reconstructs the video signal for viewing on a CCTV monitor or, if needed, to obtain printed copies of images. VCR videotape has a finite life depending on the speed with which the image was registered on the tape and the frequency on which the tape was re-recorded. Tapes that have been used beyond their normal lifetime may be useless for identification purposes. The advent of new digital storage media may enhance the ability of an image to be retained for long periods of time, retrieved when needed, and to do so at low cost.

VCRs and other storage technology have made immense improvements in the utility of security systems. In particular, countless crimes have been resolved by the visual evidence collected by image-storage devices. These are further enhanced since contemporary systems often have built-in time/date/camera number generators. For prosecution purposes, the evidence collected by such images can be persuasive. Similarly, descriptive inserts on the screen can pinpoint where the images are being recorded.

CCTV systems can include hundreds of cameras, if needed, within a single configuration. A video multiplexer allows the same system to encode, decode, or view live multiple scenes at the same time.

Technical Features. Selecting the right lenses for CCTV systems is an integral part of a high-performance system. Most lenses used for security purposes are fixed focal length (FFL) and produce a single focal length. A focal length (FL) is the distance from the lens center to a location in space where the image of a distant scene or object is focused. FFL and FL are expressed in millimeters or inches. FFL lenses must be matched with the image sensor size or with the smaller sensor size. They cannot be used with a larger sensor size than that for which they are designed. For example, a 1/2-inch sensor formatted camera will require a 1/2-inch or larger formatted lens. The image size of the picture is determined by the FL of the lens and has nothing to do with format size. Lenses for wide-angle and telephoto viewing may also be selected. Zoom lenses are variable focal length (VFL) lenses that allow a smooth, continuous change in the angular field of view so that the view can be made narrower or wider depending on the setting. This is generally accomplished by a motorized adjustment that can be directed remotely. For covert surveillance or privacy purposes, pinhole lenses are widely available.

CCTV is increasingly integrated with other technologically advanced resources. For example, video motion detection (VMD) is a software-based hardware device that detects intrusion and generates an alarm condition set by the parameters of the security system. CCTV images may also be used to confirm alarms from intruder detection systems, combining the CCTV with another intruder technology, and using the alarm of this technology to establish a video link to a remote monitoring center.[18]

Intrusion Detection Systems

Intrusion detection systems deter and detect potential entry to a protected area by unauthorized means. The security planner has an extensive choice of sensors that can identify such incursions to a protected location, each of which has advantages and disadvantages (see Box 10.4).

Sensors to detect possible intrusion may be used at outer or inner perimeters, within interior spaces (volumetric), and for particular objects or at spots requiring protection. The sensors discussed in this section are electronic. The principle of all of them is simple: a normal system is disturbed; it then goes into an alarm state. An audio alarm may be sounded at the site or at a distant monitoring station where security personnel evaluate the circumstances.

Box 10.4. Intrusion Detection Systems

Intrusion detection systems are desirable for external perimeter and internal detection. No one system is ideal, and many security planners employ two or more different kinds of sensors to protect the same area. Each type of intrusion detection system listed below has numerous advantages and disadvantages beyond those noted.

Type	Advantage	Disadvantage
Underground[1]	Hard to detect	Costly to install
Fence[2]	Increases deterrent value of fence	Does not detect tunneling
Photoelectric	Indoor, outdoor beams	Can be spotted and avoided
Microwave (Exterior)	Cheap, easy to install	Requires line of sight
Microwave (Interior)	Detects movement in area	Prone to some false alarms
Passive infrared	Reliable; inexpensive	Susceptible to defeat by covering lenses
Active infrared	Can protect oddly shaped areas	May require repeated adjustments
Ultrasonic	Covers large, diffuse area	False alarms from traffic, machines
Sound	Highest robbery detection	Privacy concerns
Capacitance	Triggered by weight	Avoidable if known
Vibration	Identifies burglar tool use	False alarms from nearby motors
Door/window switches	Widely used, inexpensive	Magnets can defeat
Metallic foil	Detects window attacks	Cracks with age; unaesthetic
Glass breakage	Identifies breakage sounds	False alarms from street noise

[1] Such as buried microphones, underground sensor tubes, and buried seismic sensors.

[2] Such as taut wires, leaky cable, and microphone. Taut wires signal an alarm condition when someone places tension on the wire; leaky cable sends and receives an electrical signal, which alarms when someone absorbs transmitted energy; and microphones collect audible signals of possible intrusion attempts.

Numerous environmental and other factors need to be taken into consideration in order to determine which sensors are to be selected for desired security applications. More reliable sensors and systems are constantly being created.

The workhorse for interior (volumetric) motion detection is the passive infrared (PIR) sensor. This operates on the principle of heat detection. It is widely used in non-security applications, such as to open doors in buildings. PIRs are sufficiently sensitive so that they do not cause false alarms from a wide ambient temperature range or from the heat of a small animal. However, PIRs may cause false alarms from hot spots caused by lights, reflections, and solar and mechanical heat sources. To overcome the possibility of false alarms, sensor manufacturers provide dual technology sensors, incorporating PIR with either microwave or ultrasonic technology.

Access Control Systems

Access control systems control persons, vehicles, and materials through entrances and exits of a protected area. (The term is also used in computer security where it has a different meaning.) Access control systems use hardware and specialized procedures to control and monitor movements into, out of, or within a protected area. Access to protected areas may be a function of authorization time or level, or a combination of both.

Access control depends upon the authorized person being correctly identified as part of the approval process. In a simple protective system, on-the-spot visual recognition of an unauthorized person, vehicle, or materials may suffice. However, large systems with numerous personnel and individuals with varying levels of authorization are best managed with systems that identify such persons automatically and with a high degree of certainty. Such systems typically involve use of three features:

- **Something that the person knows.** This can be an access code or password supposedly known only to the individual.
- **Something that the individual possesses.** For example, an approved identification (ID) card or a token that cannot be easily counterfeited.
- **Something physical and unique about the individual.** This could be a biometric feature such as a fingerprint, iris or retinal signature, writing dynamics, or a person's voice.

These characteristics can be designed into manual, semi-automatic, or automatic systems. Discriminatory levels can be set depending on the level of security desired.

Identification Numbers and Passwords. Individuals may select or be assigned an ID number for use in access control devices such as keypads. In higher-level security systems, individuals may gain access by using a keyboard in which letters and symbols can be combined with numbers. Systems that use numbers only may be compromised; hence, the ID number or a supposedly private password must be regarded as a minimum effort at system reliability. A password must be combined with other identifiable means to achieve a higher level of confidence.

ID Cards and Tokens. A wide variety of ID cards and tokens are available that vary regarding ease of use, degree of security, ease of automatic and personal identification, and cost. Badges may be permanent documents with a lifetime measured in years or they may be designed intentionally to expire within a defined period of time. Some disposable self-expiring badges self-void from ink that migrates from the back of the badge to the front, indicating that the time for its use has expired.

Widely used cards and badges may have one or more features, including visual images in color or black and white; logotypes; signature panels; key personal information; encrypted data; magnetic stripes; computer chips; and redundant features to make counterfeiting difficult or impossible. These ID cards may be visually identifiable or machine-readable. Each factor affects use, image, and cost. Security operations planners often seek to commit to systems that may be used for extensive periods of time so that any capital costs can be amortized.

While the control of visitors and employees is emphasized, such systems have numerous capabilities that relate to other management operating concerns. For example, such systems may be linked to time and attendance procedures whereby individuals' payroll data can be created from normal badge use. Also, some systems indicate where in an organization the individual may be found at a particular time. Above all, however, such systems have remarkable flexibility in allowing or denying personal access to defined locations. Records of such activities may be easily retained and consulted concerning access patterns. Lost cards may be replaced, while anyone who finds the lost ID and tries to use it could be shut out and called to the attention of security personnel. A single card could allow an individual into numerous facilities, including parking lots and at locations in other parts of the world if the organization is managed by a single integrated system.

Biometric Features. A password can be learned by another person who could misuse it. An ID card, badge, or token may be lost or stolen and used by another until the card is no longer system-accessible. However, biometric features, such as fingerprints and iris or retinal information, rarely alter over a lifetime. Therefore, in theory, systems that use these features have a more reliable means of identification.

Biometric systems have been made more user-friendly in recent years. They are no longer as expensive and now rely on simple hardware interfaces. In earlier years, some biometric systems used long data signatures that organizations had to store and then sort through for identification purposes. The cost of storing data signatures and searching among an extensive data file for a match is no longer a significant economic issue. New commercial biometric applications—like facial recognition—have intriguing possible applications.

Like any system, biometric systems have limitations. Systems operate by first enrolling people into the system, often taking several recordings of the physical feature crucial to the system. This analog information is transferred by an algorithm into a digital number according to a proprietary algorithm for the system. When a person seeking access presents a physical feature to be identified, the digital identifier will not be absolute, but will have some variability. Sensitivity can be adjusted by the systems operator. Hence, biometric systems usually identify the person with the closest approximation to that found in the file.

False-negative (also called Type I or A) errors occur when an unauthorized person gains access to a restricted facility when he or she should have been denied. This is the more serious type of error. Obviously, security planners want this type of error to be rare or non-existent. False-positive (Type II or B) errors deny admission to someone who rightfully should be admitted but is not. This type of denial often occurs because the subject was hasty at entering his or her physical feature. A repeat attempt often confirms identification. In using biometric systems, a tradeoff may exist between ease of use with faster throughput from the system and a corresponding increase in false-positive rates.

Alarm Systems

Mechanical alarms were first used in the mid-19th century.[19] Today, alarm systems are predominantly electronic, although numerous types of alarms are available to meet a variety of needs. Alarm systems were created to deter, delay, and detect burglary, and that remains the main purpose of such systems. However, these alarm devices can detect and monitor other actions, including robbery (through a panic switch), smoke and heat signals, and requests for specific services. Such alarms may sound locally or be monitored by police, a proprietary system, or a commercial central station.

Underwriters Laboratories (UL), a voluntary non-profit organization, sets widely accepted and respected standards for alarm systems and vets the reliability of monitoring stations that receive the signals and act upon them. UL standards focus on burglary deterrence and detection capabilities. Types of UL burglar alarm certificates are shown in Table 10.2. Customers with systems meeting UL standards often receive a certificate from the UL-listed alarm monitoring service. This certificate may be required for insurance purposes. The central monitoring service may perform a number of services depending upon the agreement with the customer and the capacity of the alarm service.

In addition to receiving and verifying alarm conditions, operators at the monitoring service may call and request that police be dispatched; call designated persons and inform them of the alarm condition; send security personnel to the premises; dispatch someone to reset the alarm; call the fire department or an ambulance; direct maintenance staff to check a machine or process stoppage or irregularity; and perform other desired actions.

Table 10.2. Types of UL Burglar Alarm Certificates

Type of Alarm	Operation	Maintenance	Standard
Local Mercantile Police Connected	Outside sounding device. Grade based on equipment. Sounding device and remote connection to police or listed central station.	If called in the A.M., same day. If called in the P.M., next working day. One annual operational inspection.	Local Alarm Units UL 609
Central Station Burglary	Supervision of openings and closings and guard investigation of alarms. Response time and equipment used is shown on the certificate. One-hour response for trouble at closing.	One annual operational inspection.	Central Station Alarm Units UL 1610
Limited Mercantile	Supervision of opening and closing signals. Sounding device required. Guard response optional; 45 minutes if provided.	One-hour response for trouble at closing. Same as Central Station.	Central Station Alarm Units UL 1610
Local Bank Police Station Bank	Same as Local Mercantile. Same as Police Connected.	Service no later than the second day. One annual operational inspection.	Local Alarm Units UL 1610
Residential BA	Sounding device required. No Grade.	Same as Local Mercantile.	Household Alarm Systems UL 636

Table 10.2. *(Continued)*

Type of Alarm	Operation	Maintenance	Standard
Hold Up Alarms	Manual or Semi-Automatic.	Service no later than the second day. One annual operational inspection.	Hold Up Alarm Units UL 636
Central Station Proprietary	Supervision of openings, closings, and alarms by the subscriber UL 1076.	If called in the A.M., same day. If called in the P.M., next working day.	Proprietary Alarm Units UL 1076
Defense Industrial Security Systems	Supervision of alarms, openings, closings (alarms only at police station). Remote connection to a central station; defense contractor monitoring station or police station.	Service within four hours. One annual operational inspection.	Central Station Alarm Units UL 1610 Proprietary Alarm Units UL 1076

Because of the perennial concern with false alarms, alarm installation and monitoring businesses endeavor to select systems with low likelihood of inaccurate signals. Training users how to avoid false alarms is emphasized. Nonetheless, false alarms remain a problem. Therefore, many systems will verify the alarm before calling police or taking other action. This may be accomplished through a telephone call, real-time CCTV, or other means.

An inherent weakness of most alarm systems is that their signals travel over wires or cables that can be cut, intentionally or accidentally, leaving the alarmed premises without services. At the least, a system should identify that a connection has been broken, possibly informing the customer in the process. The loss of a primary method of communication may be backed up with a radio frequency system that does not depend upon wires or cables.

Robotic Systems

Security officers frequently patrol offices, making observations and checking the safety and security of the premises as they do so. The same activity, theoretically, could be accomplished by robotics. A robotic device may be mounted with a variety of sensors and alarms. CCTV on the mobile device can observe real-time activities; remote two-way communications can inject an immediate connection with the scene from a distant monitoring post. Robots can follow a fixed or random patrol and can climb or descend stairs, avoid unexpected obstacles, and either confront or retreat from a dangerous situation. Robotics have an intriguing potential to enhance the efforts of security personnel. However, the cost versus benefit ratio or using robotics to replace or supplement security officers has not been attractive to date. While a robotic system will work 24-hours a day, seven days a week without complaining, frequent and costly service requirements have further deterred wide use.[20] (In law enforcement, robotics have a secure place in confrontational circumstances such as bomb threat analysis.)

Communications

Effective security operations must allow seamless communication among managers, supervisors, staff personnel, and others. This is a requirement during normal operations. During an emergency, this requirement is even more important. Because a single system might be compromised or incapacitated due to an emergency, security planners think in terms of multiple means by which personnel can stay in touch during such times.

Typically, security planners rely on commercial telephone service as the basis for communications. However, some applications will merit the use of a dedicated system that only serves a single organization or network. Dedicated lines to local law enforcement authorities, fire, or ambulance services are common features at larger central monitoring stations. In the event that hard-line communications are down, contact with significant parties by two-way radio or cellular telephones is important.

Many security managers have different layers of personal communications. They will have available ordinary telephone service and, in high-security applications, a separate encrypted communications system. They may also carry with them a two-way radio, personal pagers, and cellular telephones.

Information Security Systems

Protection of data systems is an important and complex topic. The nature of cyber-threats continues to grow as networks play a larger role in everyday operations and as new vulnerabilities from the Internet and e-commerce emerge. Information security is covered competently in other books;[21] for the sake of this discussion, physical security and systems protection will be considered.

Physical Security for Information Systems. Data facilities are usually considered one of the most restricted and sensitive areas in an organization. Unauthorized visitors are not welcome. Extensive measures are taken to protect hardware and software of the central processor, as well as file storage areas, other processors, switchers, and communications lines. Many of the highest applications of access control are applied to the computer environment. In addition to access restriction, attention is given to fire risks within such a facility.

Protection against loss or disturbances in electrical power is generally a protective activity. An uninterruptible power supply (UPS) is one way in which normal operations can be maintained at least temporarily when power fails. UPS systems may be provided with batteries and supplemented with a solid-state rectifier that continually charges a battery bank. Additionally, an emergency alternator, such as a diesel engine or gas turbine, may be available to drive the alternator. Such UPS systems may provide emergency power for a few minutes to a few hours.

A more common problem with computer systems relates to power irregularities that cause momentarily spikes, surges, and drops in voltage levels. In the event power loss exceeds the capacity of the UPS system, the system should be backed up and, if possible, activities transferred to another facility not likely to be affected by the power failure. Data must be backed up, preferably on a real-time basis, though batch or computer-run backups will be adequate for less than critical applications. Backup may be via teleports, physical records archival procedures, and by other means.

Systems Security. Hardware and software used to protect local area networks (LANs) and wide area networks (WANs) are supplemented by numerous procedural factors to enhance file service security. These measures concentrate on login, password, trustee, directory, and file attribute security. Other factors such as directory and user creation must be safeguarded from being easily compromised by those outside the system. Firewalls, encryption, and traffic management systems also play important roles in reducing the possibilities of successful attack from those outside the LAN.

Other Considerations. Preventing sabotage, vandalism, and theft are high priorities of data security. While a process center may be protected internally by good physical security, other considerations must be included in planning. Printed records, diskettes, printer ribbons, and tapes may be destroyed by a shredder or turned over to a bonded destruction service. Electronic collection of information is also possible. In high-security federal systems, "Tempest" programs prevent electronic emanations from leaving the immediate environment and being collected and analyzed by a hostile power. Tempest-enclosed components are more expensive than devices without such protection and must be certified by the Department of Defense.

Contraband Detection

Any articles or materials that are illegal for the public to possess and carry into one's protected area might be screened by specialized processes. A physical pat-down may identify the presence of the weapon on an individual carrying one onto a commercially scheduled airline, for example. However, a pat-down process is slow, uncertain, objectionable to some, and costly. By contrast, automated systems that screen for such illegal objects are rapid, more reliable, non-intrusive, and cost-effective when large numbers of persons must be screened. Like other types of preventative technology, contraband detection systems are constantly evolving.

X-ray. Packages, garments, and baggage may be inspected by X-ray technology for contraband, including explosives and illegal drugs. Computer-enhanced and analyzed images increase the accuracy of contraband identification. Agents must monitor the enhanced images to determine whether a physical search is indicated. The principle of X-ray technology is that pulsed energy that penetrates most objects (lead and some alloys are exceptions) is absorbed by a plate, which then intensifies the image of materials programmed to be highlighted by intense, distinctive coloring. The images are projected for analysis on a color monitor.

X-rays have some disadvantages. Some explosives and bomb-making materials may mimic items normally found within packages or luggage. Similarly, some firearms have plastic parts and are not easily identifiable in enhanced X-ray imaging. Much depends on the skills of the agent who interprets the images.

Explosive and Drug Detection. While X-rays may detect some explosives and illegal drugs, other technology may be specifically devoted to such detection. The physical principle is that explosive compounds and illegal drugs may be identified by either sniffing for telltale molecules of the contraband materials or bombarding a container with energy that will "excite" materials such that they can be identified automatically. This technology permits screening packages or luggage without opening the contents. The accuracy and utility of such systems vary greatly. A bomb-detection system may be highly reliable, but it could also be stationary, costly to operate, subject to variable service problems, slow to process items, and expensive to purchase. Other devices are lighter, portable, and less expensive, but with an equivalent lower detection rate.

A drawback is that some materials may be prepared with such awareness of the detection technology and its limitations that their contraband contents are undetected.

Metal Detectors. Most metal detectors operate by transmitting a time-varying magnetic field, which is monitored by a receiver. When a metallic object is introduced into the electromagnetic field, reception of the signal is disturbed; this is then reported by a light, an audio signal, or both. Walk-through metal detectors may also indicate where on the body metal has been identified. A security officer may then use a handheld metal detector to identify precisely where under the clothing the metal is hidden.

A disadvantage of metal detectors is that they can provide false alarms from metal within the body, such as from a prosthetic device or from safety shoes with metal parts. Also, the sensitivity of metal detectors may be changed at the discretion of the operator. For example, sensitivity can be so acute that a single coin or a metal button weighing less than one-half an ounce can be detected. In such cases, however, the throughput is slow. Yet the same instrument can be set less sensitively so that small weapons will escape detection.

Fire Detection and Life Safety

About 550,000 fires occur in the U.S. each year, producing losses in the range of $11 to $13 billion.[22] Of these, about 150,000 are non-residential structural fires. Yet the trend for such events has been a steady decline. For example, in the decade ending in 1998, structural fires declined 26 percent. Tragic multiple-death fires of the past are less likely to occur, particularly in commercial, industrial, or institutional buildings. Many reasons might be cited to explain this trend: more extensive use of fire-resistant and noncombustible materials; better design; and more frequent imposition of standards and enforcement over time. One additional factor is that the detection and response to smoke and fire conditions continues to improve.

Smoke and heat detectors. Two technologically different smoke detectors are widely available. The ionization type uses a small amount of two radioactive materials (usually Americium-241 and Radium-226) to make the air electrically conductive—or to ionize it. Smoke from flaming fires contains carbon particles that are electrically conductive. When they enter the ionization chamber of the smoke detector, an alarm is activated. Hence, flaming fires with darker combustion products tend to trigger this type of detector better than other types of fire.[23] This type of smoke detector is susceptible to low temperature, high humidity, and dirt or dust which may interrupt the current and cause a false alarm.

Photoelectric smoke detectors operate on a light principle in which smoke entering a chamber either obscures the beam's path to a photocell receptor or reflects light into a photocell, causing an alarm condition. Visible particles common in smoldering fires are apt to cause an alarm with such a detector faster than an ionization type.

Ionization and photoelectric detectors require some maintenance relative to heat detectors, but are more sensitive and provide an earlier warning of fire. Since the type of fire or fuel that may affect an area often is unknown, the best strategy may be to use both types of smoke detectors in the same system. While smoke detectors may be battery operated, such devices are usually found in residences. Industrial and commercial applications require that smoke and heat detection systems be hardwired and monitored by a system.

Another type of fire-detection technology measures heat. These detectors have the lowest false alarm record, but are slow to respond to incipient fire conditions. One type of sensor is a fixed-temperature detector, which uses a bimetallic strip thermostat possessing a different coefficient of expansion for two metals. When the detector is heated, the strip bends in one direction in a way that an electrical circuit is completed, causing an alarm.

A rate-of-rise detector goes into an alarm state when the temperature increase exceeds a stated rate, usually 12 to 15 degrees Fahrenheit per minute. These detectors may trigger a false alarm when temperatures increase rapidly but not because of a fire. They also may not respond when a fire propagates slowly with a gradual temperature rise.

In industrial applications, still other specialized smoke and heat detectors are available. Security planners often elect to use a combination of detectors to increase the possibility of early detection of fires.

Life safety considerations are embodied in design and construction methods. The National Fire Protection Association promulgates the *Life Safety Code* (NFPA #101), which is revised every three years. The code contains detailed provisions and requirements relating to structural occupancy; wall openings and door fire resistance; emergency lighting; smoke and fire detection; and other topics.

One life safety trend has been the increased use of automatic sprinkler systems. A national standard (NFPA #13) relates to the design and installation of such systems, which mandates that only qualified contractors should install these systems. Six main types of sprinkler systems exist, and numerous models of sprinkler heads are available for different applications. Systems currently available break water into a fine mist, thus extinguishing the fire with greater efficiency and causing less property damage than past types in which large water droplets were discharged.

DESIGNING SECURITY SYSTEMS

Security operations managers tend to follow a formal process when a significant new security system is required or when an existing system faces an upgrade. Once management determines that the capital investment for such a system must be considered, the following steps generally are taken:

1. **Preliminary design.** The manager and the project team assume various tasks. Necessary physical information is collected, such as exist-

ing facility drawings and security documents. Often with the help of an architect, engineer, or consultant, project scheduling and initial cost estimates are produced relative to the design requirements.

2. **Design approach.** After approval of the initial design requirements, preliminary drawings, including site plans and system block diagrams, are produced. A cost analyzer will review the requirements and produce a cost estimate range. Each component in the system will be specified. Estimators may use worksheets—paper or electronic—to determine project costs, an example of which is shown in Figure 10.3. The cost analyst will identify possible job cost and time-completion uncertainties at this time. The design will consider such issues as crime prevention strategy, human factors including ergonomics, and the rapid change of technology likely to affect the project. Such issues as operational factors, communications, lighting, power sources, terrain, emergency possibilities, and year-round weather variability will also be considered. A more detailed cost estimate will be created next, including design cost, hardware (including some extras), installation, construction supervision and testing, security during construction, and other costs such as contractors' taxes, profit, bonding, and a contingency fee to be held pending completion of the project.
3. **Bidding or negotiation.** The project will be placed up for bid for local and regional contractors. These will be evaluated by management and its consultant prior to awarding the contract. While management seeks to achieve the best price, total overall value is the primary objective, and often the lowest bidder is not awarded the contract. Point systems are often used to help select the winning contractor.
4. **Construction phase.** This is the period and process in which the materials and services are integrated and the work is undertaken to construct, assemble, and install the system.
5. **Testing and training phase.** Integrated security systems are complex. As portions of the system are completed, they need to be tested repeatedly to make sure they perform as intended. Similarly, operations personnel may be trained by installers and manufacturers on using the new systems. Operating manuals, instructional materials, and possibly graphic user interfaces may be created for use by operating personnel.
6. **Operational phase.** The system is fully installed and operational. Unexpected adjustments are made. The final payments to the engineers and contractors are authorized pending approval and acceptance of the work completed.

SUMMARY

In this time of high-tech innovation, security systems constantly draw upon new advances, incorporating them into a cohesive, integrated system.

Security Alarm System Planning Estimate Form					
Product		Brand, Type	Cost	Quantity	Total
Control Panel	Burglar Alarm				
	Burglar-Fire				
	Residential				
	Commercial				
	Digital Communicator				
Remote Stations	Digital Keypad				
	Keyswitch				
	Electronic Guard Monitoring				
Perimeter Detection	Magnetic Contracts				
	Glass Breakage				
	Fence Sensors				
	Outdoor Beams				
	Shock, Vibration				
Interior Detection	Ultrasonic				
	PIR (Volumetric)				
	PIR (Wide Angle)				
	PIR (Long Range)				
	Microwave				
	Photoelectric Beams				
	Passive Audio				
	Switch Mats				
	Verified Sensors				
Fire	Smoke Detectors-Ionization				
	Smoke Detectors--Photoelectric				
	Pull Stations				
	Heat Sensors--Rate-Of -Rise				
	Heat Sensors--Fixed Temperature				
Audible visible	Alarm				
	Bells, Horns				
	Strobe Lights				
Reporting Back-up	Digital Communicator				
	Communications Back-up				
	Power Back-up				
Other	Hold-up/Panic				
	Safe/Vault Sensor				
	Total				

Figure 10.3. Security Alarm System Planning Estimate Form

Increasingly, management seeks technical resources to improve the quality of protection operations while decreasing operational costs. The Internet offers management an extraordinary tool for flexibly, reliably, and economically controlling operations from disparate locations and under different circumstances.

DISCUSSION AND REVIEW

1. How does situational crime prevention differ philosophically and practically from the traditional objectives of policing?
2. Why must security planners be involved early in facility design? What is the expected payoff from such involvement?
3. What is the reasoning behind security signage? What are its drawbacks?
4. Explain the limitations of cylinder-type locks.
5. Why have cylinder-type locks been the mainstay for commerce and industry for almost a century? Why have electronic proximity locks become preferred for some installations?
6. Why is CCTV almost invariably an important part of an integrated security system?
7. Describe capacities and qualities of a central alarm monitoring station.
8. Provide examples of non-invasive technological methods to detect contraband.
9. What critical stages occur before a security contractor is selected to undertake a major new installation or retrofit?

ENDNOTES

[1] J. Young (1994). "Incessant Chatter: Recent Paradigms in Criminology," in M. Maguire, R. Morgan, and R. Reiner (Eds.), *The Oxford Handbook of Criminology.* Oxford: Clarendon Press, p. 86.

[2] O. Newman (1972). *Defensible Space: Crime Prevention through Urban Design.* New York, NY: Macmillan.

[3] P. Mayhew (1979). "Defensible Space: The Current Status of a Crime Prevention Theory." *Howard J. of Penology and Crime Prevention.* Vol. 18, pp. 150–59.

[4] P.J. Brantingham and P.L. Brantingham (1981). *Environmental Criminology.* Beverly Hills, CA: Sage Publications; and P.J. Brantingham and P.L. Brantingham (1984). "Surveying Campus Crime: What Can Be Done to Reduce Crime and Fear?" *Security J.*, 5(2):160–71.

[5] Examples from the literature include R.V. Clarke (1992). *Situational Crime Prevention.* Albany, NY: Harrow and Heston; R.V. Clarke (1997). *Situational Crime Prevention: Successful Case Studies.* Albany, NY: Harrow and Heston; U.S. Department of Housing and Urban Development (1997). *Creating Defensible Space.* Washington, DC: Criminal Justice Department, Office of Policy and Research.

[6] R. Gigliotti and R. Jason (1984). *Security Design for Maximum Protection.* Boston, MA: Butterworth-Heinemann.

[7] An example is the Cost-of-Risk Analysis (CORA) system, which provides estimates of expected loss or annualized loss expectancy at an average rate in dollars per year from International Security Technology, New York City.

[8] A.A. Hopkins (1928). *Lure of the Lock.* New York, NY: General Society of Mechanics and Tradesmen, pp. 29–31.

[9] J.J. Fay (1987). *Butterworth's Security Dictionary.* Boston, MA: Butterworth-Heinemann, p. 142.

[10] Castles were designed for security. Defenders operated from parapets atop the walls during an attack. The concept of concentric circles of protection is clear enough in construction of some castles, which have two or three walls, a moat, and protected places within the central structure.

[11] A. de Sélincourt (Trans.) (1960). *Livy: The Early History of Rome.* Baltimore, MD: Penguin Books, pp. 376–77.

[12] Geese may be more alert than sleeping dogs at detecting nocturnal intruders.

[13] R.S. Eden (1993). *K9 Officer's Manual.* Calgary, ALBA: Detselig Enterprises.

[14] Note that the term "fence" also refers to a receiver of stolen goods. It is also a metal pin that extends from the bolt of a lever lock, preventing retraction of the bolt unless it is aligned with the gates of the lever tumblers.

[15] R. Gigliotti and R. Jason (1992). "Physical Barriers," in L.J. Fennelly (Ed.), *Effective Physical Security.* Boston, MA: Butterworth-Heinemann, p. 77.

[16] *Barrier Technology Handbook* (1980). Albuquerque, NM: Sandia Laboratories.

[17] H. Kruegle (1995). *CCTV Surveillance: Video Practices and Technology.* Boston, MA: Butterworth-Heinemann.

[18] J. Keble (1998). "Crest of a Wave." *Security Surveyor,* Vol. 28, pp. 11–13.

[19] R.D. McCrie (July 1988). "Development of the U.S. Security Industry," *Annals, AAPSS,* 498: 24–25.

[20] Robotics for security purposes should not be confused with applications for law enforcement, particularly in high-risk situations to evaluate bombs or contraband and to enter dangerous areas where lives are at risk. Also see J.J. Harrington et al. (1989). "Sandia National Laboratories Proof-of-Concept Robotic Security Vehicle." *Proceedings of the 5th Annual Symposium and Technical Displays on Physical and Electronic Security.* Philadelphia, PA: Armed Forces Communications and Electronic Association, p. B3–16.

[21] For example, see J.M. Carroll (1996). *Computer Security.* 3rd edition. Boston, MA: Butterworth-Heinemann; C.P. Pfleeger (1997). *Security in Computing.* 2nd edition. Upper Saddle River, NJ: Prentice Hall PTR; and K.S. Rosenblatt (1995). *High-Technology Crime.* San Jose, CA: KSK Publications.

[22] *The Fact Book 1999* (1998). New York: Insurance Information Institute, pp. 63–70.

[23] T.H. Ladwig (1991). *Industrial Fire Prevention and Protection.* New York, NY: Van Nostrand Reinhold, pp. 148–54.

Additional References

D.G. Aggleton (March 1991). "Security Up Front." *Security Management*, 35(3):71.

J. Barry and P. Finnegan (1997). "System Integration," in J. Konicek and K. Little. *Security, ID Systems and Locks*. Boston, MA: Butterworth-Heinemann.

M. Felson (1998). *Crime in Everyday Life*. 2nd edition. Thousand Oaks, CA: Pine Forge Press.

T. Harpole (July 1997). "Security Glazing Options." In *Construction Specifier.*

R. Homel (Ed.) (1996). *The Politics and Practice of Situational Crime Prevention.* Monsey, NY: Criminal Justice Press.

N.S. Levy (1998). *Managing High Technology and Innovation.* Upper Saddle River, NJ: Prentice Hall.

S. Lyons (1992). *Lighting for Industry and Security.* Oxford: Butterworth-Heinemann.

Proceedings of the IEEE International Carnahan Conference on Security Technology. 1986–95.

U.S. Department of Housing and Urban Development (April 1997). *Creating Defensible Space.* Washington, DC: Office of Policy and Research.

11

LEADERSHIP FOR OPTIMAL SECURITY OPERATIONS

> Leadership is a matter of intelligence, trustworthiness, humaneness, courage, and sternness.
>
> —Sun Tzu, *The Art of War*

Organizations require leadership in order to commence, grow, survive adverse circumstances, and adjust to constant change. Leaders can be entry-level workers, contract employees, full- or part-time employees, managers, executives, or members of the board. That is, leadership can come from anyone with a position to understand and influence organizational outcomes. Arthur Jago defines leadership as "both a process and a property." He states:

> The process of leadership is the use of non-coercive influence to direct and coordinate the activities of the members of an organized group toward the accomplishment of group objectives. As a property, leadership is the set of qualities or characters attributed to those who are perceived to successfully employ such influence.[1]

This chapter will pursue the topic of leadership from four perspectives. First, we will look at the aspects of leadership that are supported by widespread experience, empirical research, or both. Next, we will look at distinctive characteristics of security functions and how they differ from other management disciplines. Third, we will look at the discrete tasks that a 21st-century leader of security programs may be most fully engaged in. And, finally, we will look at the future of security operations.

LEARNING ABOUT LEADERSHIP

Many writers on leadership generalize about styles that lead to high performance, in which vision, trust, listening, and participatory skills are delineated.[2] Positive and effective leadership has always been understood as a critical component for organizations to succeed. The search for such qualities—particularly at managerial and executive levels—is unyielding. Insight into the traits of good leaders may come from numerous sources, including literature.[3] Further, looking at past military strategies can provide insight into the behavior of current organizations.[4] Meanwhile, the use of humor to lead organizations and improve management has received attention, especially for its ability to overcome personal inhibitions and to improve communications and marketing objectives.[5]

While books that interpret past behavior and apply it to contemporary situations are valuable, research on the nature of leadership has also emerged as a valid topic. For example, researchers on leadership have studied the lives of elite MBA recipients to understand their personal qualities and how they relate to the functioning of organizations.[6] Until recent years, most criteria for leadership selection and training have not been adequately validated by empirical methods. However, leadership has moved from being anecdotal, personal, and inspirational to being based more on principles that use scientific methods based on industrial and organizational psychology and sociology. Fred E. Fiedler comments:

> The most important lesson we have learned over the past 40 years is probably that the leadership of groups and organizations is a highly complex interaction between an individual and the social and task environment. Leadership is an ongoing transaction between a person in a position of authority and the social environment. How well the leader's particular style, abilities, and background contribute to performance is largely contingent on the control and influence the leadership situation provides.[7]

Leadership is changing because the nature of organizations and their requirements continue to evolve to create new needs. According to Peter R. Scholtes, the *old* competencies needed to survive and excel in the organization included:[8]

1. **Forcefulness.** Part of a manager's responsibility was to control the workforce, making people do what they may be otherwise inclined to ignore. Good managers could look their people square in the eye and get them to respond.
2. **Motivating.** The "softer" side of forcefulness was the ability to inspire people to do great work. The judicious combination of carrots and sticks, of inspiration and exhortation, was the manager's stock-in-trade.

3. **Decisiveness.** To make quick decisions in the absence of information was routinely expected of the old-style manager.
4. **Willfulness.** Good bosses knew what they wanted and were dogged in their pursuit of it.
5. **Assertiveness.** A good boss was outspoken. Old-style leaders could not show weakness or ignorance lest their people run all over them.
6. **Results- and bottom-line-oriented.** Bosses held people accountable for meeting quotas and standards and achieving measurable goals. Maximizing ever-increasing profits each quarter and minimizing ever-diminishing costs were the manager's goals.
7. **Task-oriented.** Managers kept everyone busy and occupied. There was no slacking off or socializing. Managers believed that people don't really want to work and, left to themselves, they will screw off. Therefore, it was their job to be their conscience and taskmaster.
8. **Integrity and diplomacy.** Good bosses covered toughness with tact and amiability. They were honest, fair, and respectful while letting their employees know that they knew what to do when things got out of hand.

These competencies are still the prevailing expectations of managers, Scholtes asserts. However, the changing nature of organizations has affected management structure. The needs and expectations of contemporary organizational leaders have changed. Newer competencies emphasize collaboration, education, and worker development so goals may be achieved with less hierarchical intervention than in the past. A newer competency addresses the importance of understanding systems and how to lead them, while another concept recognizes the nature in which the traditional relationships of hierarchy are counterproductive. Some of this thinking stems from the research and management practices advocated by W. Edwards Deming, who devised these practices over a career lasting several decades (see Box 11.1).

Leadership and Power

The military or police command-and-control method does not work well, or for an extended time, in non-military or policing organizations, even those concerned with security services. The screaming, tyrannical boss of past generations is gone (mostly). The new generation of leaders, however, has a different kind of power. According to John Kotter, who created a course on the topic at the Harvard Business School, five basic types of power exist:[9]

1. Power to reward by providing a promotion, raise, better working conditions, or direct approval.

Box 11.1. W. Edwards Deming and the Quest for Quality Improvement

In the 20th century, quality control of manufactured products increasingly came under analysis in a structured way. One pioneer was W. Edward Deming, who determined that about 85 percent of manufacturing errors were related to work structure and manufacturing design, not to the performance of individual workers. Deming created statistical quality control concepts for Bell Telephone Laboratories in the 1920s. Deming's precise way of creating a new philosophy of the workplace was adopted by Japanese manufacturers in the years following World War II and helped that nation transform itself into a quality leader in only a few years time.

Deming's philosophy called for teamwork in which small groups of workers would devote themselves to an assigned task, with all the teams working as part of a collective effort to achieve quality and productivity targets. Deming wrote: "Cease dependence on mass inspection to achieve quality. Eliminate the need for inspection on a mass basis by building quality into the product in the first place."

The concepts of constant product improvement argued by Deming are mostly thought of in manufacturing terms. However, the same emphasis on quality over quantity is important to security services as well. Deming's concept produced *quality circles*, which sought to improve quality, enhance productivity, and encourage employee involvement. Ideally functioning quality circles provide an enriched job experience, greater customer satisfaction, and bottom-up improvement.

Source: M. Walton (1986). *The Deming Management Method*. Putnam, NY: Peregee.

2. Power to punish by invoking progressive disciplinary methods or coercing or terminating the individual's employment.
3. Power of authority to approve or reject the quality of work presented or plans for the future.
4. Power to assert expertise; the leader insists that she or he really does know best and makes a particular decision based on personal experience or knowledge.
5. "Referent power" is the quality of a leader that leads to admiration and compliance by others.

Kotter sees contemporary managerial leaders drawing more from this last category.

The workforce in the 21st century is highly mobile. Contemporary leaders endeavor to provide authority when delegating tasks to others. When leaders give up power to give subordinate workers more authority, it may seem as if they are transferring power away from themselves. In effect, the leader gains by the efficiencies from decentralization and subordinate empowerment.

Leadership Traits

Personal qualities, including ethics and psychology, matter enormously in the success of persons who influence others to achieve a common goal. The image of a symphony conductor—managing diverse talents to produce a complete arrangement—is sometimes used to explain such leadership traits. In other circumstances, the military model of leadership provides guidance because of its broad relevance. Patrick L. Townsend and Joan E. Gebhardt have codified leadership traits in *Fundamentals of Marine Corps Leadership (MCI 03.3m)*, a correspondence course for noncommissioned officers.[10] The following are personality traits which the Marine Corps uses to teach leadership with private security in mind. Although the Marine Corps states that possession of these traits does not guarantee success, it also says that these traits "are a good guide for determining the desired personality to be developed as a leader."

1. **Integrity.** The authors comment: "Attempts to practice integrity part-time are hypocritical and forfeit any chance of engendering trust in seniors or subordinates." Honesty must be practiced with oneself as well as with others. The manual urges: "Don't tell your superiors only what you think they want to hear. Tell it as it is—but tactfully."
2. **Knowledge.** The leader may be a generalist, but in some aspects, for example, in investigations or technical security, he or she is likely to possess a high degree of information and understanding. This knowledge is retained at an advanced level through reading, attending conferences, obtaining new experiences, and consulting with experts in the topics.
3. **Courage.** The leader looks for and readily accepts new responsibilities. She or he never blames others for personal mistakes. In some circumstances, fear is natural and should be recognized, while one's emotions meanwhile are to be controlled.
4. **Decisiveness.** Good ideas can come from anywhere in an organization. The leader should consider all points of view for every problem, take a stand, and then determine whether the decision is sound.
5. **Dependability.** This quality is similar to reliability, which in turn is akin to professionalism. Dependable leaders tend to be prompt and complete all tasks to the best of their ability. They are careful about making promises, but build a reputation for keeping them once they are made.

6. **Initiative.** A leader with initiative looks for tasks to be performed and then completes them without being asked by a superior.
7. **Tact.** In the air force leadership manual, General George C. Marshall is quoted as saying: "A decent regard for the rights and feelings of others is essential to leadership." Tactful leaders are considerate of others. They are tolerant and patient. Similarly, the Corps manual urges: "Let no Marine, superior or subordinate, exceed you in courtesy and consideration for the feelings of others."
8. **Justice.** A leader searches his or her mental attitudes to determine what prejudices may exist and then seeks to rid them from their thinking and behavior. A leader recognizes those subordinates worthy of praise, not just those who merit punishment.
9. **Enthusiasm.** This word is derived from the Greek *enthousiamos*, meaning "inspiration" or "to be inspired by a god within one." Enthusiastic people believe in what they are doing and what they can accomplish. Their behavior is contagious.
10. **Bearing.** This means both looking good but also conducting oneself in a positive manner. The Marine Corps manual advises: "Frequent irritation, loss of temper, and vulgar speech indicate a lack of self-control or self-discipline and should be avoided."
11. **Endurance.** The Marine Corps manual advises readers to keep fit physically by exercise and proper diet and to avoid excesses that lower physical and mental stamina. Security practitioners demand much the same stamina. Often, the hours are long. Planned days off or vacations may be jettisoned because of changing circumstances or an emergency. To accomplish the goal, fatigue must be successfully fought.
12. **Unselfishness.** The Marine Corps manual states: "Put the comfort, pleasures, and recreation of subordinates before your own. In the field, your Marines eat before you do." An unselfish leader focuses on subordinates. The dispatch office of a quality regional security services business sought to make this emphatic. On the wall, a large sign was directed to the managers and dispatchers who worked there. It read: "What have you done to help security officers on their posts today?" Leaders who care for subordinates retain them in their service even when tempting opportunities for them elsewhere beckon.
13. **Loyalty.** This trait pertains to an ideal or custom, or the feeling of faithfulness to a cause or activity. Loyalty today, like yesterday, goes two ways: employees have a loyalty to the structure of the organization, and vice versa. Loyalty is interdependent. While the leader is loyal to the interests of subordinates, he or she expects loyalty in return.
14. **Judgment.** Leaders anticipate situations that require decisions. This helps them be ready when the need for action arises.

The manager faces a long list of expectations. To this list is added another quality: the manager as a coach and as a teacher of coaches. Coaching includes instructing, training, or guiding others in a particular activity or endeavor. Coaches make their personnel perform better, which translates into superior productivity and profit (see Box 11.2). The image of the leader as a coach has taken its place next to the image of the leader as a symphony conductor or military field commander to stimulate thought about behavior.

The Importance of Discretion

The formidable list of personality traits of leaders catalogued by Townsend and Gebhardt may be supplemented by another trait: discretion. Discretion is similar to loyalty in that organizational and personal matters are not discussed with those who do not have a right to such information. While the leader is discrete, he or she also expects discretion from anyone who has a right to know proprietary information about the organization. This is certainly the case among security practitioners and their associations.

Box 11.2. Should You Be a Coach? Do You Need a Coach?

Management strategies of the past sought to *control* others. Management practices of the present seek to *empower* others. Coaching is one technique by which managers can demonstrate a committed partnership to help the worker exceed previous levels of accomplishment.

A coach is someone who aids players and performers. A coach may be retained to help improve individual executive or managerial skills. Coaching is different from other supportive roles, such as instructors or trainers, because of the high degree of mutual trust required between the coach and those being coached. Results occur more when there is an interrelated commitment than when a hierarchical authority forces such an arrangement.

Some managerial coaches may themselves benefit from a coach. Coaches help high-performance managers achieve even greater results. Some coaches are independent management consultants who have insight into an individual's strengths and weaknesses and guide them to see patterns of behavior differently and to improve upon their current performance.

Source: R.D. Evered and J.C. Selman (Autumn 1989). "Coaching and the Art of Management." *Organizational Dynamics*. pp. 16–32; D. Benton (1999). *Secrets of a CEO Coach*. New York, NY: McGraw-Hill.

For example, contract security officers in the leader's organization may draw their paychecks from the security services firm that has selected them for the position. Yet the leader correctly expects these persons not to reveal proprietary information to anyone without a bona fide need to know. Similarly, when an organization invites architects, engineers, consultants, and others to bid for a contract, the leader expects that all parties with access to formal data presented in the RFP—and informal insights obtained during the process—to respect such information as proprietary, even if they do not receive a contract assignment.

The Problems with Leadership

Leaders are critical for the establishment, growth, and adjustment of an organization to changing times. The term "leader" is equated with the term "innovator," while the term "manager" relates to the term "program operator." Clearly, any operation—large or small—requires persons in positions of responsibility to possess some qualities of leadership. Leaders often reach heroic status, while managers are considered humdrum drones who don't so much lead as they supervise the creation of the innovator-leader. Yet it is important to note that the visionary leader can also sink an organization.[11] Managers often must step in to save the enterprise from the leader's misdirection or errors.

According to Jay A. Conger, leaders who have helped their organizations achieve also may lead these same entities into free-fall. Congers cites four reasons for this:

1. The leader's vision reflects his or her internal needs rather than those of the market or constituents.
2. The resources needed to achieve such vision have been seriously underestimated.
3. An unrealistic assessment or distorted perception of the market and constituent needs may hold sway on the leader's vision.
4. There has been a failure to recognize environmental changes that should redirect the vision.

Conger also notes that leaders may be so absorbed with the big picture that they fail to understand critical details of operations, except in terms of pet projects that involve them. Another problem with the true leader is that he or she often fails to develop an effective successor. Conger also notes: "(U)nder charismatic leadership, authority may be highly centralized around the leader—and this is an arrangement that, unfortunately, weakens the authority structures that are normally disbursed throughout an organization."[12]

WHAT IS DISTINCTIVE ABOUT LEADERSHIP FOR SECURITY OPERATIONS?

The essence of security management traditionally relates to the appropriate creation, imposition, and implementation of controls over personal behavior. Other management disciplines market and sell, finance, make, move, and administrate products or services. While security management relates to all of these, the discipline is primarily oriented in manipulating behavior to reduce or eliminate loss. To accomplish this, the operations security manager considers appropriate controls that can achieve the desired objectives within the context of the organization's total operations.

Security operations managers concentrate on internal and external controls because such controls are more amenable to change than are other factors. For example, genetic factors seem to play a role in crime causation. Men are disproportionately more responsible than women for serious violent and property crime. Yet some women commit such crimes, and it is not within the scope of a security manager's power, say, to urge that mostly women be hired because they are less likely to be involved in crime. Men and women both are needed in workplace. Similarly, employers do not select workers based on any narrow set of environmental preconditions. Most flourishing workplaces thrive when employees represent diverse social, ethnic, and national backgrounds. This means that managers must direct loss prevention while working with all types of people.

Control theory emanates from the work of Emile Durkheim, a suicidologist who concluded that the control and discipline of one's desires and the subordination of inclinations to the expectations of others stem from group integration and its intensity of involvement over behavior.[13] Those prone to suicide lose this control. Durkheim's work influenced Travis Hirschi's seminal work, *Causes of Delinquency*, which assumed that antisocial acts occur when an individual's bonds to society are weak or broken.[14] Hirschi's work centered on violence and property crime. Another theorist has concentrated not on crime in the streets, but in the suites. Edwin H. Sutherland first named and described "white collar crime," which is crime of a substantially different nature than street crime.[15]

Yet although these prescient observers' theories added to the understanding of deviance in the workplace, they provided little in the way of guidance to managers who sought to reduce loss. Other researchers would later fashion practical measures to deal with situational incidents that could be anticipated and controlled or resolved.

The security operations manager plans to respond to actual or reasonably possible situations by establishing situationally appropriate control measures. Since such measures generally cannot guaranty certainty of success, the manager also must be prepared to respond with alacrity once normative violations occur in the workplace.

CRITICAL LEADERSHIP ISSUES FOR SECURITY OPERATIONS MANAGERS

There are many distinct leadership issues facing protective program decision-makers. The priority of such tasks is as dynamic as the changing nature of the organization. New significant issues constantly emerge. Meanwhile, matters which once captured the full attention of a manger decline in importance as control measures are devised and implemented, reducing their urgency.

Critical issues will differ according to the type, size, geographical location, and financial resources of the industry. However, a survey of security executives responding to a *Fortune* 1000 list, cited earlier in this volume, identifies 22 issues of highest concern (see Table 2.3). The survey, conducted by a large security services firm, includes responses from about one-quarter of the identified security directors among the largest manufacturers, business services, and retailers.

The security operations manager normally oversees a number of ongoing programs and services. These include facilities management, which requires the services of personnel and technology. Superimposed over the organizational routine are issues that direct the security executive's time and concerns. These complex issues are covered in other books and by specialty organizations. The discussion below seeks only to explore aspects of the 22 most frequently cited threats in the order in which they are mentioned; changes in rank of such risks are to be expected.

Workplace Violence

This issue has been at or near the top of every survey conducted among *Fortune* 1000 security directors. This discussion will concern workplace violence as it applies to employees in general and to security personnel in particular.

General Workplace Violence. The drama and significance of a violent act in the workplace is undeniable. In the last decade, violence in the workplace received exceptional attention from a variety of organizations, including the Office of Safety, Health and Working Conditions of the Bureau of Labor Statistics; the Occupational Safety and Health Administration; and the Centers for Disease Control and Prevention. In addition, organizations like the American Society for Industrial Security and the Society for Human Resource Management have regarded workplace violence as a critical topic for their membership. Numerous insurance companies, trade organizations, and other state and local groups also have weighed in on the issue. Audiovisual materials have been produced and numerous books on the topic have been published.[16] Workplace attacks that result in death or serious injury are reported in the media often.

Thus it is no great surprise that security directors organize comprehensive plans to mitigate the chances of fatal or injurious incidents in their organizations. Workplace violence mitigation involves the consideration of a broad number of issues, including employee vetting, procedures, technical measures, and post-incident response. Another aspect of operations management is to reduce the level of fear among employees by presenting available facts on workplace violence. The purpose of such information is not to lull workers into a false sense of security, but rather to counter any fears they have. Meanwhile, reasonable measures must be taken to reduce risks of those most vulnerable to an incident.

An analysis of federal studies concerning occupational injury and death presents a more nuanced portrait of the workplace violence problem. The Bureau of Labor Statistics (BLS) periodically reports data on nonfatal assaults resulting in lost workdays. Reports on workplace fatalities emanate from the National Institute for Occupational Safety and Health (NIOSH), Centers for Disease Control and Prevention, BLS, and other sources.

- **Frequency of fatalities.** In a period lasting five consecutive years, occupational fatalities remained constant. The frequency of homicide by shooting, stabbing, and other incidents, including bombings, claimed over 1,000 lives each year. Substantial as that number is, a person's chances of being killed in a transportation accident are almost two-and-a-half times greater than being a victim of homicide in the workplace. The number of deaths caused from being struck, caught, or crashed by equipment or objects is about the same as those caused by homicides, as shown in Table 11.1.
- **Risks for female workers.** Three times as many male workers were murdered as female workers. However, homicide was the leading cause of job-related fatalities for women, accounting also for nearly half of women's work injuries.
- **Race and ethnicity.** Because of their occupations, not race or ethnicity per se, homicide was the leading cause of job-related deaths for blacks, Hispanics, Asians, and Pacific Islanders.
- **Risks for the self-employed.** Homicide was also the leading cause of job-related deaths for the self-employed.
- **Circumstances of job-related homicides.** In a study of 1,024 job-related homicides that occurred in a single year, most (71 percent) of the incidents resulted from robberies or robbery attempts.[17] Typically, robbery victims were store personnel, gas station attendants, or taxicab drivers. Several workers were killed during carjackings, muggings, and robberies of goods and services. The frequent targets for robbery in this study are beer truck drivers,

Table 11.1. Fatal Occupational Injuries

Event or Exposure	Number	Percent of Total
Transportation Incidents	2,560	40
Homicides	1,024	17
Struck with Objects and Equipment	915	15
Falls	643	10
Exposure to Harmful Substances/Environments	598	10
Suicides	215	4
Fires and Explosions	208	3
Other Events or Exposures	24	--
Totals	6,187	100

Source: G. Toscano and J. Windau (1996). "National Census of Fatal Occupational Injuries, 1995." Washington, DC: Bureau of Labor Statistics, p. 35.

store personnel, gas station attendants, and cab drivers who are killed for their cash receipts.

- **Victimization by work associates.** One-tenth of workplace homicide victims were killed by a current or former work associate. Twenty-five customers or clients were responsible for job-related homicides. These incidents were most likely to be featured by the media.
- **Personal acquaintance of the victim.** Several workers—primarily women—were killed as a result of domestic disputes that entered into the workplace.
- **Police and security victimization.** Eighty-one police officers were killed in the line of duty, while 59 security guards were killed.

In addition to the harshest violence, homicide, other safety issues occur in the workplace. In 1994, 20,000 incidents involving assaults and other violent acts by persons that resulted in lost workday injuries and illnesses were reported.[18] Women were the victims in almost three-fifths of these assaults, much higher than their one-third share of lost workday cases in private industry. This suggests that the level of more serious, nonfatal violent acts was higher among men than among women (see Table 11.2 for more statistics). Service workers were at high risk of nonfatal assaults, accounting for more than two-fifths of the lost workday cases resulting from assaults. Nursing aides and orderlies accounted for more than half of all assaulted service workers. The exceptional risk to workers in healthcare pro-

Table 11.2. Percent Distribution of Nonfatal Assaults Resulting in Injuries and Illnesses with Time Off from Work

Violent Act	Percent Total*
Hitting, Kicking, Beating	43
Squeezing, Pinching, Scratching, Twisting	12
Biting	3
Threat or Verbal Assault	3
Shooting	2
Stabbing	1
Other	37
Sex of Injured	**Percent Total***
Women	58
Men	41
Industry of Injured	**Percent Total**
Services†	67
Retail Trade	16
Transportation, Utilities	5
Finance, Insurance, Real Estate	4
Manufacturing	4
Other	4
Workdays Lost	**Percent Total**
1 to 5 days	54
6 to 10 days	14
11 to 20 days	10
21 days or more	22

*Does not total 100 due to rounding.
†Includes health, social, business, and educational services.
Source: "Characteristics of Injuries and Illnesses Resulting in Absences from Work, 1994" (1996). Washington, DC: Bureau of Labor Statistics. Based on 20,438 nonfatal assaults.

fessions has resulted in guidelines for these workplaces being published by the Occupational Safety and Health Administration.[19]

The strategy of mitigating possible workplace violence continues to evolve according to the type of industry. The highest danger of work-related homicide is faced by clerks at sales counters; managers and supervisors of food and lodging establishments; sales personnel and small business proprietors; and cashiers. These particular workplaces merit ongoing research and improved working conditions.

However, more than half the nation works in offices. The specter of innocent workers being gunned down in their offices, say, by a disgruntled current or former employee, is sure to seize the nation's attention (see Box 11.3). For that reason—in addition to the possibility that such incidents might produce copycat acts of violence—security operations managers must seek to improve vetting measures, identify workers who have previously made harsh or threatening comments about others in the workplace, support the Employee Assistance Program (EAP) initiatives, improve access control measures, and generally enhance the feeling of security.

Violence Against Security Personnel. In a recent Bureau of Labor Statistics study, security personnel ranked fifth in terms of risk of work-related homicides, as shown in Table 11.3. However, upon further analysis, this ranking

Box 11.3. Workplace Violence: Warning Signs Sometimes Missed

Often, signs of workplace violence are present before an events occurs. Frequently, these signs are ignored. This was the case in the 1998 killing of four employees of the Connecticut State Lottery offices in Newington, Connecticut.

Throughout the late 1990s, a frustrated computer worker employed by the lottery had shown repeated signs of stress. The worker had been transferred from the accounting department to the data processing department two years earlier, and, as a result, he filed a grievance to obtain his old job back and demanded extra pay for doing computer work that, he claimed, was not part of his job. The worker had given no previous indication that he could be dangerous to others, although he had attempted to commit suicide twice, the more recent incident occurring just one year before his homicidal rampage.

In 1997, the worker was given a five-month medical leave for depression and stress. On March 6, 1998, just days before he was scheduled to resume work, he returned to his office. After working calmly at his desk for about a half-hour, he stormed into the offices of his intended victims. In the next few minutes, the worker killed the lottery's chief executive officer, chief financial officer, a vice president, and the organization's top computer specialist. Finally, the gunman ended the bloodbath by taking his own life.

Source: *Security Letter*, March 16, 1998, p. 1.

Table 11.3. Top-Ranked Occupations at Risk for Work-Related Homicide (Based on 1,063 Incidents)

Occupation	Homicides	Rate[1]	Relative Error (in Percent)[2]
Taxi Drivers/Chauffeurs	97	43.1	6.83
Gas Station Attendants	22	11.1	7.28
Sales Counter Clerks	22	11.3	7.34
Police and Law Enforcement	57	11.2	4.53
Security Guards and Officers	54	6.5	3.54
Managers (Food and Lodging)	59	4.9	2.95
Sales, Supervisors, Proprietors	178	4.4	1.60
Cashiers	94	3.5	2.00

[1] Experimental measure using Current Population Survey (CPS) employment data. This rate represents the number of fatal occupational injuries per 100,000 employed and was calculated as follows: N/W x 100,000, where N = number of fatal work injuries and W = employment based on the 1993 CPS.

[2] The CPS employment data used to calculate rates are estimates that are based upon a sample of persons employed rather than on a complete count. Therefore, the employment estimates and fatality rates have sampling errors expressed in percentages; that is, they may differ from figures that would have been obtained if it had been possible to take a complete census of employed persons. The relative standard error is used to calculate a confidence level of 90 percent.

Source: "Characteristics of Injuries and Illnesses Resulting in Absences from Work, 1994" (1996). Washington, DC: Bureau of Labor Statistics. Data are for 1993 (see also Table 5.1).

fails to make clear that almost half of the work-related homicides occurred to armored car service personnel and private investigators. These two categories employ far fewer workers than the watch, guard, and patrol division of the industry does. Thus, the chances of a security officer being a victim of homicide—while probably higher than other work categories on average—is lower than it would appear. At the same time, the risks to armored car services personnel and private investigators are higher when the category is analyzed by constituents.[20]

Crisis Management/Executive Protection

This category, ranking second among the 22 most important security threats, may be criticized because of lack of specificity. The two issues are not necessarily related. Therefore, they will be discussed separately.

Crisis Management. This is the management process whereby potential emergencies or disasters are systematically identified and assessed for their frequency and criticality. Table 11.4 provides an overview of such processes and their consequences.

This is a formal process involving a comprehensive search for the likely crises that an organization actually or probably could experience.[21] Management usually seeks to mitigate such vulnerabilities—called risk reduction—with appropriate measures. Additionally, some small and predictable losses may be reducible by measures that are far more costly than the damages to be paid when they do occur. Therefore, management decides not to take exceptional risk reduction measures and accepts such risks. In still other cases, some risks are reduced and others are accepted, but an aggregate risk remains, such as damage from a major fire or natural disaster. This would be unacceptable should it actually occur. Such risk is transferred to an insurer. Finally, in some cases, management will conclude that the risk cannot be accepted and the actual activity is not undertaken or is terminated. This process is termed "risk elimination" because the organization disassociates itself from the venture or activity.

Crisis management is an ongoing process in which the security operations manager is constantly engaged. Despite mitigation efforts, crises will occur. Emergencies and disasters must be responded to. Operations must resume as quickly as possible if the untoward event slows or stops the pro-

Table 11.4. The Strategy of Crisis Management

Process	Consequence
Risk Reduction	Security managers identify significant risks to the organization that can be systematically reduced or eliminated. This may be achieved with appropriate programs. Sometimes, the risk can be shared with other organizations within the same entity or in the same geographic area. For example, an organization that fears a computer crash may temporarily move operations to another location within the same establishment.
Risk Acceptance	This assumes that some losses will occur as a normal consequence of operations. Management concludes that the projected losses are acceptable and will be absorbed as a routine operating cost.
Risk Transfer	The organization accepts some risk, reduces it by various means, but transfers an unacceptable risk level to an insurer.
Risk Elimination	Management determines that the risk is too great to accept and, therefore, elects to avoid the activity entirely by not commencing it, selling it off, or ceasing operations.

cess of work. Further, security operations managers generally serve as coordinators of response and recovery activities. Contingency planning aids the organization to respond quickly to incidents.

Executive Protection. The safety and security of executives, indeed all employees, is a major concern for security operations managers. Executive protection often is related to workplace violence in that risks to the executive whenever he or she is working or traveling on behalf of the organization have security implications. However, few workplaces require special protective means such as employing executive protection personnel. M.J. Braunig states in the *Executive Protection Bible*: "Far fewer corporations hire executive protection agents than you might think, and the corporations who do hire them generally do so because of reaction to a specific incident, or because of specific threats, or in some cases simply as a prudent security measure."[22]

Executive protection involves surveys of broad risk; specific evaluation of office, hotel, and residence dangers; and training for staff and family members on what to observe and what to do under threatening circumstances.[23] The executive protection professional may complete extensive relevant training; possess good interpersonal skills; be physically fit and able to adapt to a wide variety of circumstances; and has analytical ability to identify risks and avoid them.[24] Risk reduction for the executive might involve counter-surveillance operations that seek "to identify all possible attack points where the executive is vulnerable" and to eliminate or reduce exposure to them.[25]

The consequences of inadequate executive protection can be profound, leading to liquidation or substantial reorganization of the firm when a significant and creative leader is struck down. One example is the gunning down of fashion mogul Gianni Versace in Miami Beach, Florida. Robert L. Oatman observes: "Executive protection has little to do with spinning tires and knocking people over, and everything to do with threat assessment, intelligence gathering, transportation, choreography, advance work, 10-minute medicine, resources, technology, and support."[26]

Fraud/White-Collar Crime

The mantra among security practitioners is that more losses occur from fraud and so-called white-collar crime than from conventional internal or external theft. *Fraud* is the criminal offense of obtaining money or money equivalents by false pretenses; that is, the intentional perversion of truth. It usually does not involve property damage or threatened or actual physical injury. Fraud is an element in crimes such as forgery, counterfeiting, and embezzlement, in which cunning and unfairness are used to cheat people or organizations. The range of fraud is great, from petty cheating to large-scale looting of corporate or governmental assets. Frauds often are against individuals by others. This discussion, however, is limited to circumstances in which the organization is itself victimized by individuals or entities.

White-collar crime is fraud committed by persons whose occupational status is executive, managerial, entrepreneurial, professional, or semi-professional. The term, first used by Edwin H. Sutherland, was meant to encompass such business crimes as embezzlement, kickbacks, price fixing, antitrust violations, unfair labor practices, and war crimes.[27] Sutherland's work turned the attention of academics from poverty, class, and caste as factors that cause crime, to values that lie within the social system and individual malfeasance as criminogenic element. Some writers conclude that the level of crime among the wealthy and powerful can have an insidious effect on government, and vice versa, placing the social structure at risk when such depredations go unchecked.[28]

For the security operations manager, dealing with internal and external fraud and white-collar crime usually takes the form of a reactive response, whereas a preventative response would be wiser and more cost effective. Security practitioners usually are the best—or among the best—individuals in an organization to sense risks that are greater than their likely payoff or benefit to the organization. However, the nature of fraud is protean, and organizations reasonably well prepared to deter fraud may be victimized by the originality and daring of fraudsters. Further, some incidents involve offenders who appear to be making money for the organization, but in reality are doing the opposite. In such situations, the inquiring security practitioner or compliance officer will be rebuffed by managers protecting the supposed high-income producer because their bonuses and status are linked to those persons' success (see Box 11.4).

Box 11.4. Nick Leeson and the Fall of Barings Bank

Security practitioners and compliance officers often have the skills to identify risks that are too great relative to their potential payoff. However, whether management will listen to their cautionary advice is another matter. Sometimes, wild risks are taken by people who appear to be making a profit for the organization, but actually are running it into the ground. An example of this is the collapse in 1995 of Barings Bank, a London-based international bank founded in 1762. The cause was a single individual operating without adequate controls, who was protected by his supervisors and managers. In this situation, Nicholas Leeson, a 28-year-old rogue trader for Baring Futures in Singapore, racked up huge unauthorized trading losses, hiding them in a mislabeled account.

Box 11.4. *(Continued)*

Leeson appeared to be earning enormous profits for his accounts from buying and selling derivatives—future contracts linked to fluctuations in Japan's stock exchange. His reported "profits" earned him and his superiors substantial bonuses. By the time Leeson's scheme was discovered, he had racked up losses of £827 million. This exceeded the resources of the bank, which was liquidated as a result, ending the glorious history of an institution that helped finance among other things the expansion of the British Empire and the Louisiana Purchase by the U.S. in the 19th century.

Source: J.H. Rawnsley (1995). *Total Risk: Nick Leeson and the Fall of Barings Bank*. New York, NY: HarperBusiness. Also: N. Leeson with E. Whiteley (1996). *Rogue Trader: How I Brought Down Barings Bank and Shook the Financial World*. Boston, MA: Little, Brown & Company; and *Security Letter* (March 15, 1995), p. 1.

Frequently, security practitioners learn about fraud after the fact and must move quickly to conduct an internal investigation. Such an investigation may be begun the moment an indication is received that the organization may have been victimized by fraudulent means. Before calling federal agents or local prosecutors, the organization usually seeks to determine whether the indication of fraud is verifiable and, if so, how substantial the loss is to the entity.[29] In the earliest phase in which fraud is suspected, criminality cannot be assumed. An internal investigation should collect facts that can be the basis of ascertaining the truth and facilitating future interaction with the criminal justice system. Normally, security practitioners confer with internal counsel before conducting an internal investigation. Staff investigators may be delegated to conduct the interviews needed to evaluate the situation. In the event that the organization does not have skilled personnel appropriate to the task, private investigators may be retained on a project basis.

Once the investigators establish the facts and the case is confirmed by the security operations head, the internal counsel may take the necessary steps that can result in arrest and prosecution of the offenders. Security practitioners may continue their efforts even after a case is transferred to federal, state, or local authorities. In some cases, security can achieve investigative goals better than government investigators because they understand their organization better and because incentives to protect the institutional reputation and possibly recover lost assets are present.

Employee Screening Concerns

Hiring the wrong employee leads to disappointment, financial loss, missed opportunity, and a sense of inadequate screening (vetting) measures.

When employment is full and the need to fill key positions is high, the chances of failing to conduct a thorough pre-employment credentials review increase.

Employers in high-tech industries may be particularly vulnerable to theft at the workplace.[30] To achieve the desired results of a workforce with high integrity and reliability, security practitioners and Human Resource managers must work together, drawing from the skills that protection professionals know best regarding screening out candidates with undesirable prior work records or dishonest representations on résumés or application forms. (This process of pre-employment verification and selection is discussed in detail in Chapter 3.)

Computer Crime: Hardware/Software Theft

The French scholar Blaise Pascal invented the adding machine in 1642, the first example of a device in which tens were carried to the next column. This invention created the basis for finding a mechanical way to rapidly automatic computations. The first electronic computer was called the "Eniac" (for Electronic Numerical Integrator and Computer), and was created in 1945 at the University of Pennsylvania's Moore School of Electrical Engineering. In the subsequent half-century, computers grew from being esoteric scientific tools to becoming a commonplace, ubiquitous aides to most human endeavors.

As computers have grown larger in importance and smaller in physical size, they have become objects of theft. Further, the input, output, tapes, and peripheral hardware are also tempting to thieves. Meanwhile, software programs written for an organization at great cost—both in terms of money and time invested—and critical databases can be copied in a matter of minutes, often with the owner not being aware that its proprietary software or valuable records have been copied.

The strategy to protect hardware and software involves a combination of factors, mainly related to procedures, personnel, and hardware.[31] Procedural issues involve identifying what physical and electronic assets are most valuable and creating means whereby they may be identified and protected. Personnel issues include policies and procedures for those individuals having contact with the central computer facility where computers, peripheral equipment, processing units, storage units, communications equipment, and power backup are located. Hardware issues encompass physical security measures to restrict access control throughout a facility containing a computer center, particularly in the computer room itself. The physical security involves automatic access control to restricted areas, enforced by the resources of trained personnel. Software theft can be mitigated by programs that make copying difficult.

General Employee Theft

This category is presumed to involve theft of assets excluding fraud, intellectual property, and electronic assets. Quantification of employee theft is difficult since definitions can vary. The cost of employee theft, time theft, and drug abuse on the job has been estimated at $320 billion.[32] The indirect costs from such losses to the employer, government, and the general public add to the total risks (see Box 11.5). Employee theft is mitigated by measures previously discussed, including appropriate vetting (Chapter 3), adequate supervision (Chapter 5), and discipline up to discharge and prosecution in the event of dishonesty (Chapter 6). Meanwhile, electronic systems may also discourage employee theft (Chapter 10).

Computer Crime: Internet/Intranet Security

In 1990, the "World," the first commercial Internet service provider (ISP), came online. Two years later, the number of computers connected to the Net exceeded one million. By January 2000, a single ISP boasted over 20 million customers on the Net, and this supplier had scores of competitors. About 200 million people are online and about half of all American households have a personal computer.

The Internet has transformed every aspect of business, and the pace of change continues unabated. From its onset, the Net was an open system, designed so that users could enter and leave, read and write, and share and transmit data with the greatest of ease. For organizations, the Net permitted the speed of business to soar. The capacity for organizations to communicate with each other and with anyone else achieved runaway growth. Because it was designed as an open system, the Net has no real policing. Therefore, chances of a variety of abuses are high and may expose the organization to a variety of threats never anticipated during the early development of the Net.

In some ways, the security of the Internet and intranets is similar to conventional physical security. For example, access control to electronic databases can be divided into different levels of permission. Individuals may have access to databases under circumstances that can change according to the desire of the host system's management. The same way a customer's check must be trusted by a retailer, those involved in e-commerce need to be assured that the personal identification of the customer can be trusted and that the credit authorization for a purchase is genuine. At the same time, such transactions need to be protected so that unauthorized individuals will not have access to pertinent details in a transaction. Such transactions must include audit trails so that, should an irregularity occur, an investigation can determine the culprit and minimize further economic loss.

Box 11.5. Indirect Costs of Economic Crime

Determining economic losses from crime is difficult, as such offenses also produce indirect costs to the organization, including:

Increased cost effects on the organization:

- Increased security
- Internal audit activities
- Investigation and prosecution of suspects
- Reduced profits
- Increased selling prices and weakened competitive standing
- Lower employee morale
- Reduced productivity
- Damage to business reputation and image
- Deterioration in quality of service
- Higher overhead due to theft
- Lost business opportunity due to lack of needed items
- Higher insurance

Cost effects on local and national government:

- Criminal justice activities (investigating and prosecuting)
- Correctional programs
- Crime-prevention measures
- Policing and community security activities required
- Loss of tax revenue (e.g., from loss of taxes due to the untaxed income of the perpetrator and due to deductible business losses producing less tax income)

Cost effects on the public:

- More expensive consumer goods and services to offset crime losses
- Decreased investor equity
- Higher taxes to pay for criminal justice costs
- Reduced employment from business failures

Source: W.C. Cunningham, J.J. Strauchs, and C.W. Van Meter (1990). *Private Security Trends 1970 to 2000: The Hallcrest Report II*. Boston, MA: Butterworth-Heinemann, pp. 32–33.

However, computer security is different from physical security in the way it is largely electronic in nature. Thus, IT security involves a systems approach that must be comprehensive, dynamic, and flexible and suitable to the circumstances. Electronic data can be encrypted by an algorithm that protects it within a communications network from the point of origin to the final destination. The initial algorithm to protect electronic transmissions made available by the federal government was the Data Encryption Standard (DES). With the growth of powerful, cheap computing, DES alone provides modest security. The National Institute of Standards and Technology (NIST) has selected a new cryptology, Rijndael, to replace DES. Many other broadly available algorithms exist that e-security managers may consider.[33]

Other means have been created to assure trusted operating systems, database security, and protection of distributed systems. Attacks on computers by hackers, crackers, and cyber-terrorists pose a challenge to operating security practitioners.[34] Entire systems—if unprotected by filters and warning systems—can be crashed by an offender who sends a large number of messages or queries to a system at the same time. This is called a *denial-of-service attack*. The result is that IT security becomes a priority.

As William C. Boni and Gerald L. Lovacich write: "Ironically, these I-Way robbers may do more to increase demand for I-Way security than any other force in the marketplace today."[35] Some security operations managers will feel challenged by the opportunity to respond to security issues raised by the Net and e-commerce; others will be intimidated by the technical shift required to establish trusted IT security operations. But the issue cannot be ignored, due to both the ability of Net applications to enhance commerce and the vulnerabilities created at the same time. Security operations managers must move up to the challenge of e-commerce and Net security issues, or step aside for others with the interest and competence.

Drugs in the Workplace

Drug use and abuse in the workplace are issues that predate the Industrial Revolution. The media often refer to a "war on drugs"; however, alcohol and tobacco abuse are responsible for more deaths in the U.S. than all illegal drugs combined.[36] Alcohol accounts for about 100,000 premature deaths per year and tobacco is linked to 400,000, whereas illegal drugs are responsible for about 30,000 deaths. Excessive alcohol and tobacco use results in productivity losses, greater employers' health costs, and workplace accidents.

Such addictive substances as alcohol and tobacco, however, are not linked to the same extent to criminal propensities as some illicit drugs. For this and other reasons, drug use and abuse in the workplace are subject to federal programs, and policies that lead to a drug-free workplace are promoted.[37] Federal employment guidelines test for marijuana, opiates (heroin, morphine), cocaine, amphetamines and methamphetamines, and phencyclidine. Testing for such substances prior to an offer of employment or during employment was discussed in Chapter 3.

In addition to screening for drug use, security operations managers face different issues concerning workplace drug matters. Initially, a policy might be created and disseminated concerning a testing process at the workplace.[38] Legal issues, policy promotion strategy, testing methods, sample security protocols, and laboratory selection need to be considered. In the event an overt or undercover investigation must be conducted, procedures and personnel must be directed to collect facts. The investigative team, supported by protective management, will recommend subsequent actions to be taken to mitigate future incidents.

Unethical Business Conduct

Ethics are rules of conduct. They are practices that are applicable to organizations and members of a profession regarding their moral and professional obligations. As such, ethics usually have less force than legal requirements, although the organization must take them seriously. And in some circumstances, ethical breaches may overlap with criminal behavior.

Employers have ethical obligations to employees, vendors, competitors, and society at large. Nearly all organizations have a code of conduct that delineates the principles of proper behavior. At the same time, organizations expect ethical behavior from their employees, vendors, competitors, and others.

Security practitioners encounter unethical business practices in a variety of contexts. In addition to domestic issues, the Foreign Corrupt Practices Act of 1977 requires that U.S.-based for-profit businesses operate ethically when abroad. Bribes and kickbacks when offered in a seductive way for "consulting services" are certainly illegal, and possibly criminal. Organizations may, for example, ethically collect publicly available information about competitors. Thus, sales literature, Web site information, comments by personnel made at trade shows, observations offered without prompting by vendors' employees when seeking new employment, and technical papers presented at meetings may be collected, analyzed, and used to enhance an organization's marketing and general management strategy. Security personnel sometimes aid in the creation of such competitive information units. However, it is unethical for an organization to interview vendors' employees solely for the purpose of collecting desirable competitive intelligence when no job with the interviewing organization is available. It is further unethical to hire such vendors' employees as "consultants" on a project basis when the intention is to obtain critical details about such organizations rather than the employee's skills for their own sake.

Apparent internal or external ethical violations require investigations to determine their veracity.[39] Due diligence may be necessitated before domestic organizations deal with potential foreign partners. A thorough understanding of ethical obligations and risks reduces the possibility of violating ethics and legal practices in all countries involved.

Security practitioners frequently are given responsibilities for drafting and managing ethical policies and programs, and all major security organizations have a code of behavior or ethics. The American Society for Industrial Security has a Council on Business Practices and a Code of Ethics, which can punish offenders with censure or dismissal from the organization (see Appendix A).

Property crime: External Theft

External theft, damage, vandalism, and sabotage cost organizations billions of dollars each year. Their prevention depends upon a comprehensive security management program. Architectural design and space management concepts combined with the proper selection of building materials represent an important initial consideration in reducing property crime losses. This involves the concept of Crime Prevention Through Environmental Design (CPTED), which can deter criminal and uncivil behavior (as discussed in Chapter 10).[40] The appropriate use of hardware, software, and procedures aimed at reducing losses can result in further lowering the likelihood of loss. In the event that losses do occur, litigation pursuant to an investigation may recover some lost assets, or reduce the pattern of loss, or both.[41]

Sexual Harassment/Equal Employment Opportunity (EEOC) Concerns

Sexual harassment is defined as conduct directed at a specific person, usually of a different gender, causing substantial emotional distress. In 1994, 44 percent of women and 19 percent of men responding to a survey conducted by the U.S. Merit Systems Protection Board said that they had experienced some form of unwanted sexual attention during the previous two years. Similar results were seen in a survey conducted seven years earlier.[42] The survey gathered responses from employees in over 24 agencies. Nonetheless, few bothered to report the incidents. Only 6 percent of those in the 1994 survey who expressed being the target of sexually harassing behavior indicated that they took formal action in response. The unwanted attention included pressure for sexual favors; deliberate touching or cornering; sexual looks or gestures; letters or calls of a sexual nature; pressure for dates; and sexual teasing including jokes, remarks, or questions.

Such behavior violates, in part, the federal Equal Employment and Opportunity Act of 1972, which bars discrimination based on race, color, religion, sex, and national origin. It also prohibits practices caused by statistically determined adverse impact, as well as intentional unequal treatment.[43] Implications of the act encompass all private employers of 15 or more persons; educational institutions; government; employment agencies; labor unions; and joint labor-management committees established for apprenticeship and training. State regulations may not reduce federal measures, but may go further in specifically prohibiting behaviors covered in the act.

Security practitioners have an obligation to investigate apparent or actual violations of these measures. If the investigation determines that offenses have occurred, disciplinary measures, including cautioning, counseling, or dismissal, may be considered, as discussed in Chapter 6. A single act of sexually oriented harassment does not necessarily meet the definition of ongoing harassment. The exact circumstances of the case determine whether a court proceeding will occur. These include how often the conduct occurs; how serious it is; if the behavior physically threatens the other party; and if the unwanted behavior interferes with work behavior. Failure of the employer to investigate and respond to charges in a timely fashion can lead to costly sanctions (see Box 11.6).

Box 11.6. Sexual Harassment Sanctions Add Up

For any organization, having an anti-harassment policy is not enough. Organizations additionally may require customized training on sexual-harassment awareness that spells out consequences for policy violations. The complaint procedure must be made clear. The following are some incidents in which employers were forced to pay for sexual harassment at the workplace:

- Mitsubishi Motors agreed to pay $34 million in a sexual-harassment settlement brought by some 300 current and former female workers at Mitsubishi's Normal, Illinois, plant.
- The U.S. unit of Sweden's Astra Pharmaceuticals agreed to pay $9.9 million to about 80 workers in a sexual-harassment case.
- The Ford Motor Company paid $7.5 million in damages and planned to spend about $10 million more on training after 19 women at two Chicago-area plants complained to the EEOC. The women claimed that male workers often used sexually degrading words for them, placed explicit materials in their workplace, and even groped them.

Source: J.P. Miller (June 12, 1998). "Mitsubishi Will Pay $34 Million in Sexual-Harassment Settlement." *Wall Street J.*, p. B4; A. McLaughlin (September 10, 1999). "When Others Harass, Now Managers Lose Pay." *Christian Science Monitor*, p. 1.

Business Espionage/Theft of Trade Secrets

For many organizations, theft of intellectual property, including know-how and trade secrets, can far surpass losses from conventional theft. While larcenies can be overt, documented, and related to a monetary loss, the theft of information can be insidious and lead to the loss of market position and future opportunity without the victim organization perceiving the cause. According to an estimate by the American Society for Industrial Security, espionage by foreign and domestic competitors costs U.S. firms more than $300 billion a year.[44] While most saboteurs are likely to be competitors within the same industry, a few come from abroad and act with the support of their governments.[45]

Security professionals seek to protect their organizations' inherent intellectual capital by identifying what is most important and then restricting access to it. This involves broad information protection policies, requiring employees with need-to-know trade secrets to be aware of their importance for protection.[46] As more know-how becomes accessible over the Net, protection of such systems becomes a priority, lest organizations become victims of cyber-theft. A former National Security Agency expert, Ira Winkler, observes that countermeasures should include holding "fire drills" to simulate problems and their responses, and the periodic performance of vulnerability assessments with attempted penetration of systems.[47] Practitioners remain up to date about such risks by using independent consultants, surfing the Web, liaising with local sources, and studying material provided by the U.S. Department of State.

The management of competitive intelligence is linked to the protection against business espionage and the implementation of trade secret security as well as the legitimate collection of market information. The Society of Competitive Intelligence Professionals (SCIP) was founded in 1986 as a nonprofit organization that educates and provides liaison opportunities for managers of competitive intelligence programs.

Litigation: "Inadequate Security"

As described in Chapter 3, litigation may occur if employers or their agents breach their duty and if such a breach results in direct harm to others. One question to be settled is how sufficient or proportional the effective protective measures are relative to the prescribed standards or measures. "Inadequate security" covers a wide range of inadvertent, intentional, and unintentional circumstances that could lead to a civil action in which the organization is a defendant. In such plaintiffs' actions, senior executives or managers, the director of security, and individual security personnel involved may be listed individually as defendants. While the employer may provide legal counsel for any employees, individuals frequently must consult lawyers privately for guidance.

Many charges of inadequate security relate to the failure of property owners and managers to adequately protect the public from the foreseeable criminal actions of others. For example, the appellate court of Pennsylvania ruled in a case:

> While we do not consider a landlord to be an insurer of its tenants, we hold that in all areas of the leasehold, particularly in the areas under his control, the landlord is under a duty to provide adequate security to protect its tenants from the foreseeable criminal actions of third persons.[48]

A wide variety of tort actions due to inadequate security are conceivable. A fixed list of such actions cannot be created because of the changing nature of circumstances and the inventiveness of plaintiffs and their lawyers. Further, the nature of such charges varies according to the type of organization. The following are a few general examples of inadequate security charges brought against organizations:

- Alarms that are inadequate, missing, or broken
- Assaults
- Inadequate background checks (see "Negligent Hiring" below)
- Failure to prevent a crime
- Duty to protect not met
- Emotional distress
- Failure to make necessary repairs to security hardware
- Failure to anticipate crime
- Failure to warn
- Foreseeability of event not anticipated by the defendant
- Inadequate locks
- Security company and its personnel fail to protect or are negligent

Terrorism

An act of terrorism occurs through the calculated use of violence to obtain political or social goals to instill fear, intimidation, or coercion. Terrorism usually involves a criminal deed meant as a symbolic activity to influence someone unrelated to the immediate victims. The efforts of terrorists are usually against governments.[49] However, private organizations and their facilities may also be targets of or indirectly affected by a terrorist event (see Box 11.7). Acts of domestic terrorism are infrequent in the U.S. and Canada. As a result, security operations managers assess the possibilities of terrorism within a wider global context. But complacency on domestic risks is never appropriate.

Box 11.7. A Strike at the "Heartland": Terrorism in Oklahoma City

At 9:02 a.m. on April 19, 1995, a detonation outside the Alfred P. Murrah Federal Building in Oklahoma City, Oklahoma, changed the way millions of Americans thought about terrorism. The event occurred in the "heartland" of America, creating a sense of disbelief, anger, and fear unlike any other domestic disaster of its kind in memory. The explosion immediately killed 167 persons, including children attending a nursery in the building. Extensive physical damage was sustained for several blocks from ground zero.

The explosion reminded security planners of the importance of car bombs as vehicles of destruction. In Beirut in 1983, an Islamic Jihad terrorist killed 241 U.S. servicemen in a badly protected Marine barracks. (Moments later, another bomb in Beirut killed 58 servicemen at a nearby French barracks.) In London in 1982, an IRA car bomb exploded, killing eight and causing over $1 billion in property damage. In New York City at the World Trade Center in 1992, a bomb killed six and injured around 1,000. All of these instances of terrorism involved bombs composed mostly of natural materials or easily available explosives and were created with modest skill and at little cost. Perimeter control, CPTED, interior changes, and security systems enhancements are measures taken to deter these threats.

A survey by the U.S. Marshal Service completed in the late 1990s identified over 1,300 single or multi-tenant federal office buildings employing 750,000 federal employees. Such buildings, usually under the management of the General Services Administration, are divided into five categories according to number of employees, size, and criticality for federal service. Appropriate security measures are prescribed for buildings in each category. But some still wonder whether this level of security is sufficient.

At the same time, the Oklahoma City explosion made clear how uncommon domestic terrorism is. In the mainland U.S., just two major incidents occurred in the 1990s. Both were targeted against important physical symbols of capitalism and government. Would these two incidents be exceptional historical footnotes? Or would they be events to inspire others to act similarly in the years ahead?

Source: D.O. Coulson and E. Shannon (1999). *No Heroes*. New York, NY: Pocket Books; R.A. Serrano (1998). *One of Ours: Timothy McVeigh and the Oklahoma City Bombing*. New York, NY: W.W. Norton & Company; M. Dees with J. Corcoran (1998). *Gathering Storm: America's Militia Threat*. New York, NY: HarperCollins; *Security Letter* (May 1, 1995), Vol. 25, p. 1; and *Security Letter* (May 17, 1995), p. 1.

Governments prepare for the possibility of terrorism in a variety of ways. The first step is planning how to prevent or mitigate the effects of a terrorist act. Physical design, location, and construction methods influence the possibility of a particular facility or system being a target. Training, including mock exercises, help organizations respond to the emergency. Appropriate emergency supplies and continuity resources are identified in case they should be needed. In 1980, an elite antiterrorist military group was created, called the Delta Force or Blue Light, and is based in Fort Bragg, North Carolina. The troops are trained to deal with terrorist incidents on land, sea, and air.

Terrorism need not be a dramatic event seeking to bring sweeping social change. It can be a local event capable of engendering fear or inconvenience. The Model Penal Code (Section 211.3) defines terrorism as threatening "to commit any crime of violence with purpose to terrorize another or to cause evacuation of a building, place of assembly, or facility of public transportation, or otherwise to cause serious public inconvenience, or in reckless disregard of the risk of causing such terror or inconvenience." Security operations managers typically prepare for such risks as bomb threats, incidents, and kidnapping.[50] However, other criminal forms of intimidation, such as cyber-terrorism, can threaten the vitality of an organization and its people.

Litigation: "Negligent Hiring"

Employers have an obligation to conduct a reasonable investigation into a prospective employee's work experience, background, character, and qualifications. As David A. Maxwell states, "The standard of care does not vary. The greater the risk of harm, the greater degree of care necessary to constitute ordinary care."[51] The doctrine of negligent hiring and retention provides that:

> (A)n employer whose employees are brought into contact with members of the public in the course of the employer's business has a duty to exercise reasonable care in the selection and retention of his employees . . . Negligent retention . . . occurs when, during the course of employment, the employer becomes aware or should have become aware of problems with an employee that indicates his unfitness, and the employer fails to take further action such as investigating, discharge, or reassignment.[52]

The protection manager constantly evaluates personnel performance, asking whether certain behaviors have failed to meet the duty of care standard. If so, have such workers been counseled, re-trained, reassigned, placed on temporary leave, or discharged? If appropriate measures are not taken in a timely fashion, the employer faces the possibility of a difficult defense in the event a plaintiff commences an action for such negligence.

Insurance/Workers' Compensation Fraud

Security practitioners sometimes interact with an organization's risk manager; that is, the individual who directs an organization's non-health and benefits insurance coverage. When a security program reduces chances of loss, this information can be useful to the risk manager in negotiating lower premiums or improved coverage from the insurance carriers serving the property, liability, and other risks of the organization. Additionally, security directors may review or monitor insurance provided by protection-related vendors (for example, for alarm monitoring, armored car, investigations, and security patrol services).

Apart from this interaction, security practitioners frequently are called to investigate workers' compensation fraud. Workers' compensation provides cash benefits for employees who are injured or otherwise incapacitated during the course of active employment. Also, in the event of a worker's death, a surviving spouse and family members may be entitled to benefits. State workers' compensation laws dictate the terms and conditions of benefits to claimants and the obligations to employers. In the event that an individual is unable to work because of a job-related injury or circumstance, benefits may also be awarded. In such cases, the worker may not sue the employer under the doctrine of negligence unless the court rules otherwise; that is, the rights and remedies granted an employee arise out of and in the course of employment. For example, if a third party injures an employee while on the job, that employee normally may not be able to sue the employer for negligence, but will likely receive the limited recovery provided by the workers' compensation statute.[53]

Security operations managers often must investigate incidents relative to routine workers' compensation claims. In other cases, however, former employees who are enjoying such benefits may actually no longer be ill or injured and should return to work and cease receiving such benefits, which can cost the employer over an extended period. Occasionally, former workers receiving benefits because they were injured on-the-job and can no longer work may in fact have returned to active health and may be working elsewhere while continuing to receive full benefits. In such cases, investigations seek to determine the facts of such incapacitation and to determine whether the benefits should be eliminated.[54]

Cargo/Supply-Chain Theft

The cost of cargo, transportation, and distribution theft is not known with any degree of precision. Efforts made in Congress since 1970 to require that cargo theft losses be reported have failed. Many cargo thefts are reported to police and are included in larceny statistics for the area where the crime occurred, but many others are not reported. In 1976, the Office of Transportation Security of the Department of Transportation estimated that $1.5 billion per year was stolen from warehouses, shipping and receiving platforms, storage areas, depots, terminals, and piers. Of the loss, 5 percent was from highjackings, 10 percent from breaking and entering, and 85 percent from internal theft, collusion, or unexplained shortage.[55]

Emphasis on recruiting, screening, and training reliable personnel head the agenda of cargo security practitioners. Technology and containerization help to cut losses from both internal and external crime.[56] Using bar codes, radio frequency sensors, electronic seals, and asset-tracking technology helps organizations manage assets better, and also decreases the chances of merchandise being stolen or abused. Arms, cash and money equivalents, jewelry, pharmaceuticals, and microchips are low-weight/high-value cargo highly susceptible to theft, and are thus special targets for enhanced security. Cargo security practitioners learn to work with insurers, claim agents, and law enforcement—domestically and abroad—as part of their activities.

Kidnapping/Extortion

People have been seized, borne off, and held for ransom ever since the beginning of time.[57] Greek mythology is full of examples, usually of females being carried off by males in acts of lust. In fact, familial disputes are still the causes of scores of kidnappings each year. However, kidnappings-for-profit, or extortion, infrequently occur in the U.S. and Canada. Security practitioners with executives requiring protection in other nations, however, will find risks high under some circumstances. Charles P. Nemeth writes: "Kidnapping and false imprisonment actions are relevant to the security industry because of their executive protection and counter-terrorism roles."[58] Despite the unlikely event of a kidnapping, organizations rely on executive protection to make the possibility of such an attempt costly to the attackers, thus decreasing its likelihood.

Kidnap and ransom (K&R) insurance requires a number of protective measures to be taken, including the agreement of the insured not to reveal that the organization has a K&R policy on the lives of its employees. K&R insurance coverage usually includes specialized services to support family members during the time a loved one is forcefully held. The same service interacts with local and national police officials and the kidnappers to negotiate a ransom, if permitted by local laws, and to obtain freedom for the victim.

Political Unrest/Regional Instability

Commerce is global. Managing in an international environment demands sensitivity to issues such as relations with government, especially local police; sex roles in the workplace; and relations with customers and vendors. It also demands that management knows how to direct and control programs from a distance.

Sometimes, investment in foreign-based operations can be placed at risk for political or social reasons. In the worst of circumstances, executives and managers are kidnapped or assets are seized by a vengeful and corrupt foreign government. Security practitioners in global enterprises where substantial resources are allocated may be responsible for evaluating and reporting on changing political and social circumstances that conceivably could put people and capital at risk. The means by which managers stay informed of such circumstances include staying in touch with managers in distant locations; occasionally visiting such facilities and including courtesy calls to local governmental or police officials; and subscribing to services that provide information on changing political events around the world.

Product Diversion/Transshipment

Manufacturers sometimes have differential pricing policies for products. That is, the same product can cost different amounts in different countries. The reasons for this are because some governments set the price of products to be sold. Further, prices may be different due to local trading conditions. Companies may manufacture products in one country and export them elsewhere, where the price may be lower. An exporter purchases products at the lower price to market them abroad. Normally, such exporters agree not to sell products into the markets not covered by the purchase agreement. However, some exporters or shippers divert the product back to the original country and sell it into local channels at a discount. This deprives the manufacturer of conventional sales and profits.

The security practitioner learning of such a scam must be able to prove it. Investigations in the field must be conducted to ascertain whether products have been diverted from their intended destination. Fact-finders visit retailers, distributors, shippers, buyers, public markets, and elsewhere to check the reliability of the distribution program. If product diversion is confirmed, the investigators may seek to prepare an analysis of the profits lost and seek to receive lost profits from the exporters or distributors who have broken terms of the contract. Typically, audio products, computer chips and devices, new fashions, and pharmaceuticals are targets for diversionary fraud.

Product Tampering/Contamination

The security and safety of a product are sometimes elusive objectives. Generally, if a product has been contaminated, it is due to deficient production controls. On rare occasions, however, an empty container of a product may be used to store a caustic substance, and then instead of being discarded is re-used in food service. In other cases, consumers can appear to have a reaction to a product for reasons unrelated to its purity. Nonetheless, the resulting reports of victimization may draw media attention. In some cases, however, the contamination is due to deliberate tampering, which can lead to deaths, injuries, and loss of immediate sales and market share for the product involved. The stock market valuation of a large, diverse organization can be affected by a single criminal case.

In 1982, capsules of Tylenol, a popular analgesic made by McNeil Laboratories, a division of Johnson & Johnson, were laced with cyanide.[59] Seven persons died. The result was that a massive criminal investigation was launched, the product was recalled and destroyed, and the organization sought to make future packaging tamper-resistant. The attack on Tylenol is the most dramatic case history of medical product contamination, and is remembered for the way in which the corporation demonstrated its security and safety measures to the public. Yet other cases of product contamination occur with less publicity.[60] In such circumstances, security practitioners must engage in supporting investigative efforts, coping with negative publicity, collecting and protecting recalled products, and helping to design procedures to avoid a recurrence of the incident.

Organized Crime

The Federal Omnibus Crime Control Act of 1970 defines organized crime as "the unlawful activities of the members of a highly organized, disciplined association engaged in supplying illegal goods and services, including but not limited to gambling, prostitution, loansharking, narcotics, labor racketeering, and other unlawful activities of members of such organizations." In effect, organized crime is crime that is organized. It should not be associated with any particular race, ethnicity, or national origin because the types of organized deviance and their perpetrators shift over time.

In 1986, the President's Commission on Organized Crime estimated that the net income from such activity ranged from between $26.8 billion to $67.7 billion. Dennis J. Kenney and James O. Finckenauer write: "Organized crime is not unique to the United States or American society. However, the wealth, the economic, social, and political structures, and the criminal opportunities available in the United States present a unique set of circumstances that enable organized crime to achieve its highest form here."[61]

Security practitioners in banking are concerned with the possibility of organized crime using financial institutions to launder money. In 1989, the Bank of Commerce and Credit International (BCCI) paid a $50 million settlement after pleading guilty to conspiring to launder drug money. The bank subsequently ceased operations.[62] The U.S. Department of Justice believes that the vast majority of cargo thefts are the result of organized crime. Businesses also have suffered from organized crime, from product counterfeiting, product diversion, credit and insurance fraud, and labor racketeering. Each type of organized crime is somewhat different, as are the means of attacking it. Security operations managers work with law enforcement and other victims in quantifying the amount of loss and in qualifying the means by which such losses occur so that the offenses can be stopped. Despite the organization and determination of well-financed criminals, the grasp of organized crime can be loosened by concerted effort.[63]

THE FUTURE DIRECTION OF SECURITY OPERATIONS

This book has emphasized origins of various security practices and institutions. It is certain that the pattern of rapid change will continue into the future, perhaps at a still faster pace. Technologically systems will be able to provide a higher degree of security, sureness, and safety by enhanced supervision, tracking, and reporting. Management styles will continue to change. Computer systems may lessen the burden of certain procedures and enhance decision-making. Meanwhile, the role of security operations managers will also evolve. Separate security departments for individual tasks may be created, such as general patrol, investigations, competitive intelligence, IT protection, and emergency planning, response, and recovery. In other circumstances, managers may be expected to direct security programs as well as other activities not currently part of their job description. This could include human resources, compliance, and risk management.

Senior management may also seek to eliminate, downsize, or outsource security programs to contractors and consultants. The likelihood of this occurring is related to the value perceived and demonstrated by the security program initiatives. Operating security managers who demonstrate the ongoing worth of their programs, through efficient operations, measurable benefits, and reliable services, will thrive.

DISCUSSION AND REVIEW

1. While some people may be better leaders than others, all people can lead and all people can learn to lead better. Discuss some ideas of how leadership skills may be improved.
2. Can leadership exist without power? What kinds of power?
3. Why do security operations managers tempt to possess—or should endeavor to possess—exceptional discretion for workplace activities?

4. What is the philosophical essence of security as a management practice?
5. Workplace violence is a major concern among protection professionals in large industrial and commercial organizations. Why should this be so when the workplace—especially an office—is essentially one of the safest places for people to be?
6. Why does senior management expect security practitioners to head the organization's crisis management and contingency planning operations?
7. Numerous challenges for security operations at the present involve the ability to direct and use technology and investigative services. Is the same person likely to have the skills for both corporate activities? If not, what are management's options?

ENDNOTES

[1] A. Jago (March 1982). "Leadership: Perspectives in Theory and Research." *Management Science*, Vol. 28, pp. 315–36.

[2] J.D.W. Beck and N.M. Yeager (1994). *The Leader's Window: Mastering the Four Styles of Leadership to Build High-Performance Teams.* New York, NY: John Wiley & Sons.

[3] R.A. Brawer (1998). *Fictions of Business: Insights on Management from Great Literature.* New York, NY: John Wiley & Sons.

[4] S. Howarth (August 1–2, 1998). "Leadership—Fleets Ahead of Its Time." *Financial Times*, p. IV.

[5] A. Bryant (February 7, 1999). "A Rebuff to the Ministry of Silly Bosses." *New York Times*, Sec. 3, p. 1, and "Test: Can You Laugh at His Advice?" (July 6, 1998). *Fortune*, p. 203.

[6] J.P. Kotter (1995). *The New Rules: How to Succeed in Today's Post-Corporate World.* New York, NY: Free Press.

[7] F.E. Fiedler (1996). "Research on Leadership Selection and Training: One View of the Future." *Administrative Science Quarterly*, Vol. 41, p. 243.

[8] P.R. Scholtes (1998). *The Leader's Handbook: Making Things Happen, Getting Things Done.* New York, NY: McGraw-Hill, pp. 18–19.

[9] T.A. Stewart (November 6, 1989). "New Ways to Exercise Power." *Fortune*, p. 52.

[10] P.L. Townsend and J.A. Gibhardt (1997). *Five-Star Leadership.* New York, NY: John Wiley & Sons, pp. 64–75.

[11] J.A. Conger (1990). "The Dark Side of Leadership." *Organizational Dynamics.* vol 19; pp. 44–55.

[12] Ibid., p. 55.

[13] E. Durkheim (1951). *Suicide: A Study in Sociology.* J.A. Spaulding and G. Simpson, Trans. New York, NY: Free Press.

[14] T. Hirschi (1969). *Causes of Delinquency.* Berkeley, CA: University of California Press.

[15] E.H. Sutherland (1949). *White Collar Crime.* New York, NY: Dryden Press.

[16] Examples: M. Braverman (1999). *Preventing Workplace Violence.* Thousand Oaks, CA: Sage Publications; S.L. Heskett (1996). *Workplace Violence: Before, During, and After.* Boston, MA: Butterworth-Heinemann; M.D. Kelleher (1997). *Profiling the Lethal Employee: Case Studies of Violence in the Workplace.* Westport, CT: Praeger; J.A. Kinney (1995). *Violence at Work: How to Make Your Company Safer for Employees and Customers.* Englewood Cliffs, NJ: Prentice Hall; J.W. Mattman and S. Kaufer (1997). *The Complete Workplace Violence Prevention Manual.* Costa Mesa, CA: James Publishing; and M.D. Southerland, P.A. Collins, and K.E. Scarborough (1997). *Workplace Violence: A Continuum from Threat to Death.* Cincinnati, OH: Anderson Publishing.

[17] G. Toscano and J. Windau (1996). "National Census of Fatal Occupational Injuries, 1995." Washington, DC: Bureau of Labor Statistics, p. 36.

[18] "Characteristics of Injuries and Illnesses Resulting from Absences from Work, 1994" (1996). Washington, DC: Bureau of Labor Statistics.

[19] "Guidelines for Preventing Workplace Violence for Healthcare and Social Service Workers" (1996). Washington, DC: Occupational Safety and Health Administration.

[20] In 1990, armored car services personnel employed an estimated 15,000 workers, or 1.0 percent of the total private security employment. Private investigators employed 7,000 workers, or 4.7 percent of the total private security employment. Contract and proprietary security combined employed 1,048,000 workers, or 70.2 percent of the total private security employment. Source: W.C. Cunningham, J.J. Strauchs, and C.W. Van Meter (1990). *Private Security Trend 1970 to 2000, The Hallcrest Report II.* Boston, MA: Butterworth-Heinemann, p. 196.

[21] A.M. Levitt (1997). *Disaster Planning and Recovery.* New York, NY: John Wiley & Sons.

[22] P.T. Holder and D.L. Holley (1998). *The Executive Protection Professional's Manual.* Boston, MA: Butterworth-Heinemann; and M.J. Braunig (1993). *The Executive Protection Bible.* Aspen, CO: ESI Education Development Corporation, p. 221.

[23] *NAI Executive Protection Manual* (1997). Tigard, OR: Noble & Associates.

[24] R.L. Oatman (1998). *The Art of Executive Protection.* Baltimore, MD: Noble House; B. Mares (1994). *Executive Protection: A Professional's Guide to Bodyguarding.* Boulder, CO: Paladin Press; and P.T. Holder and D.L. Hawley (1998). *The Executive Protection Professional's Manual.* Boston, MA: Butterworth-Heinemann.

[25] J. Autera and M. Scanlan (April, 1999). "Seeing Through Enemy Eyes." *Security Management*, p. 35.

[26] R.L. Oatman (June 1998). "The Challenge of Protecting the Chief." *Security Management*, p. 40.

[27] E.H. Sutherland (1983). *White-Collar Crime: The Uncut Version.* New Haven, CT: Yale University Press.

[28] J. Nolan (1999). *Confidential.* New York, NY: HarperBusiness; and T.G. Poveda (1994). *Rethinking White-Collar Crime.* Westport, CT: Praeger.

[29] R.J. Magnuson (1992). *The White-Collar Crime Explosion*. New York, NY: McGraw-Hill, pp. 100–109.

[30] T.D. Schellhardt (November 12, 1997). "A Former High-Tech Thief Shares Tricks of the Trade." *Wall Street J.*, p. B1.

[31] J.M. Carroll (1996). *Computer Security.* 3rd edition. Boston, MA: Butterworth-Heinemann.

[32] R. Kuhn (1990). "The Attack on Employer's Rights." *Security J.* 1:74-75.

[33] P. Wayner (May 28, 1998). "From Toy Rings to Sophisticated Codes, a Quest for Security." *New York Times*, p. G9.

[34] C.P. Phleeger (1997). *Security in Computing*. 2nd edition. Upper Saddle River, NJ: Prentice Hall PTR.

[35] W.C. Boni and G.L. Kovacich (1999). *I-Way Robbery: Crime on the Internet*. Boston, MA: Butterworth-Heinemann, p. 152.

[36] J. Normand, R.O. Lempert, and C.P. O'Brien (Eds.) (1994). "Under the Influence? Drugs and the American Work Force." Washington, DC: National Academy Press.

[37] For example, Executive Order 12564, September, 1986, and the consequent 1998 U.S. Department of Health and Human Services' "Mandatory Guidelines for Federal Workplace Drug Testing Programs."

[38] P.B. Schur and J.F. Broder (1990). *Investigation of Substance Abuse in the Workplace*. Boston, MA: Butterworth-Heinemann. Aiso: H. Abadinsky (2001). *Drugs: An Introduction*, 4th Edition. Belmont, CA: Wadsworth/Thompson Learning.

[39] B.R. Hollstein (February 1998). "Don't Do as the Romans Do." *Security Management*, pp. 56–57.

[40] T.D. Crowe (1991). *Crime Prevention Through Environmental Design*. Boston, MA: Butterworth-Heinemann.

[41] W.F. Blake and W.F. Bradley (1999). *Premises Security: A Guide for Security Professionals and Attorneys*. Boston, MA: Butterworth-Heinemann.

[42] U.S. Merit Systems Protection Board (1995). Washington, DC: Government Printing Office, p. viii.

[43] The legislation is enforced by the U.S. Equal Employment Opportunity Commission, www.eeoc.gov. The act may be found at www.eeoc.gov/laws/vii.html.

[44] R.E. Silverman (January 4, 1999). "Stop, Thief!" *Wall Street J.*, p. R50.

[45] P. Schweitzer (1993). *Friendly Spies*. New York, NY: Atlantic Monthly Press; and J.J. Fialka (1997). *War by Other Means*. New York, NY: W.W. Norton & Company.

[46] J.H.A. Pooley (1987). *Trade Secrets*. New York, NY: American Management Association.

[47] I. Winkler (1997). *Corporate Espionage*. Rocklin, CA: Prima Publishing.

[48] J.F. Vaughan, Ed. (1999). *Avoiding Liability in Premises Security.* 4th edition. Atlanta, GA: Strafford Publications, pp. 2–3.

[49] R.A. Falkenrath, R.D. Newman, and B.A. Thayer (1998). *America's Achilles' Heel*. Cambridge: MA: MIT Press; P.B. Heymann (1998). *Terrorism and America*. Cambridge, MA: MIT Press.

[50] F. Bolz, Jr., K.J. Dudonis, and D.P. Schulz (1990). *The Counter-Terrorism Handbook.* New York, NY: Elsevier Science Publishing.

[51] D.A. Maxwell (1993). *Private Security Law: Case Studies.* Boston, MA: Butterworth-Heinemann, p. 8.

[52] J.F. Vaughan (Ed.) (1999). *Avoiding Liability in Premises Security.* 4th edition. Atlanta, GA: Strafford Publications, pp. 221–22.

[53] Ibid., p. 166.

[54] S.L. Wallace (October 1997). "Curing a Claims Crisis." *Security Management,* pp. 72–75; and R.H. Schmedlen (August 1997). "Heading Off the Liability Headache." *Security Management,* pp. 79–83.

[55] National Advisory Committee on Criminal Justice Standards and Goals (1976). Report of the Task Force on Private Security. Washington, DC: Government Printing Office, p. 59. See also G.O.W. Mueller and F. Adler (1985). *Outlaws of the Ocean.* New York, NY: Hearst Marine Books.

[56] L.A. Tyska and L.J. Fennelly (Eds.) (1983). *Controlling Cargo Theft.* Boston, MA: Butterworth-Heinemann.

[57] C. Moorehead (1980). *Hostages to Fortune.* New York, NY: Atheneum.

[58] C.P. Nemeth (1995). *Private Security and the Law.* 2nd edition. Cincinnati, OH: Anderson Publishing, p. 219.

[59] *Security Letter* (October 4, 1982), Vol. XII, p. 1; (February 15, 1986), Vol. XIV, p. 1.

[60] *Security Letter* (May 15, 1990), Vol. XX, p. 3; (July 1, 1999), Vol. XXIX, p. 1.

[61] D.J. Kenney and J.O. Finckenauer (1995). *Organized Crime in America.* Belmont, CA: Wadsworth Publishing, p. 371.

[62] Silver Lining (June 27, 1998). *Economist,* p. 74–75.

[63] J.B. Jacobs with C. Friel and R. Radick (1999). *Gotham Unbound.* New York, NY: New York University Press.

Additional References

E. Brown (September 28, 1998). "War Games to Make You Better at Business." *Fortune.* pp. 299–96.

D.K. Denton and B.L. Wisdom (July–August 1989). "Shared Vision." *Business Horizons,* 32(4):67–69.

F. Fukuyama (June 12–13, 1999). "Death of Hierarchy." *FT Weekend,* p. I, IX.

S.J. Garone (1999). "Concepts for the New Leadership." Conference Report 1231-99-CH. New York, NY: Conference Board.

D.A. Sklansky (April 1999). "The Private Police." *UCLA Law Review,* 46(4):1165–287.

A. Zaleznik (March–April 1992). "Managers and Leaders: Are They Different?" *Harvard Business Review,* 70(2): 126–35.

Appendix A

CODE OF ETHICS OF THE AMERICAN SOCIETY FOR INDUSTRIAL SECURITY

PREAMBLE

Aware that the quality of professional security activity ultimately depends upon the willingness of practitioners to observe special standards of conduct and to manifest good faith in professional relationships, the American Society for Industrial Security adopts the following Code of Ethics and mandates its conscientious observance as a binding condition of membership in or affiliation with the society:

CODE OF ETHICS

I. A member shall perform professional duties in accordance with the law and the highest moral principles.

II. A member shall observe the precepts of truthfulness, honesty, and integrity.

III. A member shall be faithful and diligent in discharging professional responsibilities.

IV. A member shall be competent in discharging professional responsibilities.

V. A member shall safeguard confidential information and exercise due care to prevent its improper disclosure.

VI. A member shall not maliciously injure the professional reputation or practice of colleagues, clients, or employers.

ARTICLE I

A member shall perform professional duties in accordance with the law and the highest moral principles.

Ethical Considerations

I-1 A member shall abide by the law of the land in which the services are rendered and perform all duties in an honorable manner.

I-2 A member shall not knowingly become associated in responsibility for work with colleagues who do not conform to the law and these ethical standards.

I-3 A member shall be just and respect the rights of others in performing professional responsibilities.

ARTICLE II

A member shall observe the precepts of truthfulness, honesty, and integrity.

Ethical Considerations

II-1 A member shall disclose all relevant information to those having the right to know.

II-2 A right to know is a legally and enforcible claim or demand by a person for disclosure of information by a member. Such a right does not depend upon prior knowledge by the person of the existence of the information to be disclosed.

II-3 A member shall not knowingly release misleading information nor encourage or otherwise participate in the release of such information.

ARTICLE III

A member shall be faithful and diligent in discharging professional responsibilities.

Ethical Considerations

III-1 A member is faithful when fair and steadfast in adherence to promises and commitments.

III-2 A member is diligent when employing best efforts in an assignment.
III-3 A member shall not act in matters involving conflicts of interest without appropriate disclosure and approval.
III-4 A member shall represent services or products fairly and truthfully.

ARTICLE IV

A member shall be competent in discharging professional responsibilities.

Ethical Considerations

IV-1 A member is competent who possesses and applies the skills and knowledge required for the task.
IV-2 A member shall not accept a task beyond the member's competence nor shall competence be claimed when not possessed.

ARTICLE V

A member shall safeguard confidential information and exercise due care to prevent its improper disclosure.

Ethical Considerations

V-1 Confidential information is nonpublic information, the disclosure of which is restricted.
V-2 Due care requires that the professional must not knowingly reveal confidential information, or use a confidence to the disadvantage of the principal or to the advantage of the member or a third person, unless the principal consents after full disclosure of all the facts. This confidentiality continues after the business relationship between the member and his or her principal has terminated.
V-3 A member who receives information and has not agreed to be bound by confidentiality is not bound from disclosing it. A member is not bound by confidential disclosures made of acts or omissions that constitute a violation of the law.
V-4 Confidential disclosures made by a principal to a member are not recognized by law as privileged in a legal proceeding. The member may be required to testify in a legal proceeding to the information received in confidence from his principal over the objection of his principal's counsel.

V-5 A member shall not disclose confidential information for personal gain without appropriate authorization.

ARTICLE VI

A member shall not maliciously injure the professional reputation or practice of colleagues, clients, or employers.

Ethical Considerations

VI-1 A member shall not comment falsely and with malice concerning a colleague's competence, performance, or professional capabilities.

VI-2 A members who knows, or has reasonable grounds to believe, that another member has failed to conform to the Society's Code of Ethics shall present such information to the Ethical Standards Committee in accordance with Article VIII of the Society's bylaws.

CENSURE, SUSPENSION, AND EXPULSION

According to the ASIS bylaws, as revised through June 1999, censure, suspension, or expulsion of a Society member is possible, regardless of when the alleged offense may have been committed and regardless of when the alleged offense came to the attention of the Society. Six conditions exist for a sanction to be imposed. These include (1) violation of the Code of Ethics established by the Society and (2) any conviction for commission of a felony or a misdemeanor which has been reduced from a felony.

If in the opinion of the majority the members of the Ethical Standards Committee find that a member has engaged in conduct enumerated as a cause, the Committee shall notify the member by registered mail that charges will be submitted to a Board of Review for consideration. The member is permitted to appear before the Board and defend himself or herself and have legal counsel or a member of the Society appointed by him also present. The Chair of the Ethical Standards Committee shall appoint at least three members including a chair to hold a hearing or take testimony in a particular case.

Source: ASIS and J. Kleinig with Y. Zhang (1993). *Professional Law Enforcement Codes: A Documentary Collection*. Westport, CT: Greenwood Press.

Appendix B
REPORT OF THE TASK FORCE ON PRIVATE SECURITY

INTRODUCTION

Private security in the U.S. employed over one million persons in 1976, and was growing at a rate of over 10 percent per year when the *Report on the Task Force on Private Security* was published. Relative to its size, little had been written about private security on a national level. Yet the report noted, "There is virtually no aspect of society that is not in one way or another affected by private security." The Task Force had the mission of improving the quality of private security. To do so, the Task Force—made up of 13 persons with a staff of 11 and supplemented by an editorial staff—produced an extensive report. Over a quarter of a century after its publication, this report contains reasonable objectives for the interest of the public—not all of which has been achieved. The following are the standards and goals identified by the Task Force:

1. Selection of Personnel

Goal 1.1: Selection of Qualified Personnel. Primary emphasis in the screening process should be placed on selecting qualified personnel who will perform efficiently and who will preferably make a career of private security.

Goal 1.2: Commensurate Salaries. In an effort to reduce the attrition rate of the industry, salaries for private security personnel should be commensurate with the experience, training and/or education, job responsibilities, and other criteria related to the job performed.

Standard 1.3: Pre-employment Screening. In order to determine whether prospective personnel are trustworthy and capable, pre-employment screening should be initiated. Pre-employment screening should include a screening interview, an honesty test, a background investigation, and other appropriate job-related tests.

Standard 1.4: Employer Exchange of Information. Employers should cooperate in exchanging information on previous work performance and other data relating to selection criteria.

Standard 1.5: Equal Employment Opportunity. Employers should comply with equal employment opportunity guidelines and other federal, state, or local guidelines that preclude discrimination based on sex, race, creed, or age.

Standard 1.6: Application for Employment. An employment application should be used to provide a basis for the screening process (and should contain comprehensive educational and work histories of the applicant and other information).

Standard 1.7: Availability of Criminal History Records. Criminal history records for offenses, specified by statute or other authority as grounds for denying employment, should be made available to employers to assist them in the screening of private security personnel.

Standard 1.8: Minimum Pre-employment Screening Qualifications. The following minimum pre-employment screening qualifications should be established for private security personnel:

1. Minimum age of 18
2. High school diploma or equivalent written examination
3. Written examination to determine the ability to understand and perform duties assigned
4. No record of conviction, as stated in Standard 1.7
5. Minimum physical standards

2. Personnel Training

Goal 2.1: Training in Private Security. The responsibilities assumed by private security personnel in the protection of persons and property require training. Training should be instituted at all levels to ensure that personnel are fully prepared to exercise their responsibilities effectively and efficiently.

Goal 2.2: Professional Certification Programs. Professional associations should study the feasibility of developing voluntary certification programs for private security managerial personnel. (ASIS established the Certified Protection Professional program in 1979. Other certification programs, including Architectural Hardware Consultant (AHE), Certified Fraud Examiner (CFE), Certified Protection Officer (CPO), and Certified Security Supervisor (CSS), have also emerged.)

Standard 2.3: Job Description. Private security employers should develop job descriptions for each private security position.

Standard 2.4: Training Related to Job Functions. Private security employers should ensure that training programs are designed, presented, and evaluated in relation to the job functions to be performed.

Standard 2.5: Pre-assignment and Basic Training. Any person employed as an investigator or detective, guard or watchman, armored car personnel or armored courier, alarm systems installer or servicer, or alarm respondent, including part-time personnel and those presently employed, should successfully:

1. Complete a minimum of eight hours formal pre-assignment training
2. Complete a basic training course of a minimum of 32 hours within three months of assignment. A maximum of 16 hours can be supervised on-the-job training.

Standard 2.6: Arms Training. All armed private security personnel, including part-time personnel and those presently employed, should:

1. Be required to successfully complete a 24-hour firearms course that includes legal and policy requirements—or submit evidence of competence and proficiency—prior to assignment to a job that requires a firearm
2. Be required to requalify at least once every 12 months with the firearm(s) they carry while performing private security duties. (The requalification phase should cover legal and policy requirements.)

Standard 2.7: Ongoing Training. Private security employers should ensure that private security personnel are given ongoing training by using roll-call training, training bulletins, and other training media.

Standard 2.8: Training of Supervisors and Managers. Private security employers should provide effective job-related training for supervisory and managerial employees. Appropriate prior training, education, or professional certification should be accepted to meet this requirement.

Standard 2.9: State Authority and Responsibility for Training. A state government regulatory agency should have the authority and responsibility to accredit training schools, approve training curriculums, and certify instructors for the private security industry.

Standard 2.10: State Boards to Coordinate Training Efforts. Appropriate state boards and agencies should coordinate efforts to provide training opportunities for private security personnel and persons interested in prepared for security employment, through utilization of physical and personnel resources of area vocational schools and colleges and universities.

3. Conduct and Ethics

Goal 3.1: Code of Ethics. A code of ethics should be adopted and enforced for private security personnel and employers.

Standard 3.2: Conduct of Private Security Personnel. Private security personnel should perform their security functions within generally recognized guidelines for the protection of individual rights.

Standard 3.3: Reporting of Criminal Violations. All felonies and serious misdemeanors discovered by private security personnel should be reported to appropriate criminal justice agencies. Private security personnel should cooperate with those criminal justice agencies in all subsequent actions relating to those crimes.

Standard 3.4: Employer Responsibilities. Employers should provide a working environment, including adequate and serviceable equipment, conducive to the efficient performance of security functions assigned.

Standard 3.5: Maintaining Data on Criminal Activities. The private security industry has a responsibility to maintain internal data on criminal activities to develop, improve, and assess effectiveness of crime reduction programs.

4. Alarm Systems

Standard 4.1: Alarm Systems Research. Appropriate research should be conducted to develop new methods and techniques to transmit alarm signals and enhance alarm systems capabilities.

Standard 4.2: Backup Power for Alarms. All alarm systems terminating at a law enforcement agency should be equipped with a standby power source.

Standard 4.3: Certified Training of Alarm Sales and Service Personnel. There should be a certified training program for alarm sales personnel and alarm service technicians.

Standard 4.4: Compatibility of Sensors. Alarm companies and alarm users should only use those sensor devices in alarm systems that are operationally compatible with the area in which the system is located.

Standard 4.5: Training and Instruction of Alarm Users by Alarm Companies. Companies and others installing alarm systems should be required to instruct or train users and their employees in the proper operation of the system and to provide continued guidance when needed.

Standard 4.6 Joint Corporation to Reduce Alarm System Costs. Governmental agencies such as the law enforcement assistance administration; federal, state, and local regulatory agencies, the alarm industry; law enforcement agencies; and telephone companies should work together to reduce the cost of alarm systems and improve the efficiency and the reliability of operation and transmission.

Standard 4.7: Special Trunklines into Law Enforcement Facilities and Automatic Dialers. Consistent with existing technology, automatic telephone dialing services that are connected to alarm systems should not be keyed or interconnected with emergency law enforcement agencies' telephone lines.

Standard 4.8: Annual Alarm Inspection. Local government should require all alarm systems users whose systems ordinarily result in a law enforcement response to have their systems inspected at least once every year.

Standard 4.9: Alarm Systems Servicing Capability. Every jurisdiction should have a disclosure law requiring persons in the business of alarm systems sales to make known prior to a sale where the alarm system can be serviced or where a service arrangement can be obtained. Proof that a servicing arrangement, such as a contract or agreement, is in existence should be submitted to a law enforcement agency by persons desiring to transmit alarm signals to that agency.

Standard 4.10: Alarm User Permit Systems and the False Alarm Problem. Local governments should establish and enforce an alarm user permit system to regulate and reduce false alarms. Verified excessive false alarming ordinarily resulting in a law enforcement response should be grounds for permit revocation, suspension, and other appropriate penalties.

Standard 4.11: Ownership and Operation of Alarm Systems. Ownership or operation of alarm systems should be the province of private enterprise, and the government ownership or operation of alarm systems should be discouraged, provided, however, that government should not be precluded from:

1. Operating such systems in temporary or emergency situations
2. Owning or operating alarm systems that are located in publicly owned or leased buildings, annunciate in the same or other government buildings, and are responded to by government employees

3. Providing private individuals and businesses with funds for the acquisition of crime prevention devices provided that such devices are purchased on the open market and remain the property of the consumer.

5. Environmental Security

Standard 5.1: Improvement of Door and Window Security. Governments should examine those standards developed for protection of doors, windows, and other openings. Those standards that provide the most economical level of effective protection and deterrence should be considered for incorporation into building codes.

Standard 5.2: Adequate Security Lighting. Where appropriate, property should be adequately lighted to discourage criminal activity and enhance public safety.

Standard 5.3: Computer Security. Possessors of computers should have a comprehensive protection plan for both physical site and data, regardless of whether the computer is used solely for their own needs or for providing computer services to others.

Standard 5.4: Crime Prevention in Design. Architects, builders, and/or their professional societies should continue to develop performance standards of crime prevention in design with advice from law enforcement agencies and the private security industry.

Standard 5.5: Development of Environmental Security Expertise. Those companies selling security services should develop the expertise necessary to offer environmental security planning services.

Standard 5.6: Environmental Security in Comprehensive Planning. Environmental security should be a part of comprehensive planning from the design phase to the completion of construction projects.

Standard 5.7: Crime Prevention Courses in School of Architecture and Urban Planning. Schools offering courses in architecture or urban planning should include in the curricula courses on architectural design for crime prevention.

Standard 5.8: Inclusion of Crime Prevention Measures in Existing Codes and the Consideration of Building Security Codes. Crime prevention measures should be an identifiable part of assisting all proposed regulatory codes. Building, fire, and safety codes should be reviewed by regulatory bodies and private security representatives to avoid conflicting with implementation of effective crime prevention measures.

Standard 5.9: Crime Impact Forecast. Crime impact statements should be included in the planning phase of all new public and private building and development projects.

Standard 5.10: Crime Prevention Courses as a Job Requirement. Architects and urban planners should be encouraged to attend seminars or classes in Crime Prevention Through Environmental Design (CPTED). Proof of successful completion of a CPTED seminar or course should then become a necessary prerequisite for employment or the obtaining of a license.

6. Law Enforcement Agencies

Goal 6.1: Interactive Policies. Effective interaction between the private security industry and the law enforcement agencies is imperative for successful crime prevention and depends to a large extent on clear and understandable published policies developed by their administrator. Policies should be developed to serve as guides for modification by appropriate agencies.

Standard 6.2: Survey and Liaison with Private Security. Law enforcement agencies should conduct a survey and maintain a current roster of those private security industry components operating in the agencies' jurisdictions, and designate at least one staff officer to serve as a liaison with them.

Standard 6.3: Policies and Procedures. For law enforcement agencies and the private security industries to most effectively work within the same jurisdiction, policies and procedures should be developed covering:

1. The delineation of working roles of law enforcement officers and private security personnel
2. The continuous prompt and responsible interchange of information
3. Cooperative actions between law enforcement agencies and the private security industry

Standard 6.4: Multilevel Law Enforcement Training in Private Security. There should be multilevel training programs for public law enforcement officials, including but not limited to the following topics:

1. The role and mission of the private security industry
2. The legal status and types of services provided by private security companies
3. Interchange of information, crime reporting, and cooperative actions with the industry
4. Orientation in technical and operating procedures

Standard 6.5: Mistaking Identity of Private Security Personnel. Title terms, verbal representations, and visual items that cause the public to mistake private security personnel for law enforcement officers should be eliminated; security employers should be ensured that their personnel and equipment are easily distinguishable from public law enforcement personnel and equipment.

Standard 6.6: State Regulation of Private Security Uniforms, Equipment, and Job Titles. Each state should develop regulation covering the use of private security uniforms, equipment, company names, and personnel titles that do not conflict with those in use by law enforcement agencies within the state.

Standard 6.7: Law Enforcement Secondary Employment. Law enforcement administrators should ensure that secondary employment of public law enforcement personnel in the private security industry does not create a conflict of interest, and that the public resources are not used for private purposes.

Standard 6.8: Law Enforcement Officer Employment as a Private Security Principal or Manager. No law enforcement officer should be a principal or manager of a private security operation where such association creates a conflict of interest.

Standard 6.9: Private Investigative Work. Law enforcement officers should be strictly forbidden from performing any private investigative work.

7. Consumers of Security Services

Goal 7.1: Consumer Responsibility for Selection of Security Services. The consumer of private security services has a responsibility to evaluate systems and services prior to acquisition in order to ensure the best crime-reduction results for himself or herself and other members of the public affected by those systems and services.

Standard 7.2: Consumer Assistance Committees. Private security professional associates and organizations should form permanent committees to develop useful guides for the evaluation and acquisition of goods and services and to provide clearinghouses for professional response to consumer inquiries.

Standard 7.3: Development of Expertise by Private and Governmental Consumer Agencies. Government consumer protection bureaus, better business associations, and private consumer groups should develop sufficient knowledge of the private security industry to enable them to intelligently evaluate complaints and advise consumers.

Standard 7.4: Private Security Advertising Standards. The private security industry should adhere to advertising standards that accurately portray to the public the nature and quality of the service to be provided.

8. Higher Education and Research

Standard 8.1: State Review of Private Security Task Force Report. Each state should provide a mechanism to review and recommend implementation, as appropriate, of the standards and goals contained in this report.

Standard 8.2: National Private Security Resource and Research Institute. The Law Enforcement Assistance Administration should encourage the development of a national private security resource and research institute. In addition, all universities, companies, organizations, associations, and individuals concerned with private security should increase their efforts in private security research.

Standard 8.3: Noncredit and Credit Seminars and Courses. Colleges and universities should develop and offer noncredit and credit seminars and courses to meet the needs of private security personnel.

Standard 8.4: Degree Programs for Private Security. The private security industry and the Law Enforcement Assistance Administration should cooperate in the encouragement and development of:

1. Certificate, Associate of Arts, or Associate of Science degree programs designed to meet local industry needs
2. Undergraduate and graduate programs designed to meet private security needs

9. Regulatory Board

Standard 9.1: State Regulation. Regulation of the private security industry should be performed at the state level with consideration for uniformity and reciprocity among all the states.

Standard 9.2: Regulatory Board for Private Security. State-level regulation should be through a regulatory board and staff responsible for the regulation of private security activities within that state. This board should have sufficient personnel to perform adequately and promptly their tasks of screening and investigating.

Standard 9.3: State Regulatory Board Membership. The state regulatory board should include, as a minimum, representatives of licensed security service businesses; businesses using proprietary security; local police departments; consumers of security services; members of the general public; and individuals who are registered with the board and are presently employed in the private security field.

Standard 9.4: Regulatory Board Hearing Procedure. The state regulatory board should establish a hearing procedure for consideration and resolution of the complaints of applicants, licensees, registrants, consumers, and the public. To assist in the implementation of this role, the board should be granted the means necessary to require appearance of witnesses and production of documents.

Standard 9.5: Regulatory Board Funding. The state regulatory board should be funded by non-confiscatory license and registration fees and such general revenue funds as may be necessary for the effective operation of the board.

Standard 9.6: Regulatory Board Access to Criminal Record Information. The State regulatory board should be granted statutory authority for access to all criminal history record information so that it can conduct the necessary criminal history record check of all applicants for licenses and registration.

10. Licensing

Standard 10.1: Licensing of Security Businesses. Appropriate licensing should be required for any person or legal entity engaged in the business of:

1. Selling, installing, or servicing alarm systems
2. Providing respondents to alarm signal devices
3. Providing secured transportation and protection of valuables from one place to anther under armed guard
4. Providing guard or patrol services
5. Providing investigative services
6. Providing detection-of-deception services for the benefit of others

Standard 10.2: License Applications. License applications should include sufficient information about the applicant to enable the regulatory agency to determine if ethical, competent, and responsible services can be provided. (The Standard proposes a verified statement of the qualifying agent's experience qualifications, form of business, and information on the applicant, each director, and shareholders holding 10 percent of equity or more equity than the qualifying agent. The information should include directory-type data and two classifiable sets of fingerprints.)

Standard 10.3: Qualifying Agents. License applicants should be required to name one individual who will act as the licensee's qualifying agent. The qualifying agent should meet the following qualifications:

1. Be at least 18 years of age
2. Be an active participant in the business of the licensee
3. Not have been convicted of any felony or crime involving moral turpitude or have any criminal charges and/or indictments pending, unless pardoned or granted a special exemption by the regulatory agency
4. Not be under any present adjudication of incompetency
5. Be experienced in some area of security relevant to the license being sought

Standard 10.4: Notification of Changes in Status of Licensee. The licensee should be required to notify the regulatory board within 14 days of any change in the status of the licensee previously reported in the licensee application.

Standard 10.5: License Renewal. The license to engage in a security service business should be renewed every year.

Standard 10.6: Display of License Certificate. The licensee should be required to display the license certificate in public view in the licensee's principal place of business as well as display a copy of the certificate in each branch office.

Standard 10.7: Bonding and Insurance. When appropriate, due to the nature of the work, the applicant should file a surety bond and proof of public liability insurance with the regulatory agency before a license is issued.

Standard 10.8: License Denial, Revocation, or Suspension. A license may be denied, revoked, or suspended if the licensee, its qualifying agent, resident manager, or any officer, director or shareholder owning a 10 percent or greater interest in the licensee (provided the licensee is not listed on a national securities exchange or registered under section 12 of the Securities and Exchange Act of 1934, as amended) does the following:

1. Violates any provisions of the regulatory act or of the rules and regulations promulgated under the act
2. Commits an act resulting in the conviction of a felony or a crime involving moral turpitude, where such conviction reflects unfavorably on fitness to engage in a security service business
3. Practices fraud, deceit, or misrepresentation
4. Makes a material misstatement in the application for or renewal of the license

5. Demonstrates incompetence or untrustworthiness in actions affecting the conduct of the security services business

11. Registration

Standard 11.1: Registration of Private Security Personnel. Every person who is employed to perform the functions of an investigator or detective, guard or watchman, armored car personnel or armed courier, alarm systems installer or servicer, or alarm respondent should be registered with the private security regulatory board.

Standard 11.2: Registration Qualifications. Every applicant seeking registration to perform a specific security function in an unarmed capacity should meet the following minimum qualifications:

1. Be at least 18 years of age
2. Be physically and mentally competent and capable of performing the specific job function being registered for
3. Be morally responsible in the judgment of the regulatory board
4. Have successfully completed the training requirements set forth in Standard 2.5

Standard 11.3: Qualifications for Armed Security Personnel. Every applicant who seeks registration to perform a specific security function in an armed capacity should meet the following minimum qualifications:

1. Be at least 18 years of age
2. Have a high school diploma or pass an equivalent written examination
3. Be mentally competent and capable of performing in an armed capacity
4. Be morally responsible in the judgment of the regulatory board
5. Have no felony convictions involving the use of a weapon
6. Have no felony or misdemeanor convictions that reflect upon the applicant's ability to perform a security function in an armed capacity
7. Have no physical defects that would hinder job performance
8. Have successfully completed the training requirements for armed personnel set forth in Standards 2.5 and 2.6

Standard 11.4: Permanent Registration Card. So that employers, consumers of security services, and the public know that an individual is registered to perform specific security job functions, armed or unarmed, a permanent registration card should be issued and strictly controlled by the regulatory board. This card should not be issued until the applicant has met the minimum qualifications for registration in an armed or unarmed capacity.

Standard 11.5: Temporary Permit. Pending the issuance of the permanent registration card, provision should be made for the issuance of a nonrenewable temporary permit to allow an applicant to perform a specific security job function, in an unarmed capacity only, for a maximum of 30 days. This permit should be issued immediately upon completion of a favorable preliminary check of the applicant with the local law enforcement agency and other available sources.

Standard 11.6: Registration Renewal. Individuals who are registered as armed security personnel should be required to renew their registrations annually. All other registrants should be required to file for renewal of registration every five years.

Standard 11.7: Suspension and Revocation. Registration cards and temporary permits may be suspended or revoked for good cause, after a hearing, when a registrant:

1. Is convicted of a misdemeanor or felony that reflects unfavorably on his or her fitness to perform a security function
2. Has been formally charged with a criminal offense, the nature of which may make him or her unable to meet the minimum qualifications of registration
3. Fires a weapon without justification
4. Engages in conduct detrimental to the public safety or welfare
5. No longer meets the requirements of registration or violates any provisions of the act

Standard 11.8: Sanctions. Non-registered persons who perform a security function requiring registration should be subject to criminal penalties. Any person authorizing or permitting a non-registered person to perform a security function requiring registration should be subject to criminal penalties.

GLOSSARY

ABC technique: A time management concept in which the adherent categorizes items to be done according to their vital importance, nominal importance, and unimportance.

Acceptance theory: A concept advanced by Chester Barnard holding that subordinates will cooperate in the goals of the organization and assent to authority when certain conditions were met.

Adverse impact: Having an opposed or contrary effect on members of the public who are protected by legislation. For example, an individual with a physical disability may be able to perform all the duties of a security receptionist or alarm monitoring agent except that the computer keyboard is difficult to reach, creating an adverse impact. Cutting a semi-circular area from the counter would be an appropriate accommodation.

Amortization: The allocation and charge to expense of the cost or other basis of an intangible asset over its estimated useful life. Intangible assets like goodwill are not amortizable. Examples of amortizable intangibles include capital costs, leasehold improvements and interests, patents, and copyrights.

Analysis: The process of separating a problem into its constituent parts or basic principles so as to determine the nature of the whole and to examine it methodically. Related to planning.

At-will employment: The concept that an employee may be fired at any time for good cause, bad cause, or no cause at all. Originated by Horace G. Wood in 1877. The at-will concept of employment was less common in the second half of the 20th century, but still is the basis of employment for most private sector and institutional employees.

Auditor: See independent auditor.

Baseline: Readings on a polygraph chart that form a point of comparison for the psychological responses to polygraph questions.

Behaviorally Anchored Rating Scale (BARS): A behaviorally oriented scale sometimes used to assess performance.

Biometric: A security identification system that uses a physical feature and measures it automatically. The information from the physical feature is translated into a digital register, which is then compared with the values found in the approved database list.

Break-even: The point at which fixed costs and variable costs equate with sales or revenues volume.

Budget: A financial statement prepared prior to an accounting period containing the financial plans to be achieved during that period.

Bureaucracy: An organization with the following traits: a chain of command with fewer people at the top than at the bottom; well-defined positions and responsibilities; generally inflexible rules and procedures; and delegation of authority from the top down. Considered the most rational form of organization by Max Weber.

Capital expenditures: The cost of purchasing or improving a fixed asset, which will be depreciated over the estimated useful life of that asset.

Cash flow: A measurement of the inflow and outflow of cash and cash equivalents over a financial period.

Certificate of good conduct: A written document that determines whether a person is law-abiding at that time in the area in which the document is produced.

Certificate of relief: An order by a civilian or military judge in which an offender's criminal record may be sealed from public scrutiny or consideration in employment circumstances. It may be lifted in cases of employment consideration by government.

Charge coupled device: A camera that uses a semiconductor microchip as the imaging device.

Chief financial officer (CFO): The senior official in an organization responsible for financial activities.

Collusion: A conspiracy between two or more persons to defraud a person or persons of their rights by fraudulent means. When one fraudster is an employee and another is outside the organization, the effects from the secret combination can be deep and costly.

Concentric circles of protection: The notion that all countermeasures to crime and disorder have their limitations; therefore, numerous countermeasures need to be designed and implemented to protect a facility or program. Also called layered security.

Contingency: Something that may or may not happen; a possibility. In emergency planning, the term refers to an emergency or casualty.

Contingency management: See crisis management.

Contracting-out: The process of transferring responsibility for certain tasks or duties to another party or organization.

Crisis management: The process of dealing with an event or set of circumstances that can lead to loss to persons or an organization; a critical point that demands resolute action.

Cross-training: When an employee in one primary job task is trained in another or other tasks.

Cyber-crime: A variety of offenses related to information technology. They include extortion, boiler-room investment and gambling fraud, and fraudulent transfers of funds. The term was coined by Donn B. Parker.

Delegation: The giving of authority by one person to another; for example, a manager delegates responsibility to employees.

Denial-of-service attack: An attack on a Web site that usually crashes the system. The attacker arranges to trigger a large number of simultaneous demands on a system that crashes as a consequence.

Depreciation: The cost of the economic benefits of a tangible fixed asset that has been consumed over a financial period.

Deputizing: The process of selecting a personnel to manage or direct an operational plan.

Diligence: A measure of prudence or activity that is to be expected from a reasonable and prudent person under the particular circumstances; it is not measured by an absolute standard, but is dependent upon circumstances of a particular case. Also referred to as due diligence.

Director: A person appointed or elected according to law and authorized to manage and direct the affairs of a corporation or company. Inside directors are employees, like the CEO or CFO; outside directors are not employees, but may be officers of other corporations and possess skills and experience believed to be valuable in directing the affairs of the corporation.

Discipline: The process of controlling behavior (for example, in a workplace).

Disguised purpose test: A test instrument that seeks to determine characteristics of deviance or unreliability without asking direct questions. Also called a covert test instrument.

Dividends: The distribution of current or accumulated earnings to the shareholders of a corporation pro-rata based on the number of shares owned.

Embezzlement: The willful taking or converting to one's own use the assets of another which the wrongdoer acquired possession of lawfully during the course of office, employment, or reasons of trust. Embezzlement differs from larceny in that the original taking of property was lawful or with the consent of the owner; in larceny, the felonious intent must have existed at the time of the initial taking.

Equity: The extent of a stockholder's proportionate ownership interest in the corporation's capital stock and surplus; the extent of ownership interest in a venture. Also refers to injustice as a concept of fairness as contrasted with strictly formulated rules of common law.

Ethics: Relating to moral action, conduct, motive or character; professionally right or befitting conduct, conforming to professional standards of conduct.

False negative: An erroneous decision that a person is not deceptive when she or he actually is deceptive. Also called Type I or A.

False positive: An erroneous decision that a person is being deceptive when she or he actually is being truthful. Also called Type II or B.

Firewalls: A network security system used to monitor and restrict external and internal traffic.

Fraud: The criminal offense of intentionally deceiving another person or persons to obtain financial or monetary gains. Fraud usually does not involve property damage or threatened or actual physical injury to another.

Functional foremanship: A concept introduced by Frederick W. Taylor that when different responsibilities are understood, managerial or supervisory employees can change without affecting production by workers.

Generally accepted accounting principles (GAAP): Standards and concepts followed by accountants in measuring, recording, and reporting transactions.

Graphic user interface (GUI): A system whereby a user may direct a computer program by touching or clicking on symbols located on the monitor screen.

Grievance: A complaint filed by an employee, or by his or her union representative, regarding working conditions to seek resolution for which procedural measures are available in the union agreement.

Gross Domestic Product (GDP): The economic activity of a country during a particular calendar year; an international convention aggregating economic activity.

Gross National Product (GNP): GNP includes the GDP plus net income transfer from international economic activity.

Halo effect: The tendency of some evaluators to judge workplace performance of subordinates as consistently superior, despite evidence to the contrary in some situations.

Hawthorne investigations: Investigations conducted by Elton T. Mayo and others at the Western Electric company's plant in Hawthorne, Illinois. The research seemed to establish that paying attention to workers affected production output more than extrinsic factors did.

Hierarchy of human needs: Proposed by Abraham Maslow, this hierarchy identifies five levels of needs, with a lower one having to be satisfied before the next higher one can become important. Also called Maslow's pyramid or ladder.

Independent auditor: Usually a firm or corporation retained by a corporation to check and attest to the accuracy, fairness, and general acceptability of accounting records and statements. Usually performed by a certified public accountant.

Informed consent: A person agrees in writing to allow something to happen (such as a check of credit records) based on disclosure of relevant facts relating to the procedure.

Inspector-general: Government personnel whose primary function is to conduct and supervise audits and investigations relating to programs and operations of the particular agency. May investigate allegations by whistler-blowers.

Job description: A statement of facts about a particular job that can be used to determine job requirements (skill, knowledge, physical and mental efforts), responsibility, and working conditions. It sets out the requirements that applicants should be able to meet.

Knowledge workers: A term coined by Peter F. Drucker referring to executives, managers, and individual professionals who, through their knowledge or positions, make decisions that have a significant impact on the performance and results of the whole group or organization.

Local area network (LAN): A collection of computing resources (such as PCs, printers, mini-computers, and mainframes) linked by a common transmission medium such as a coaxial cable.

Larceny: The unlawful taking and removal of another's property with the intent to convert it or deprive the owner of its use.

Liabilities: An obligation to transfer economic benefit as a result of a passed transaction.

Likert scale: The scale used in certain tests that ask test-takers to judge a question on a scale of 1 to 7, in which 1 and 7 represent the extremes and 4 represents the midpoint, and the numbers between the extremes represent degrees of difference.

Line-item: An expense planned on an ongoing basis, such as a position.

Management by Objectives (MBO): A popular management strategy in which employees and supervisors set mutually agreed-upon goals and endeavor to reach them. Proposed by Peter F. Drucker and Douglas McGregor in the 1950s.

Middle management: Managers who supervise first-line managers and some non-management personnel. They coordinate tasks and do some planning to achieve organizational goals.

Model penal code: A codification of the principles of criminal law published by the American Law Institute in 1962. It served to unify state codes following its publication.

Moonlighting: Working at a second job after regular working hours. Some police officers moonlight as private security officers.

National Labor Relations Act: This 1935 law requires organizations to recognize and bargain with the union if that union has been legally established by the organization's employees. Section 9(b)(3) permits employers the right to terminate voluntary recognition of non-guard unions to perform security services. Also called the Wagner Act.

Negligence: The failure to do something that a reasonable person, guided by ordinary considerations, would do; the failure to use such care as a reasonably prudent and careful person would use under similar circumstances. The opposite of diligence. Negligence varies in significance from slight negligence—a failure to exercise great care—to gross negligence—the intentional failure to perform a manifest duty in reckless disregard of consequences affecting the life or property of others.

Organizing: The process of amassing critical resources needed in a plan so that the action aspect of the plan may get under way.

Outsourcing: See contracting-out.

Overt integrity test: An employment test that raises specific questions about the test-taker's past deviant behavior as well as attitudes toward such behavior in others. See Disguised-purpose test.

Pareto principle: A rule that posit that 80 percent of business activity comes from about 20 percent of the customers or clients. Named for Vilfredo Pareto, an Italian economist.

Peace Officer Standards of Training (POST): Training requirements whereby someone is trained following the statute of a particular state in the legal basis of making arrests. Some security officers are trained as peace officers with supplemental training that requires an average of 40 hours of classroom instruction.

Performance standards: These standards compare the actual work of an employee with a standard rate of work.

Peter principle: This idea stipulates, semi-seriously, that employees rise to the level of their incompetence. Proposed by Lawrence J. Peter and Raymond Hull.

Piece-rate: A form of financial incentive, proposed by Frederick W. Taylor, that provided for higher compensation to workers who produced at rates above the expected level.

Plaintiff's action: Such cases are usually in civil cases in which a complainant seeks redress for alleged grievances and harms. Negligent hiring and retention may be the basis for such a case.

Planning: The process of determining how a problem or opportunity may be responded to. Involves identifying problems or opportunities, analyzing relevant characteristics of the circumstances, organizing the formal response, deputizing a leader to head the response effort, and supervising the person(s) selected.

Privatization: The process whereby a public entity, usually a government, contracts-out for services or materials to the private sector that the public sector might formally have managed for itself.

Profit: A reward to that factor of production known as enterprise; the residual figure in payment for risk-bearing; the money remaining after all expenses, amortization, taxes, and other charges have been subtracted.

Proprietary: Refers to property ownership; possessorship of assets or opportunity. In a proprietary security department, all employees are normally full-time, not contract, employees.

Proximate cause: A natural and continuous sequence that produces a direct result.

Publicly-held corporations: Corporations whose stock is held by and available to the public. Such shares are usually traded on a securities exchange or over-the-counter.

Pygmalion effect: Named for the George Bernard Shaw play in which a simple flower girl was transformed into a refined socialite; a kind of self-fulfilling prophecy.

Quality circles: Employees' committees that analyze and solve quality problems. An outgrowth of the work conducted by W. Edwards Seming.

Relevant/irrelevant technique: An examination technique that uses two types of questions—relevant questions and neutral questions—to assess the subject's baseline response.

Relevant questions: Polygraph text questions about the topic or topics under investigation.

Reliability (in pre-employment tests): Implies that the same test will produce about the same results if administered at a later time.

Request for Proposal (RFP): The process by which an organization formally requests that bidders indicate how they will provide the services required by a client and their proposed fee.

Respondeat superior: The doctrine that states that the master (employer) is liable in certain cases for the wrongful acts of servants (employees) and is a principal for these agents.

Return on equity (ROE): An accounting ratio in which the net income is expressed as a percentage of capital employed.

Return on investment (ROI): An accounting ratio in which the net income is expressed as a percentage of capital employed plus cost of capital.

Risk management: The process of identifying hazards of property insured; the casualty contemplated in a specific contract of insurance; the degree of hazard; a specific contingency or peril. Generally not the same as security management, but may be related in concerns and activities. Work is done by a risk manager.

Security design: The process by which a new facility or retrofit is designed and engineered with protective principles considered, usually from the earliest stages of planning.

Senior management: Usually refers to headquarters staff officers.

Shareholder: A person who owns shares or stocks in a corporation or joint-stock company. Also called stakeholder, stockholder, or stockowner.

Shareholders' equity: The asset value in the organization belonging to shareholders, plus any reserves.

Supervising: The process whereby a manager responsible for an operation ascertains the progress of the intended plan, including ongoing evaluation of the persons specifically responsible for carrying out the plan. See deputizing.

Tempest: The technology and processes of illuminating undesirable electronic emanations.

Theory X and Theory Y: Two management theories described by Douglas McGregor that reflect opposite ways management has of viewing the workforce. The Theory X manager favors authoritarian leadership with centralized decision-making and close supervision of work activities. The Theory Y manager favors participatory decision-making, a decentralized authority structure, and less emphasis on coercion as a motivator.

Theory Z: Proposed by William Ouchi at a time when market share in some industries was lost to Japanese competitors. Theory Z incorporates aspects of American and Japanese styles of management.

Uniform Crime Reports (UCR): A national report of "serious crime," or Type I incidents, compiled by the Federal Bureau of Investigation. Includes four violent incidents (non-negligent homicide, rape, robbery, and aggravated assaults) and four property crimes (burglary, larceny, vehicular theft, and arson).

Uninterrupted power supply (UPS): A device that stores energy during normal operations so that it can provide backup energy if power fails.

Validity (in pre-employment test): Implies that the test measures what it is supposed to.

Vetting: The process of ascertaining the accuracy and completeness of information. Often refers to pre-employment screening.

Whistle-blower: An individual who informs on an employee's or employer's misconduct. In federal and state statutes, public employees are protected from retaliation for such disclosures. Some states also protect private sector whistler-blowers.

White-collar crime: Unlawful, nonviolent conduct by corporations and individuals, including theft or fraud and other violations of trust occurring during the time of the offender's occupation. White-collar crime is a frequent focus of internal investigations within the organization. A term devised by the sociologist Edwin Sutherland. See also Embezzlement.

INDEX